VOLUME ONE HUNDRED AND TWENTY SIX

ADVANCES IN
HETEROCYCLIC CHEMISTRY

EDITORIAL ADVISORY BOARD

A. T. Balaban *Galveston, Texas, United States of America*
A. J. Boulton *Norwich, United Kingdom*
M. Brimble *Auckland, New Zealand*
D. L. Comins *Raleigh, North Carolina, United States of America*
J. Cossy *Paris, France*
J. A. Joule *Manchester, United Kingdom*
P. Koutentis, *Cyprus*
V. I. Minkin *Rostov-on-Don, Russia*
B. U. W. Maes *Antwerp, Belgium*
A. Padwa *Atlanta, Georgia, United States of America*
A. Schmidt *Clausthal, Germany*
V. Snieckus *Kingston, Ontario, Canada*
B. Stanovnik *Ljubljana, Slovenia*
C. V. Stevens *Ghent, Belgium*
J. A. Zoltewicz *Gainesville, Florida, United States of America*

VOLUME ONE HUNDRED AND TWENTY SIX

Advances in
HETEROCYCLIC CHEMISTRY

Editors

ERIC F. V. SCRIVEN
Department of Chemistry,
University of Florida,
Gainesville, FL, USA

CHRISTOPHER A. RAMSDEN
Lennard-Jones Laboratories,
Keele University, Staffordshire,
United Kingdom

Academic Press is an imprint of Elsevier
50 Hampshire Street, 5th Floor, Cambridge, MA 02139, United States
525 B Street, Suite 1650, San Diego, CA 92101, United States
The Boulevard, Langford Lane, Kidlington, Oxford OX5 1GB, United Kingdom
125 London Wall, London, EC2Y 5AS, United Kingdom

First edition 2018

Copyright © 2018 Elsevier Inc. All rights reserved.

No part of this publication may be reproduced or transmitted in any form or by any means, electronic or mechanical, including photocopying, recording, or any information storage and retrieval system, without permission in writing from the publisher. Details on how to seek permission, further information about the Publisher's permissions policies and our arrangements with organizations such as the Copyright Clearance Center and the Copyright Licensing Agency, can be found at our website: www.elsevier.com/permissions.

This book and the individual contributions contained in it are protected under copyright by the Publisher (other than as may be noted herein).

Notices
Knowledge and best practice in this field are constantly changing. As new research and experience broaden our understanding, changes in research methods, professional practices, or medical treatment may become necessary.

Practitioners and researchers must always rely on their own experience and knowledge in evaluating and using any information, methods, compounds, or experiments described herein. In using such information or methods they should be mindful of their own safety and the safety of others, including parties for whom they have a professional responsibility.

To the fullest extent of the law, neither the Publisher nor the authors, contributors, or editors, assume any liability for any injury and/or damage to persons or property as a matter of products liability, negligence or otherwise, or from any use or operation of any methods, products, instructions, or ideas contained in the material herein.

ISBN: 978-0-12-815209-6
ISSN: 0065-2725

For information on all Academic Press publications
visit our website at https://www.elsevier.com/books-and-journals

Publisher: Zoe Kruze
Acquisition Editor: Jason Mitchell
Editorial Project Manager: Shellie Bryant
Production Project Manager: Vignesh Tamil
Cover Designer: Alan Studholme

Typeset by SPi Global, India

CONTENTS

Contributors vii
Preface ix

1. **The Preparation and Properties of Heteroarylazulenes and Hetero-Fused Azulenes** 1
   Taku Shoji and Shunji Ito

   1. Introduction 2
   2. Azulene Derivatives With 6-Membered Ring Heterocycles 4
   3. Azulene Derivatives With 5-Membered Ring Heterocycles 23
   4. Azulene Derivatives With Azoles and Benzoazoles 46
   5. Concluding Remarks 50
   References 51
   Further Reading 54

2. **Recent Developments in the Chemistry of Pyrazoles** 55
   Andrew W. Brown

   1. Introduction 56
   2. Synthesis 57
   3. Properties and Applications 93
   4. Concluding Remarks 101
   References 102
   Further Reading 107

3. **Recent Developments in the Chemistry of 1,2,3-Thiadiazoles** 109
   Yuri Shafran, Tatiana Glukhareva, Wim Dehaen, and Vasiliy Bakulev

   1. Introduction 110
   2. Synthesis of 1,2,3-Thiadiazoles 110
   3. Ring Transformations of 1,2,3-Thiadiazoles 144
   4. 1,2,3-Thiadiazoles in Medicine and Agriculture 162
   5. Concluding Remarks 163
   Acknowledgment 164
   References 164

4. **The Literature of Heterocyclic Chemistry, Part XVI, 2016**   173
Leonid I. Belen'kii and Yulia B. Evdokimenkova

1.	Introduction	174
2.	General Sources and Topics	175
3.	Three-Membered Rings	207
4.	Four-Membered Rings	208
5.	Five-Membered Rings	208
6.	Six-Membered Rings	215
7.	Rings With More Than Six Members	220
8.	Heterocycles Containing Unusual Heteroatoms	222
	Appendix	223
	References	224

Index   *255*

CONTRIBUTORS

Vasiliy Bakulev
Technology for Organic Synthesis, Ural Federal University, Ekaterinburg, Russia

Leonid I. Belen'kii
Russian Academy of Sciences, Moscow, Russian Federation

Andrew W. Brown
Sygnature Discovery Ltd, Biocity, Nottingham, United Kingdom

Wim Dehaen
Molecular Design and Synthesis, KU Leuven, Belgium

Yulia B. Evdokimenkova
Russian Academy of Sciences, Moscow, Russian Federation

Tatiana Glukhareva
Technology for Organic Synthesis, Ural Federal University, Ekaterinburg, Russia

Shunji Ito
Graduate School of Science and Technology, Hirosaki University, Hirosaki, Japan

Yuri Shafran
Technology for Organic Synthesis, Ural Federal University, Ekaterinburg, Russia

Taku Shoji
Graduate School of Science and Technology, Shinshu University, Matsumoto, Japan

PREFACE

The volume begins with a chapter entitled "The preparation and properties of heteroarylazulenes and hetero-fused azulenes" by Taku Shoji (Shinshu University, Japan) and Shunji Ito (Hirosaki University, Japan). Azulene derivatives have unique properties with potential applications in material sciences. This chapter describes recent developments in the synthesis of heteroarylazulenes and hetero-fused azulenes, together with an account of their reactivity and physical properties. "Recent developments in the chemistry of pyrazoles" are then reviewed by Andrew W. Brown (Sygnature Discovery Ltd., UK). Pyrazole derivatives continue to have wide-ranging applications not only in pharmaceuticals but also in areas such as metal organic frameworks, liquid crystals, and electroluminescence. The review covers advances in the synthesis and application of pyrazoles in the past two decades.

In Chapter 3 Yuri Shafran, Tatiana Glukhareva, and Vasiliy Bakulev (Ural Federal University, Russia) and Wim Dehaen (University of Leuven, Belgium) describe "Recent developments in the chemistry of 1,2,3-thiadiazoles." The chapter covers significant advances in well-known synthetic routes, such as a redox organocatalytic version of the Hurd–Mori reaction, together with new methods for the synthesis of 1,2,3-thiadiazoles based on oxidative heterocyclization of hydrazines. The discovery of new thermal transformations and rearrangements of 1,2,3-thiadiazoles is discussed. In Chapter 4 Leonid I. Belen'kii and Yulia B. Evdokimenkova present "The literature of heterocyclic chemistry, Part XVI, 2016." The chapter provides a systematic survey of reviews and monographs published in 2016 on all aspects of heterocyclic chemistry.

Eric Scriven and Christopher A. Ramsden
March 2018

CHAPTER ONE

The Preparation and Properties of Heteroarylazulenes and Hetero-Fused Azulenes

Taku Shoji*,[1], Shunji Ito[†,1]

*Graduate School of Science and Technology, Shinshu University, Matsumoto, Japan
†Graduate School of Science and Technology, Hirosaki University, Hirosaki, Japan
[1]Corresponding authors: e-mail address: tshoji@shinshu-u.ac.jp; itsnj@hirosaki-u.ac.jp

Contents

1. Introduction	2
2. Azulene Derivatives With 6-Membered Ring Heterocycles	4
2.1 Synthesis and Reactions of Pyridylazulene Derivatives and Related Compounds	4
2.2 Azulene Derivatives With Oxygen-Containing 6-Membered Heterocycles	20
3. Azulene Derivatives With 5-Membered Ring Heterocycles	23
3.1 Synthesis and Reactions of Thienylazulene Derivatives and Related Compounds	23
3.2 Synthesis and Reactions of Pyrrolylazulene Derivatives and Related Compounds	35
3.3 Synthesis and Reactions of Furylazulene Derivatives and Related Compounds	39
3.4 Synthesis and Reactions of Porphyrin- and BODIPY-Substituted Azulene Derivatives	43
4. Azulene Derivatives With Azoles and Benzoazoles	46
4.1 Synthesis and Reactions of 1-(Azolyl)- and 1-(Benzoazolyl)Azulenes and Related Compounds	46
4.2 Synthesis and Reactions of Other Azolyl- and Benzoazolylazulenes and Related Compounds	48
5. Concluding Remarks	50
References	51
Further Reading	54

Abstract

Azulene derivatives having heteroaryl substituents have been prepared, for the purpose of application to material sciences, by utilizing their unique properties. These derivatives have been synthesized by various methods including transition metal-catalyzed coupling, electrophilic substitution, and cyclization reactions. In this review, we describe recent developments in the synthesis of heteroarylazulenes and hetero-fused azulenes, as well as their reactivity and physical properties.

Keywords: Azulene, Heteroarylazulene, Cross-coupling reaction, Electrophilic substitution, Cyclization reaction

1. INTRODUCTION

Azulene (**1**) is a nonalternant 10π-electron nonbenzenoid aromatic hydrocarbon, which has a fused structure of five- and seven-membered rings. Although the structural isomer naphthalene is a colorless compound, azulene (**1**) shows a deep blue color. Therefore, its compound name originates from "azur" and "azul," which mean "blue" in Arabic and Spanish, respectively. Since the ionic cyclopentadienide and tropylium structures contribute to the azulene skeleton, electrophilic substitution reactions take place at the 1- and 3-positions, and nucleophilic addition occurs at the 4-, 6-, and 8-positions. The reason for the large dipole moment of 1.08 D and a blue color of azulene, despite the hydrocarbon structure, is explained by the intramolecular charge transfer (ICT) derived from the polarized resonance structure as shown in Scheme 1.

The first synthesis of azulene (**1**) was reported by Plattner and Pfau in 1937 (37HCA224). However, the product yield is low, because their method includes a troublesome dehydrogenation process of a hydroazulene derivative at high temperature in the last stage (Scheme 2). Therefore, this method is not a practical synthesis of azulene.

In the 1950s, efficient and practical methods for the synthesis of azulene (**1**) and its derivatives were discovered by Ziegler–Hafner and Nozoe et al. (55AG301, 04EJOC899). The Ziegler–Hafner's method, in which either pyridinium salt **2** or pyrylium salt **3** is used as a starting material in a reaction with cyclopentadienide ion (**4**$^-$), can be applied to a large-scale synthesis of azulene (**1**) and its alkyl and aryl derivatives in a short process (Scheme 3)

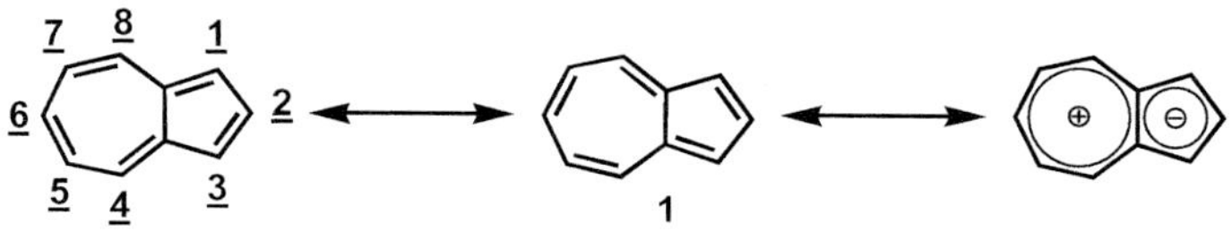

Scheme 1 Numbering and resonance structure of azulene (**1**).

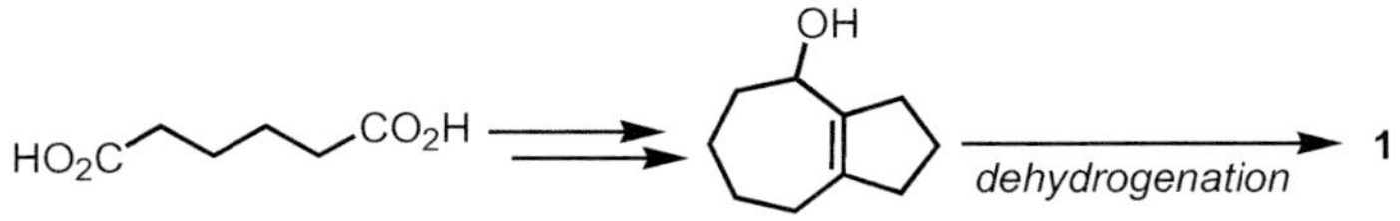

Scheme 2 Synthesis of azulene (**1**) reported by Plattner and Pfau.

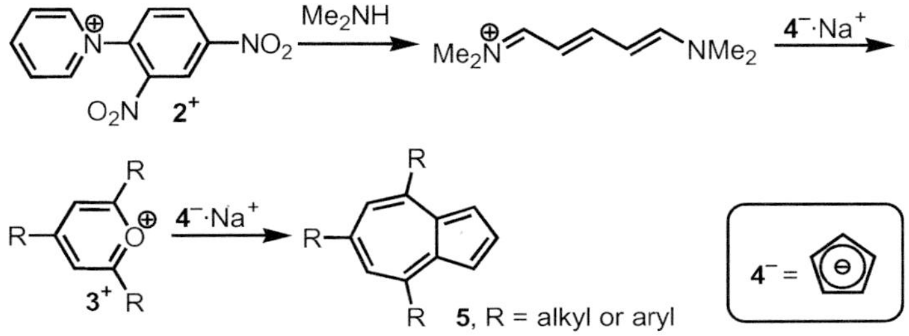

Scheme 3 Synthesis of azulene (**1**) reported by Ziegler and Hafner.

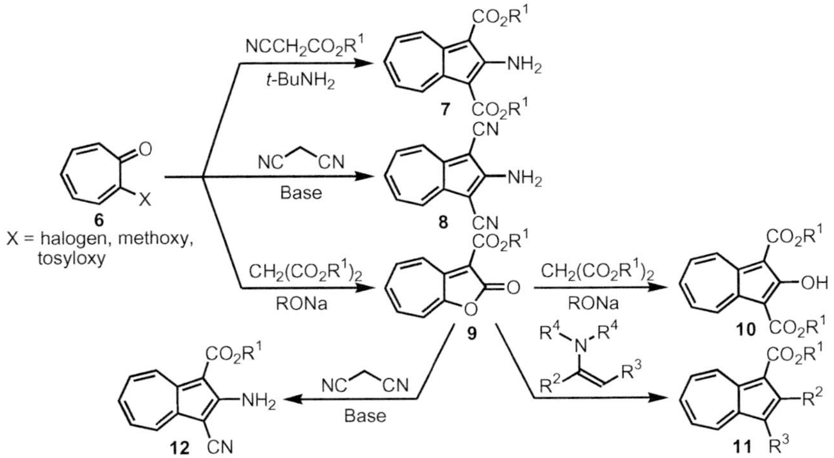

Scheme 4 Synthesis of azulene derivatives from tropolones.

(55AG301). Their synthetic method is highly effective for the preparation of azulene derivatives having the substituents on the seven-membered ring.

In the synthetic procedure reported by Nozoe et al., 2-amino- and 2-hydroxyazulene derivatives **7**, **8**, and **10** can be prepared in excellent yields by the reaction of tropone derivatives **6** having a leaving group (e.g., a halogen, methoxy, or tosyloxy group) at the 2-position with an active methylene reagent (e.g., cyanoacetate ester, malonic acid ester, and malononitrile) in the presence of a base (Scheme 4) (04EJOC899). This method is potentially useful for the preparation of azulene derivatives having a functional group, such as amino and hydroxy groups, at the 2-position.

Yasunami and Takase developed a synthetic method for 1-alkylazulene derivatives **11** by the reaction of 2*H*-cyclohepta[*b*]furan-2-one derivatives **9** with enamines (Scheme 4) (81JSOCJ1172). This procedure is very effective for the preparation of azulene derivatives bearing alkyl groups or a fused-ring structure at the five-membered ring. 2*H*-Cyclohepta[*b*]furan-2-one derivatives **9** are also utilized in a reaction with malononitrile to afford the 2-amino-1-cyanoazulenes **12** in excellent yields (71T3357, 12H305).

As well as the synthesis of the large number of azulene derivatives, heteroarylazulenes have been prepared for the purpose of application to material sciences, such as organic light-emitting diodes (OLEDs), organic field-effect transistors (OFETs), and solar cells, by utilizing their unique optical and electrochemical properties. Therefore, many researchers have studied the synthesis, reactivity, and physical properties of these derivatives as described in the later sections (11SL2279). However, despite the recent progress in the synthetic methods of heteroarylazulenes and their related compounds, there are no reviews that systematically summarize the synthesis and reactivity of these compounds so far.

In this review, we describe the developments in the synthesis of heteroarylazulenes and hetero-fused azulene derivatives, as well as their reactivity and physical properties.

2. AZULENE DERIVATIVES WITH 6-MEMBERED RING HETEROCYCLES

2.1 Synthesis and Reactions of Pyridylazulene Derivatives and Related Compounds

2.1.1 1-(Pyridyl)-, 1-(quinolyl)-, and 1-(isoquinolyl)azulene Derivatives

In 1996, Yasunami et al. discovered a method leading to 1-(dichloropyridyl)azulenes during an investigation into the fluorination of azulene (**1**) and its derivatives (Scheme 5) (96BCJ1645). By this investigation, the 1-(dichloropyridyl)azulene derivatives **13**, **14**, and **15** were obtained by the reaction of azulene (**1**) with *N*-fluoro-3,5-dichloropyridinium triflate, which is known as a fluorination reagent; the expected 1-fluoroazulenes were not obtained. The formation of these compounds is considered to be the consequence of the decomposition of a charge transfer (CT) complex between azulene and the pyridinium salt followed by a dehydrofluorination reaction to furnish the substitution products.

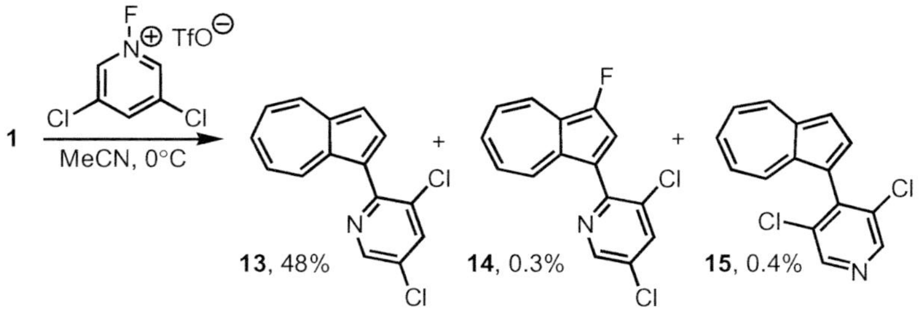

Scheme 5 Reaction of azulene (**1**) with N-fluoro-3,5-dichloropyridinium triflate.

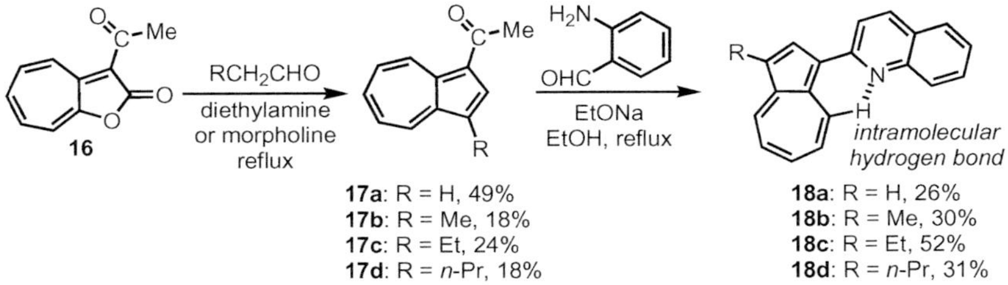

Scheme 6 Synthesis of 1-(2-quinolyl)azulenes **18a–18d** by Friedländer reaction of **17a–17d**.

In 1996, Imafuku and coworkers reported the synthesis of 1-(2-quinolyl) azulenes **18a–18d** by Friedländer reaction of 1-acetylazulenes **17a–17d** with o-aminobenzaldehyde (Scheme 6) (96JHC841). This reaction tolerates substituents on both the azulene and the o-aminobenzaldehyde moieties, and the 1-(2-quinolyl)azulenes **18a–18d** have been produced in low to moderate yields (26%–52%). In the ^{1}H NMR spectra, the proton signals of 1-(2-quinolyl)azulenes **18a–18d** at the 8-position of the azulene moiety show a remarkable down-field shift ($\delta_H = 9.80$–9.97 ppm), compared with that of the parent azulene (**1**). The down-field shift is considered to be due to intramolecular hydrogen bonding between the nitrogen atom of quinoline and the ring proton of the azulene as represented in Scheme 6.

Imafuku et al. have also demonstrated the preparation of 1-(4-pyridyl- and 3-pyridyl)azulene derivatives **22a–22d**, **23a**, and **23b** with aryl, cyano, and methoxy groups on the pyridine moiety (Scheme 7) (00JHC1019). Chalcone analogues **20a–20d**, **21a**, and **21b** with a 1-azulenyl group, which are prepared by aldol condensation of 1-formyl- and 1-acetylazulenes **19a** and **19b** with aryl ketones and aldehydes, react with malononitrile in the presence of MeONa to afford the 1-(4-pyridyl- and 3-pyridyl)azulenes **22a–22d**, **23a**, and **23b** in moderate to good yields. The reaction takes

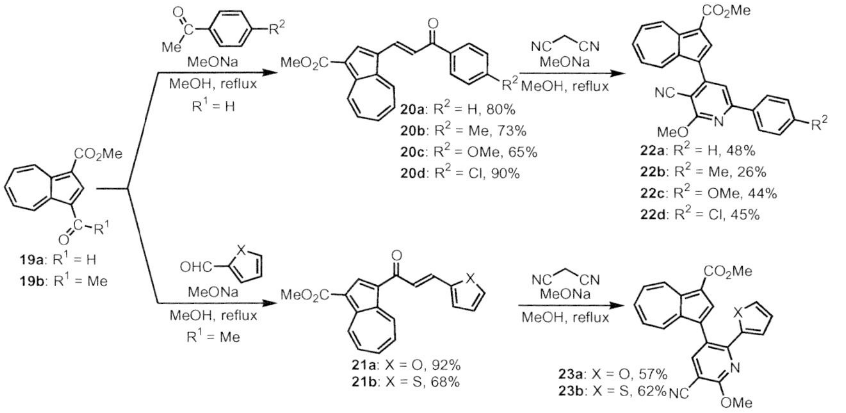

Scheme 7 Synthesis of 1-(4-pyridyl- and 3-pyridyl)azulenes from chalcone analogues **20a–20d**, **21a**, and **21b**.

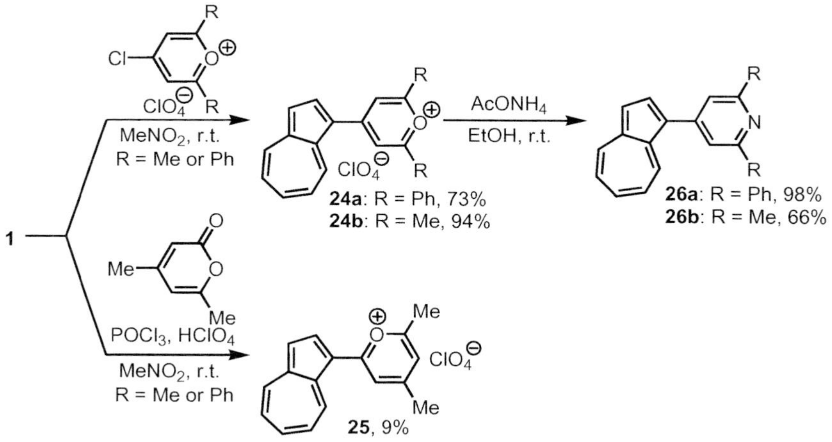

Scheme 8 Synthesis of 1-(2,6-diphenyl- and 2,6-dimethyl-4-pyridyl)azulene derivatives **26a** and **26b** from (1-azulenyl)pyrylium salts **24a** and **24b**.

place via Michael-type 1,4-addition of malononitrile to the chalcone analogue in the presence of a base to give a 2-amino-3-cyano-4H-pyran intermediate, followed by the ring opening of the 4H-pyran to result into the pyridine ring.

Preparation of 1-(2,6-diphenyl- and 2,6-dimethyl-4-pyridyl)azulene derivatives **26a** and **26b** has been demonstrated by Razus et al. via (1-azulenyl)pyrylium salts **24a** and **24b** as synthetic intermediates (Scheme 8) (07JHC245). The pyrylium salts **24a** and **24b** with 1-azulenyl group are readily prepared by the reaction of azulene (**1**) with 4-chloropyrylium salts, which can be generated in situ by the reaction of the corresponding 4-pyranones with PCl$_5$ or POCl$_3$ in the presence of HClO$_4$. 4,6-Dimethyl-2-pyranone also reacts with azulene (**1**) to afford the corresponding pyrylium salt **25** in low yield (9%), although this pyrylium salt has not been examined for transformation to the corresponding pyridyl derivative (Scheme 8). The (1-azulenyl)pyrylium salts **24a** and **24b** are converted into the corresponding 1-(4-pyridyl)azulenes **26a** and **26b** by treatment with AcONH$_4$. 1-[2,6-Bis(heteroaryl)- and 2,6-bis(heteroarylvinyl)-4-pyridyl] azulenes can also be synthesized by the same method (17T2488).

As shown in Scheme 9, 1-(4-pyridyl)azulenes without any substituents on their pyridine ring have been synthesized by several methods. 1-(4-Pyridyl) azulene (**28a**) is prepared by the electrophilic substitution reaction of azulene (**1**) with triflate of pyridinium salt, which is readily prepared by the reaction of pyridine with trifluoromethanesulfonic anhydride (Tf$_2$O), to give 1-(*N*-trifluoromethanesulfonyl-4-dihydropyridyl)azulene, followed by aromatization with KOH. The yield is 71% in two steps from azulene (**1**) (07TL1099). This procedure is also applicable to other N-containing heterocycles, such as quinoline, isoquinoline, acridine, and 1,10-phenanthroline, to afford the corresponding 1-(heteroaryl)azulenes. Furthermore, the two-step reaction of 1-methylthioazulene derivatives also takes place to give the

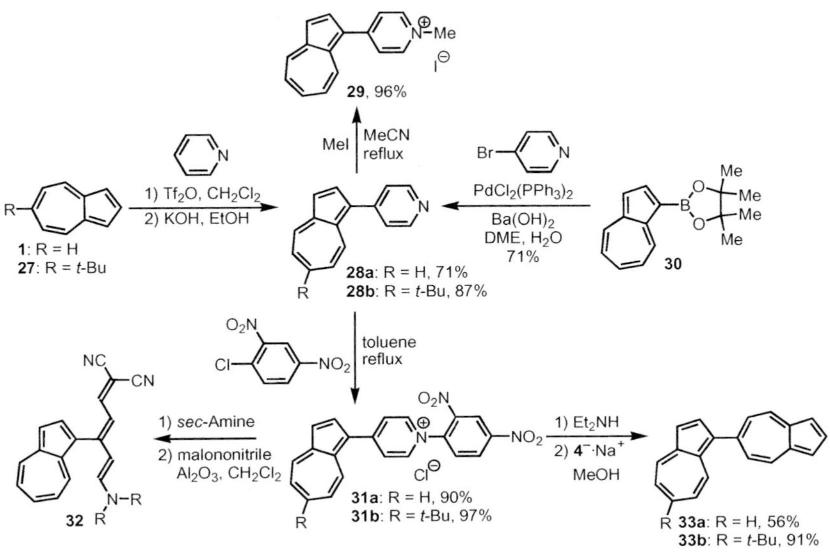

Scheme 9 Synthesis and reaction of 1-(4-pyridyl)azulenes **28a** and **28b**.

corresponding 1-methylthio-3-(heteroaryl)azulenes, which show a reversible oxidation wave on their cyclic voltammograms (CVs), in good to moderate yields (08EJO5823). In a similar manner to the reaction of azulene (**1**), the reaction of 6-*tert*-butylazulene (**27**) in the presence of a large excess of pyridine and Tf$_2$O followed by treatment with KOH affords the corresponding 1-(4-pyridyl)azulene **28b** in high yield (87%).

Wakabayashi and coworkers have also reported the preparation of 1-(2-, 3-, and 4-pyridyl)azulenes in good yields by Suzuki–Miyaura cross-coupling reactions of 1-azulenylboronic acid ester **30** with 2-, 3-, and 4-bromopyridines (Scheme 9) (07JOC744). They also demonstrated the color dependence of these compounds on the addition of metal ions and acid.

The pyridine moiety of 1-(4-pyridyl)azulenes **28a** and **28b** reacts with iodomethane or 2,4-dinitro-chlorobenzene to afford the corresponding pyridinium salts **29**, and **31a** and **31b**, respectively (Scheme 9) (11EJO5311, 12CL1644). Pyridinium salt **31a** is convertible to the donor–acceptor-type polymethines **32** with a 1-azulenyl substituent by treatment with secondary amines, e.g., diethylamine, pyrrolidine, morpholine, and piperidine, followed by the Al$_2$O$_3$-catalyed Knoevenagel condensation with malononitrile. The polymethines **32** exhibit intense ICT absorption bands in the visible region in their UV/vis spectra, along with the solvatochromic features by changing the solvent polarity. For instance, the polymethine **32** with a diethylamino substituent (R = Et) shows an ICT absorption band at λ_{max} 505 nm in CH$_2$Cl$_2$. On the other hand, the longest wavelength absorption maximum of **32** (R = Et) exhibits a hypsochromic shift to λ_{max} 470 nm in 10% CH$_2$Cl$_2$/*n*-hexane mixed solvent.

1,6′-Biazulene derivatives **33a** and **33b** can also be obtained from the pyridinium salts **31a** and **31b** by Ziegler–Hafner's method (Scheme 9) (13CL638). The reaction of **31a** and **31b** with diethylamine affords Zincke-type intermediates, which are treated with cyclopentadiene (CPD) in the presence of MeONa as a base, to yield the 1,6′-biazulenes **33a** and **33b**. The ICT between the two azulene rings of the 1,6′-biazulenes **33a** and **33b** is revealed by UV/vis spectra and computational chemistry using density functional theory at the B3LYP/6-31G** level.

Dependence of the amount of pyridine employed for the reaction with the pyridinium salt has been examined by utilizing 1,6-di-*tert*-butylazulene (**34**) to give a variety of products (Scheme 10) (09EJO1554). When excess pyridine (5 equiv.) compared with Tf$_2$O is used in this reaction, a 1-(dihydropyrid-4-yl)azulene intermediate is obtained as the main product, which is readily converted into the corresponding 1-(4-pyridyl)azulene **35**

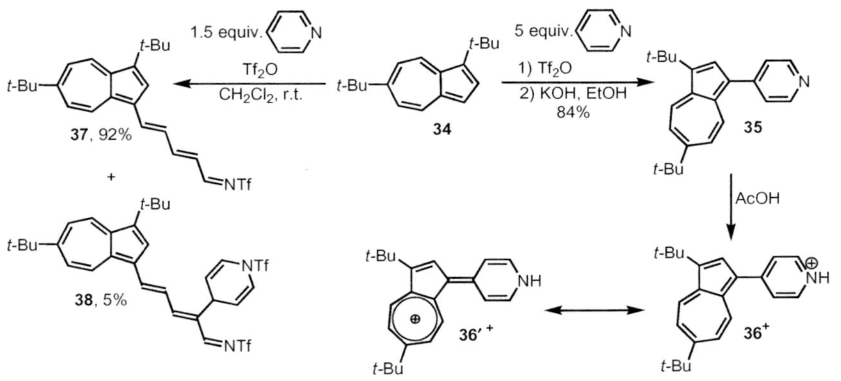

Scheme 10 Reaction of 1,6-di-*tert*-butylazulene (**34**) with pyridinium salt.

by the treatment with KOH. In the UV/vis spectra, the compound **35** shows a weak absorption band at λ_{max} 594 nm in CH_2Cl_2, whereas the strong absorption band is observed at λ_{max} 438 nm in acetic acid due to the protonation of the pyridine moiety to produce the cationic species **36$^+$**. On the other hand, when pyridine (1.5 equiv.) and Tf_2O are reacted with **34** using substantially the same amount, a polymethine **37** is obtained as a main product, along with **38** as a by-product. Formation of the polymethine **37** is explained by an electrophilic substitution reaction of the azulene **34** with the pyridinium salt at its 2-position, followed by the ring opening of the pyridine ring.

Oda et al. demonstrated the stepwise preparation of 1-(2-pyridyl)azulene (**42**), as well as 1,3-di(2-pyridyl)azulene (**43**) (Scheme 11) (99S1349, 07H1413). Nazarov cyclization of the substituted ethyl 3-(cyclohepta-1,3,5-trien-1-yl)-2-methylene-3-oxopropionate (**40**), which is prepared from 1-acetylcyclohepta-1,3,5-triene (**39**) in two steps, gives 1,2,3,8-tetrahydroazulen-1-one **41** with a 2-pyridyl group at the 3-position. Reduction of **41** with $NaBH_4$ followed by aromatization affords 1-(2-pyridyl)azulene (**42**) in 45% yield. Nucleophilic addition of 2-pyridyllithium to ketone **41** and a subsequent dehydration/oxidation sequence with palladium–carbon in refluxing diphenylether provide 1,3-di(2-pyridyl)azulene (**43**) in 30% yield. The same compound **43** is also obtained by a palladium-catalyzed Negishi coupling reaction of 1,3-diiodoazulene (**44a**) with 2-pyridylzinc bromide in 85% yield (Scheme 11) (07H1413). The reaction of 1,3-dibromoazulene (**44b**) with 2-(tri-*n*-butylstannyl)pyridine in the presence of CuO also gives product **43** in 66% yield under Stille coupling conditions, but in lower yield than that under the Negishi coupling conditions. The synthesis of 1-(pyridyl)azulenes by

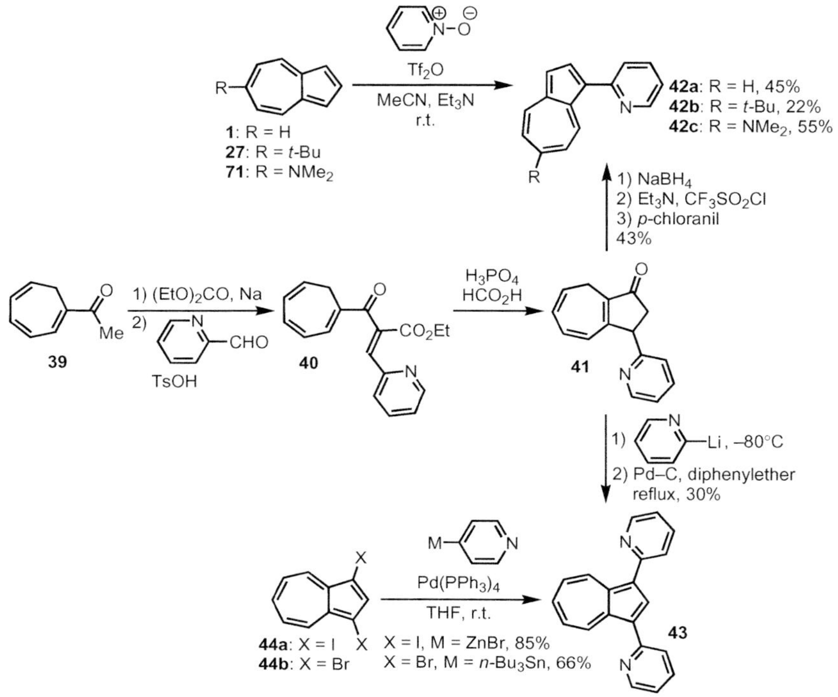

Scheme 11 Synthesis of 1-(2-pyridyl)azulenes **42a–42c** and 1,3-di(2-pyridyl)azulenes **43**.

a Negishi coupling reaction of 1-azulenylzinc reagent with aryl halides or 1-haloazulenes with arylzinc reagents was also reported by Bredihhin et al. (15S2663, 15S538).

1-(2-Pyridyl)azulene (**42a**) and its derivatives **42b** and **42c** can be prepared in 22%–55% yields by Reissert–Henze-type reactions of the corresponding azulene derivatives with pyridine N-oxide in the presence of Tf$_2$O (Scheme 11) (10TL5127). This approach is also applicable to substituted pyridine N-oxides and quinoline N-oxide, and these N-oxides react with azulene (**1**) in the similar manner to produce the corresponding 1-(2-pyridyl)- and 2-quinolyl)azulenes. The unsymmetrically substituted product, 1-(2-pyridyl)-3-(4-pyridyl)azulene, has been obtained in 55% overall yield by the reaction of compound **42a** with pyridine/Tf$_2$O, followed by the base-induced aromatization.

The 1,3-di(4-pyridyl)azulenes **45a** and **45b** are available by a similar synthetic procedure as for **28a** and **28b** (Scheme 12) (07TL1099). The reactions of **1** and **27** with excess pyridine are carried out by using excess Tf$_2$O,

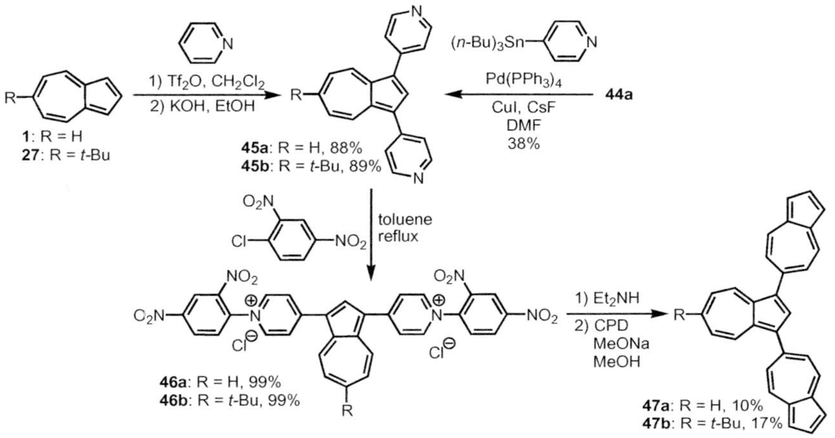

Scheme 12 Synthesis of 1,3-di(4-pyridyl)azulenes **45a** and **45b** and conversion to 1,6′:3,6″-terazulenes **47a** and **47b**.

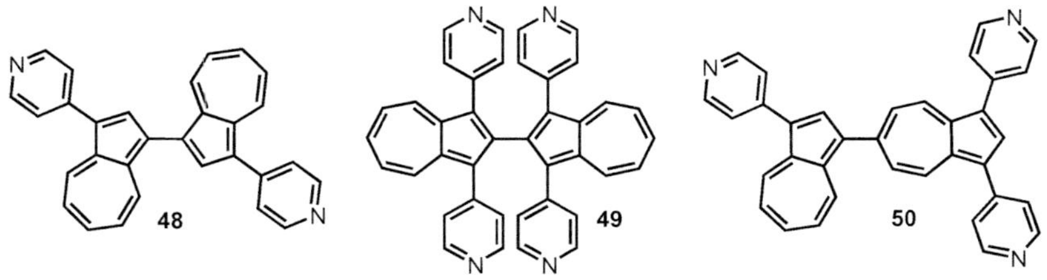

Fig. 1 Structures of di, tri, and tetra(4-pyridyl)biazulenes **48–50**.

followed by base-induced aromatization to afford **45a** (88%) and **45b** (89%), respectively. Tri(4-pyridyl)- and tetra(4-pyridyl)biazulenes **48–50** can also be synthesized by similar reactions (Fig. 1).

Stille coupling of 1,3-diiodoazulene (**44a**) with 4-(tri-n-butylstannyl)pyridine also generates the 1,3-di(4-pyridyl)azulene **45a**, as well as 1,3-di(2-pyridyl- and 3-pyridyl)azulenes, from the corresponding (tri-n-butylstannyl) pyridines (Scheme 12) (07H1413). 1,3-Di(4-pyridyl)azulenes **45a** and **45b** are convertible to 1,6′:3,6″-terazulenes **47a** and **47b** by the Ziegler–Hafner's method, in a similar manner to the preparation of 1,6′-biazulenes **33a** and **33b** (Scheme 9). 1,6′:3,6″-Terazulenes **47a** and **47b** exhibit an anomalous ICT absorption band at around λ_{max} 430 nm, which is attributable to the transition from 1-azulenyl to 6-azulenyl groups. 1,3-Di(4-pyridyl)azulene (**45a**) undergoes coordination to a cobalt salt to give the corresponding cobalt complexes, which acts as a single-ion magnet (15IC16).

As mentioned earlier, 1-(pyridyl)azulenes can be synthesized by palladium-catalyzed cross-coupling reactions, but unstable 1-haloazulenes must be used as a starting material in these reactions. Lewis and colleagues demonstrated the synthesis of 1-(2-pyridyl)- and 1-(3-quinolyl)azulenes **52** and **53** by the cross-coupling reaction of 1-azulenylsulfonium salt **51** instead of 1-haloazulenes (Scheme 13) (16AGE2564). By their method, Suzuki–Miyaura cross-coupling reactions of the 1-azulenylsulfonium salt **51** with boronic acid or boronic acid esters of heterocycles in the presence of a palladium catalyst give the corresponding 1-(heteroaryl)azulenes. The salt **51** can be prepared by the reaction of azulene (**1**) with tetramethylene sulfoxide in the presence of trifluoroacetic anhydride.

Preparation of 2,2′:6′,2″-terpyridine-substituted azulene derivatives under Kröhnke-type reaction condition was reported by Nica et al. (Scheme 14) (16BJOC1812). The Claisen–Schmidt reaction of 1-formylazulene (**54**)

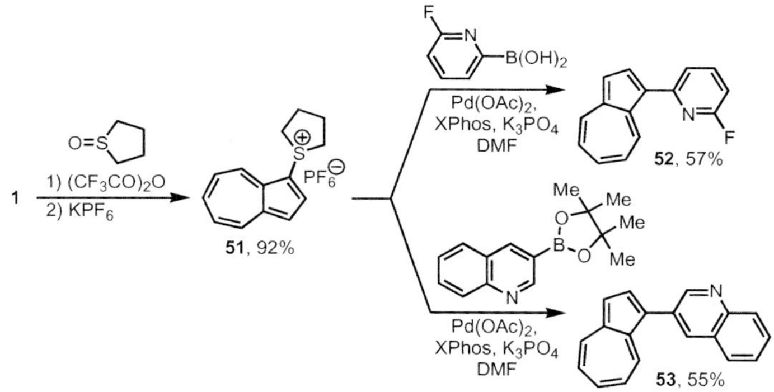

Scheme 13 Synthesis of 1-(2-pyridyl)- and 1-(3-quinolyl)azulenes **52** and **53** by cross-coupling reaction of 1-azulenylsulfonium salt **51** with boronic acids.

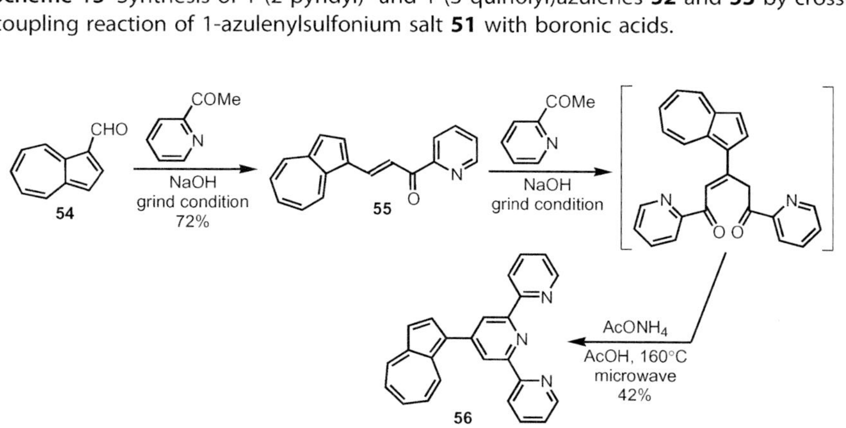

Scheme 14 Synthesis of 2,2′:6′,2″-terpyridine-substituted azulene derivative **56** by Kröhnke-type reaction.

with 2-acetylpyridine under the "grind" conditions yields the corresponding azulene-substituted chalcone **55** (72%), which reacts with further 2-acetylpyridine under grind conditions, followed by microwave irradiation with excess ammonium acetate in acetic acid to give 4-(1-azulenyl)terpyridine **56** in 42% yield. Azulene-substituted chalcone **55** can also be obtained by stirring in ethanol solution or by microwave irradiation in almost the same yields as that under the grind conditions. The molecular structure of 4-(1-azulenyl) terpyridine **56** has been revealed by single-crystal X-ray structural analysis (16BJOC1812). Furthermore, compound **56** shows dual fluorescence emission at λ_{FL} 435 and 530 nm in CH_2Cl_2 when excited at λ_{EX} 375 nm.

2.1.2 2-(Pyridyl)-, 2-(quinolyl)-, and 2-(isoquinolyl)azulene Derivatives

The literature on the synthesis of 2-(pyridyl)azulenes and their derivatives is not as extensive as that of 1-(pyridyl)azulene derivatives. As one of the reasons, introduction of the substituent by electrophilic substitution reactions is difficult at the 2-position of azulenes, unlike that at the 1- or 3-position. Furthermore, the synthesis of 2-haloazulenes, which have become good precursors for cross-coupling reactions, is not straightforward, even though 1-haloazulene derivatives can be easily prepared by reaction with N-halosuccinimides (53JACS4980). For example, the synthesis of 2-iodoazulene (**58**) is achieved by a halogen exchange reaction of 2-chloroazulene (**57a**), which can be produced by the Sandmeyer reaction of 2-aminoazulene with a large excess potassium iodide (Scheme 15) (62BCJ1990).

Synthesis of 2-(pyridyl)- and 2-(quinolyl)azulene derivatives is accomplished by the cross-coupling reaction of 2-iodoazulene **58** with lithium heterocyclic magnesium ate complexes, prepared in situ in the presence of $PdCl_2(PPh_3)_2$ as a catalyst (Scheme 15) (05H91). This method is also applicable to the synthesis of 6-heteroarylazulenes, but details of the synthesis of these derivatives are described in Section 2.1.4. The cross-coupled 2-(pyridyl)- and 2-(quinolyl)azulene derivatives are converted into 1,3-di (4-pyridyl)-2-(pyridyl)- and -2-(quinolyl)azulenes by reaction with pyridine in the presence of Tf_2O and subsequent aromatization with KOH (14H2588). When the ^1H NMR spectrum of compound **60** was measured in CF_3CO_2D, not only the chemical shifts of the pyridine moieties but also the ring proton signals on the azulene moiety exhibited remarkable downfield shifts. This phenomenon can be explained by the protonation of the three pyridyl groups to produce the tricationic species **61**$^{3+}$ and the seven-membered ring of the azulene moiety takes on cationic character by the contribution of resonance structure **61'**$^{3+}$.

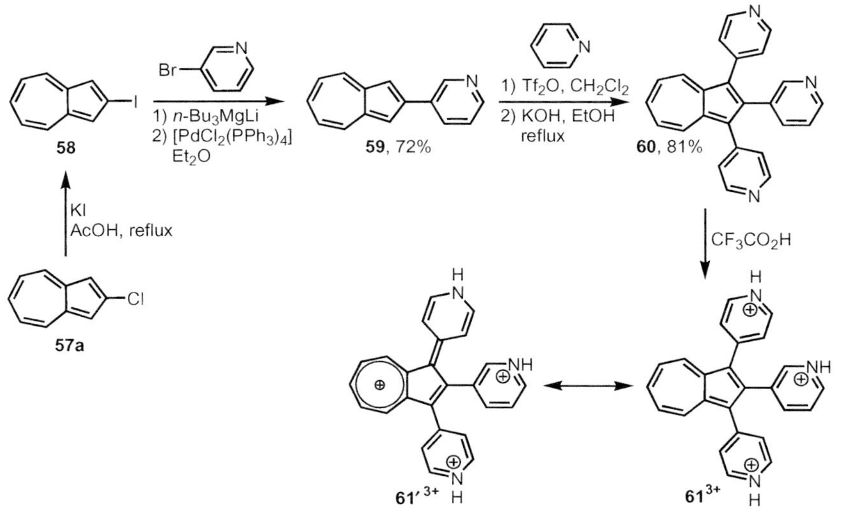

Scheme 15 Synthesis of 2-(3-pyridyl)azulene **59** by cross-coupling reaction with lithium magnesium ate complex.

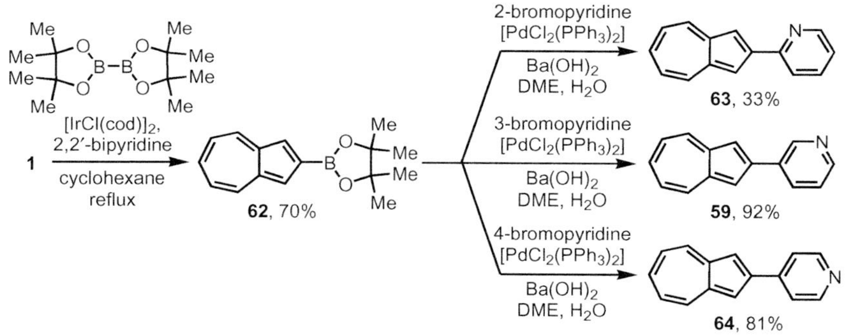

Scheme 16 Synthesis of 2-(pyridyl)azulenes **59**, **63**, and **64** by Suzuki–Miyaura coupling of **62**.

The synthesis of a series of 2-(pyridyl)azulenes **59**, **63**, and **64** has been achieved by Wakabayashi et al. by utilizing the Suzuki–Miyaura cross-coupling between 2-azulenylboronic acid pinacol ester (**62**) and the corresponding bromopyridines (Scheme 16) (07JOC744). Compound **62** was synthesized by the Ir-catalyzed direct borylation reaction of azulene in 70% yield (02AGE3056, 03EJO3663). Palladium-catalyzed cross-coupling reaction of **62** with bromopyridines affords 2-(2-, 3-, and 4-pyridyl)azulene derivatives (**63**: 33%, **59**: 92%, and **64**: 81% yields).

These 2-(pyridyl)azulenes **59**, **63**, and **64** retain their original blue color in solution by the addition of trifluoroacetic acid, although 1-(pyridyl) azulenes show the remarkable change in their colors from blue to red in their solution.

Fujimori et al. have reported the synthesis of pyrazines, quinoxalines, and their analogues substituted by two 2-azulenyl groups in good to excellent yields (54%–97%) (Scheme 17) (09H1079). Preparation of the precursor 2-formylazulene (**67**) is achieved by the reaction of lactone **65** with an enamine, followed by the hydration with aq. HCl (16TL4514). The diketone derivative **68**, which is prepared by benzoin condensation of aldehyde **67** and subsequent oxidation of the initially formed α-hydroxyketone with MnO_2, undergoes condensation with o-phenylenediamine to form the quinoxaline derivative **69** with two 2-azulenyl substituents. A Scholl-type oxidative dehydrogenation reaction of the quinoxaline **69** using MnO_2 gives the polycyclic aromatic compound **70** with two fused azulene rings.

There is only one example of the synthesis of 2-heteroarylazulenes via an electrophilic substitution reaction reported to date. The reaction of 6-dimethylamino-1,3-bis(methylthio)azulene (**72**), which is derived from 6-dimethylaminoazulene (**71**), with pyridine in the presence of Tf_2O, followed by treatment with a base gives the corresponding 2-(4-pyridyl) azulene **73** in moderate yield (Scheme 18) (12H35, 14BCJ141). In this series of reactions, quinoline, acridine, and 1,10-phenanthroline can be used in place of pyridine to give the corresponding 2-(heteroaryl)azulene derivatives, but in the case of the reaction using isoquinoline, the aromatization

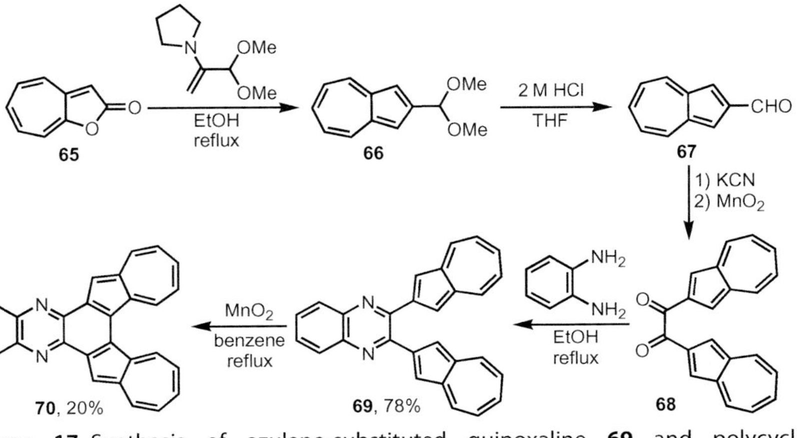

Scheme 17 Synthesis of azulene-substituted quinoxaline **69** and polycyclic derivative **70**.

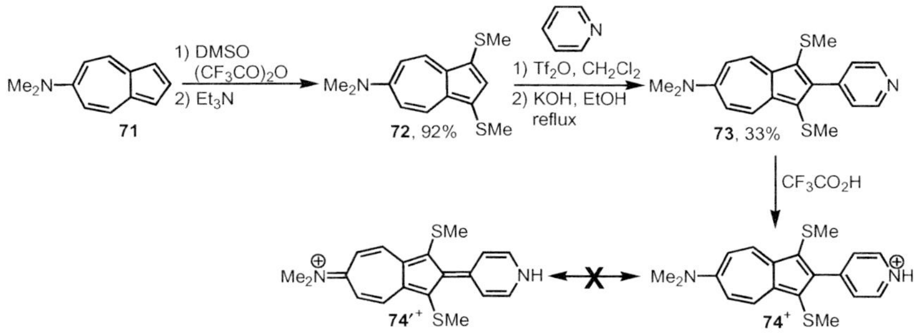

Scheme 18 Synthesis of 2-(4-pyridyl)azulene **73** by reaction of **72** with pyridinium salt.

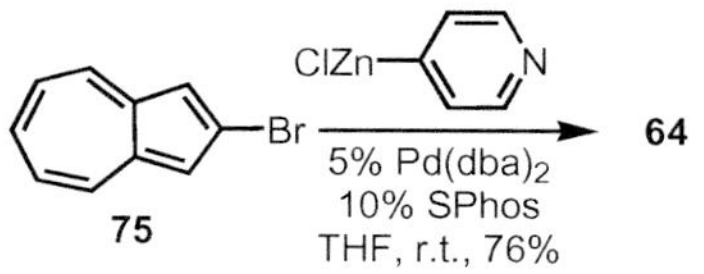

Scheme 19 Synthesis of 2-(4-pyridyl)azulene (**64**) by Negishi coupling reaction.

does not proceed to give the corresponding 2-(1-isoquinolyl)azulene derivative. The chemical shift of the proton signals of product **73** in CF_3CO_2D exhibits a down-field shift in the pyridine moiety. On the other hand, the chemical shifts of the proton signals of **73** on the azulene ring display almost the same values as those in $CDCl_3$, unlike those of 1,3-di(4-pyridyl)-2-(3-pyridyl)azulene (**61**). These results reflect the small contribution of the quinoid structure **74'$^+$** resulting from the protonation of the pyridine moiety.

Bredihhin et al. have achieved the synthesis of heteroarylazulenes by Negishi coupling in two ways. One of them is the reaction of the heterocyclic zinc reagent with bromoazulenes. For example, Negishi coupling between 2-bromoazulene (**75**) and 4-pyridylzinc reagent at room temperature gives 2-(4-pyridyl)azulene (**64**) in 76% yield (Scheme 19) (15S538).

The other method is the cross-coupling between the azulenylzinc reagent and the halogenated heterocyclic compound (Scheme 20) (15S2663). The azulenylzinc reagent **78** is synthesized by the halogen–metal exchange reaction of 1,3-dichloro-2-iodoazulene (**77**), which is prepared from 1,3-dichloroazulene (**76**) with turbo Grignard reagent (*i*-PrMgCl·LiCl), followed by transmetalation with zinc chloride. The reagent **78** is the first successful example of the preparation of an azulenylzinc reagent. The 2-azulenylzinc reagent **78** reacts with 3-bromo-5-methoxycarbonylpyridine at room temperature to yield the corresponding 2-(3-pyridyl)azulene derivative **79** in

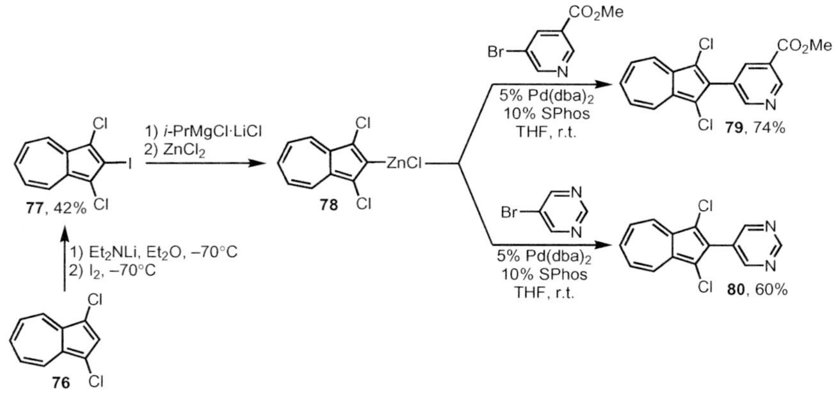

Scheme 20 Synthesis of 2-(3-pyridyl and pyrimidinyl)azulenes **79** and **80** by Negishi coupling reaction.

74% yield. In the presence of a palladium catalyst, the cross-coupling between 3-bromopyrimidine and **78** under the same condition gives the desired pyrimidine derivative **80** bearing a 2-azulenyl moiety in 60% yield.

2.1.3 5-(Pyridyl)-, 5-(quinolyl)-, and 5-(isoquinolyl)azulene Derivatives

Although the 5- and/or 7-positions of azulene have the next highest reactivity toward the electrophilic substitution to that for the 1- and/or 3-positions, there are few examples of reactions at these positions. Thus, the substituents at the 5- and 7-positions of azulene are usually introduced when the azulene ring is formed.

In 1976, Houk and coworkers found that the [6 + 4] cycloaddition of 6-phenylfulvene with diethylaminobutadiene forms the hydroazulene derivative, which can be converted to 4-phenylazulene after oxidation with chloranil (76JA7095). Inspired by Houk's results, Messmer et al. accomplished the synthesis of 5-(2-quinolyl)azulene (**83**) by a modified approach (Scheme 21) (88M1113). The [6 + 4] cycloaddition of fulvene **81** bearing p-nitrobenzoyloxy group with quinoline-substituted dienamine **82** generates the desired product **83** in low yield (12%).

In the early ages of azulene chemistry, electrophilic substitution reactions (i.e., Friedel–Crafts and Vilsmeier reactions) of 1,3-dialkylazulenes were reported by Hafner and Moritz (62LA40). In these reactions the main products are *ipso*-substituted compounds at the 1- and/or 3-positions, although substitution at the 5- and/or 7-positions of azulene takes place slightly.

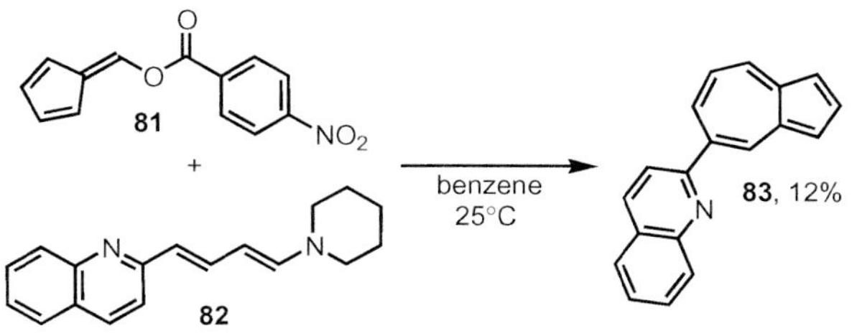

Scheme 21 Synthesis of 5-(2-quinolyl)azulene **83** by the [6+4] cycloaddition of fulvene **81** with dienamine **82**.

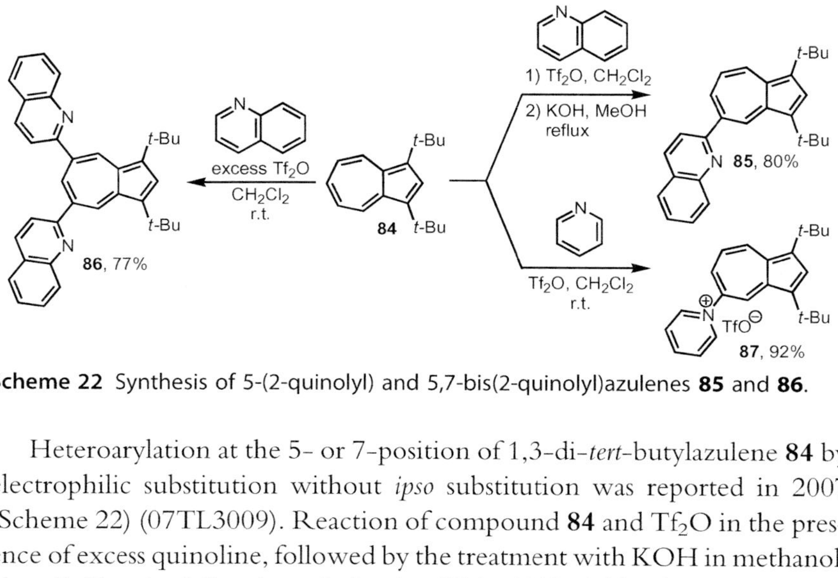

Scheme 22 Synthesis of 5-(2-quinolyl) and 5,7-bis(2-quinolyl)azulenes **85** and **86**.

Heteroarylation at the 5- or 7-position of 1,3-di-*tert*-butylazulene **84** by electrophilic substitution without *ipso* substitution was reported in 2007 (Scheme 22) (07TL3009). Reaction of compound **84** and Tf$_2$O in the presence of excess quinoline, followed by the treatment with KOH in methanol, gives 5-(2-quinolyl)azulene derivative **85** in 80% yield. The use of excess Tf$_2$O leads to further substitution to generate the 5,7-bis(2-quinolyl)azulene (**86**) in 77% yield (10EJO1059). Similar reactions can be applied with isoquinoline, acridine, 1,10-phenanthroline, and benzothiazole, but *N*-(5-azulenyl)pyridinium salt **87** is formed when the pyridine is used in this reaction.

2.1.4 Synthesis and Reactivity of 6-(pyridyl)-, 6-(quinolyl)-, and 6-(isoquinolyl)azulene Derivatives

Although there are few examples of the preparation of 6-(pyridyl)azulene derivatives, some methods including cross-coupling using 6-bromoazulenes have been reported in the literature.

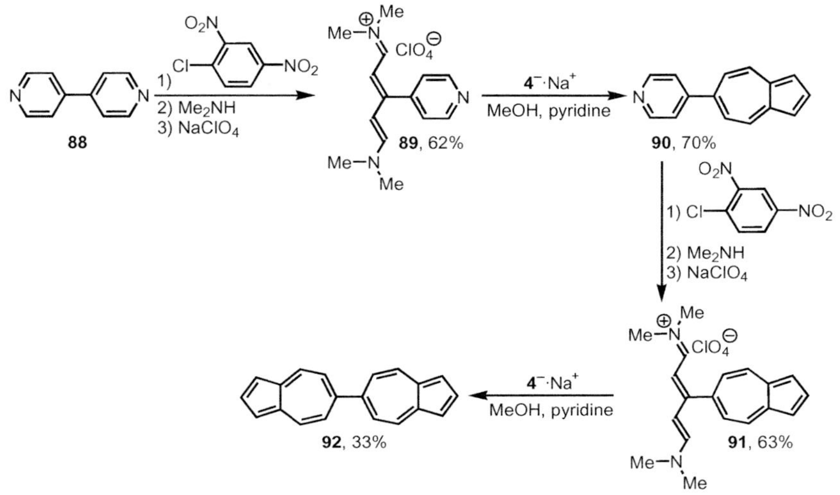

Scheme 23 Synthesis of 6-(4-pyridyl)azulene **90** and conversion to 6,6′-biazulene **92**.

In 1980, Hanke and Jutz successfully synthesized 6-(4-pyridyl)azulene by Ziegler–Hafner's method using 4,4′-bipyridine (**88**) as starting material (Scheme 23) (80S31). In this method, compound **88** is converted to a pyridinium salt by reaction with 1-chloro-2,4-dinitrobenzene and then treated with dimethylamine to give the Zincke salt **89**. The salt **89** reacts with cyclopentadienide ion (**4**), produced by the reaction of CPD and MeONa, to give 6-(4-pyridyl)azulene (**90**) in 70% yield. This method is one of the few examples for the synthesis of 6-(pyridyl)azulene derivatives, which are not accessible by cross-coupling reactions. 6-(4-Pyridyl)azulene (**90**) is transformed into 6,6′-biazulene (**92**) via the Zincke salt **91** by repeating the same reactions.

Murafuji and coworkers reported the synthesis of a 2,6-di(6-azulenyl)pyridine derivative by Suzuki–Miyaura cross-coupling (Scheme 24) (02S1013). They prepared the precursor 6-azulenylboronic acid pinacol ester **94** in 73% yield by a Miyaura–Ishiyama borylation reaction of the 2-amino-6-bromoazulene derivative **93** with bis(pinacolato)diboron (95JOC7508). Suzuki–Miyaura cross-coupling reaction of diester **94** with 2,6-dibromopyridine gives 2,6-di(6-azulenyl)pyridine **95** in 85% yield. The 6,6′-biazulene derivative **96** can also be synthesized in 24% yield by a similar cross-coupling reaction.

In a manner similar to the preparation of 2-(pyridyl- and quinolyl)azulene derivatives, palladium-catalyzed cross-coupling reaction of 6-bromoazulene

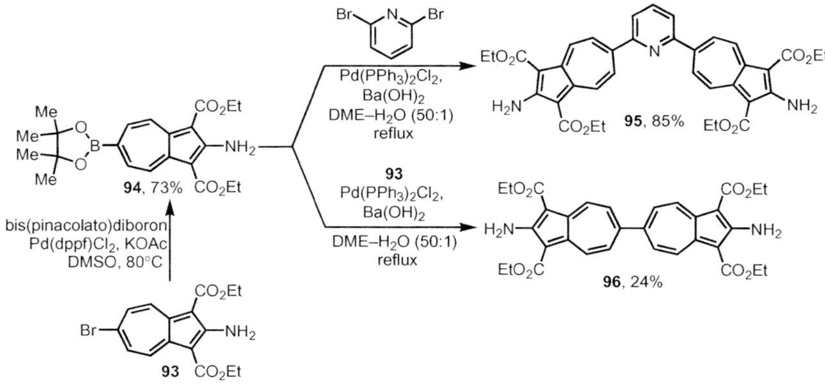

Scheme 24 Synthesis of 2,6-di(6-azulenyl)pyridine **95** by Suzuki–Miyaura cross-coupling reaction.

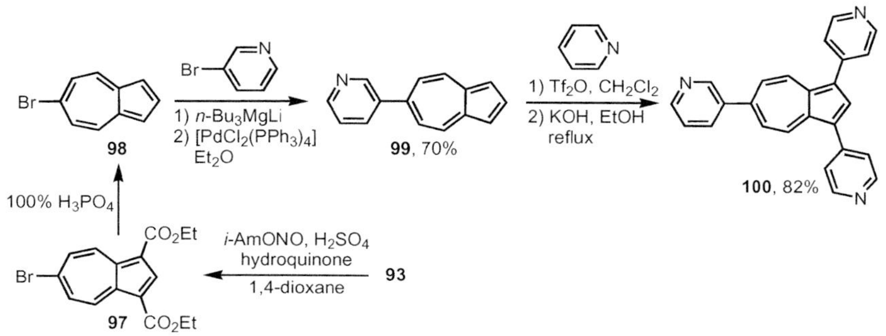

Scheme 25 Synthesis of 2-(3-pyridyl)azulene **99** by cross-coupling reaction with lithium magnesium ate complex.

(**98**), which is prepared from **93** by a two-step procedure, with lithium heterocyclic magnesium ate complexes can be applied to the synthesis of 6-(3-pyridyl)azulene (**99**), as well as 6-(2-pyridyl)- and 3-(quinolyl)azulenes (Scheme 25) (76JOC1811, 05H91). 1,3-Di(4-pyridyl)-6-(3-pyridyl)azulene (**100**) can be prepared by the reaction of **99** with pyridine in the presence of Tf$_2$O and subsequent aromatization with KOH (14H2588).

2.2 Azulene Derivatives With Oxygen-Containing 6-Membered Heterocycles

Coumarin and isocoumarin derivatives are found in many natural products as their partial structures, and their applications to pharmaceuticals and organic materials have grown in recent years. There are some reports for the syntheses of azulene analogues of these derivatives (i.e., pyrone-substituted and -fused azulene derivatives).

In 1992, Nozoe et al. reported the reaction of the 2*H*-cyclohepta[*b*]furan-2-one derivatives **101a** and **101b** with various furan derivatives at high temperature (160–190°C) via an [8+2]cycloaddition affording the azulenopyranone derivatives **102a** and **102b** in low to excellent yields (10%–90%) (Scheme 26) (92H429).

Pyrone-fused azulene derivatives **104a–104c** are also obtained by the $AlCl_3$-mediated cyclization reaction of 2-hydroxyazulenes **103a** and **103b** with activated methylene compounds in low to moderate yields (10%–52%) (Scheme 27) (14T2796). Compound **104a** can be converted into the bromides **105** and **106** by reaction with *N*-bromosuccinimide (NBS). When 1 equiv. of NBS is employed, **105** is obtained in 42% yield along with **106** in 31% yield. On the other hand, the dibromo derivative **106** is selectively generated in 91% yield by reaction with 2 equiv. of NBS. The bromine moieties in **105** and **106** can be transformed by palladium-catalyzed cross-coupling reaction.

Formation of the pyrone-fused azulene derivative **108** by an intramolecular nucleophilic reaction between alkyne and ester moieties was discovered during the investigation of hydration reactions of azulenylalkynes (Scheme 28) (16RSCA78303). When **107**, which is prepared by a

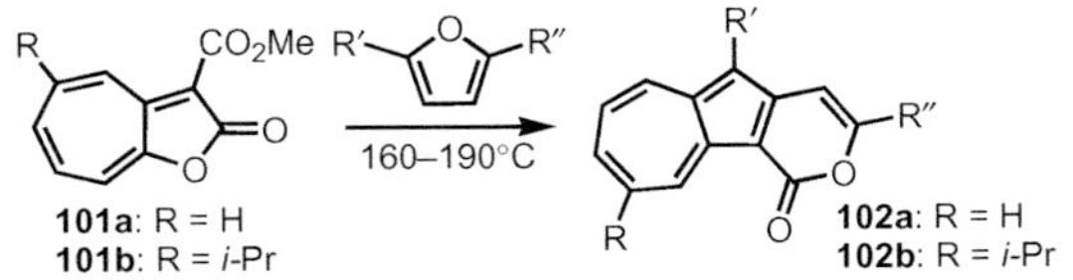
R
CO2Me
R′
O
R″
O
O
160–190°C
R′
R″
R
O
O
101a: R = H
101b: R = i-Pr
102a: R = H
102b: R = i-Pr

Scheme 26 Synthesis of azulenopyranones **102a** and **102b** by [8+2]cycloaddition with **101a** and **101b** with furan derivatives.

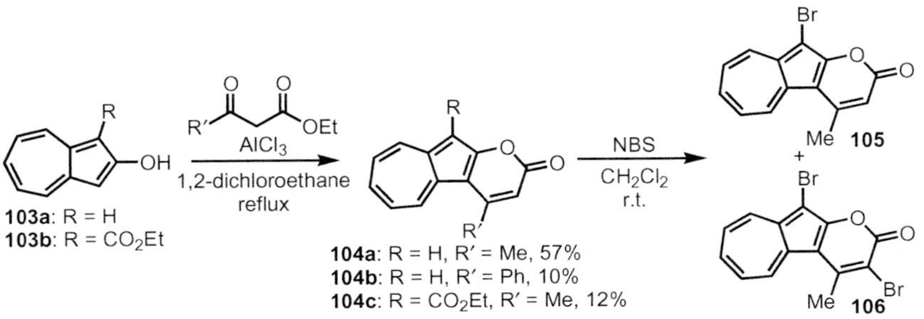
R
OH
O
O
R′
OEt
AlCl3
1,2-dichloroethane
reflux
103a: R = H
103b: R = CO2Et
R
O
O
R′
104a: R = H, R′ = Me, 57%
104b: R = H, R′ = Ph, 10%
104c: R = CO2Et, R′ = Me, 12%
NBS
CH2Cl2
r.t.
Br
O
O
Me
105
+
Br
O
O
Br
Me
106

Scheme 27 Synthesis of azulenopyranones **104a–104c** by $AlCl_3$-mediated cyclization reaction.

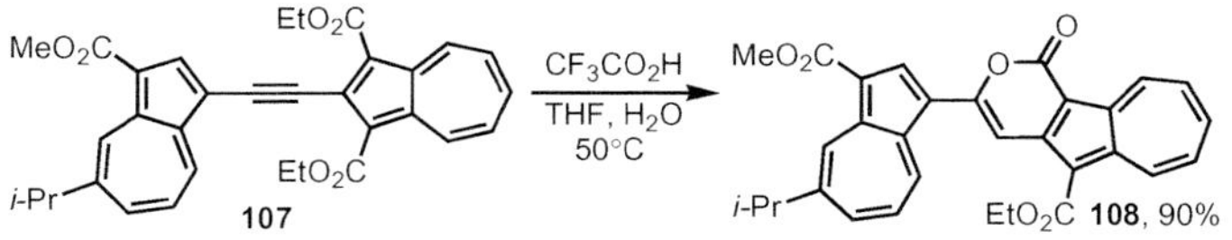

Scheme 28 Synthesis of pyrone-fused azulene derivative **108** by Brønsted acid-mediated intramolecular nucleophilic reaction.

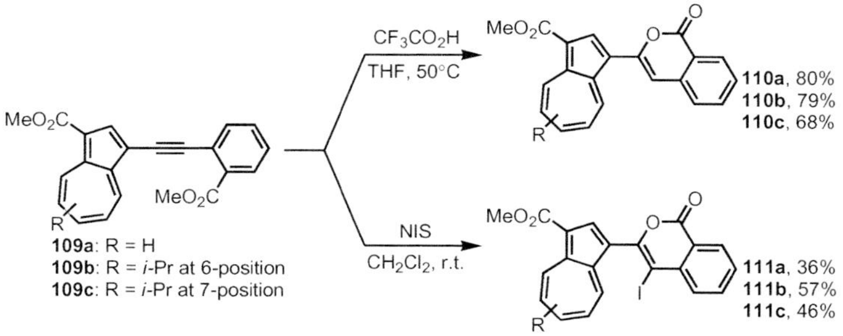

Scheme 29 Synthesis of azulene-substituted isocoumarins **110a–110c** and **111a–111c** by acid-mediated and iodocyclization reactions.

Sonogashira–Hagihara reaction of the corresponding 1-ethynylazulene with 2-chloroazulene, is treated with trifluoroacetic acid in a mixed solvent of THF and water, **108** is produced in 90% yield, instead of hydration of the alkyne moiety (13JOC12513).

Azulene-substituted isocoumarin derivatives **110a–110c** and **111a–111c** have also been prepared by acid-mediated and iodocyclization reactions, respectively (Scheme 29) (18OBC480). Intramolecular cyclization of alkynes **109a–109c** in the presence of trifluoroacetic acid produces the isocoumarin derivatives **110a–110c** in good yields (68%–80%). The alkynes **109a–109c** also react with N-iodosuccinimide (NIS) to give the corresponding iodocyclization products **111a–111c** in 36%–57% yields. Although there are many reports of the preparation of the iodine-substituted heterocycles by iodocyclization reactions of alkyne with an electrophilic iodine source, this is the first example of iodocyclization in azulene chemistry.

X-ray crystal structure analysis demonstrates that isocoumarin **110b** is an almost planar molecule with a dihedral angle of 0.9 degree between the azulene and isocoumarin moieties. In contrast, 4-iodoisocoumarin **111b** has a structure in which the azulene and isocoumarin moieties are twisted by 48.9 degree, owing to the steric repulsion between the iodine substituent, introduced at the 4-position of the isocoumarin, and the azulene ring.

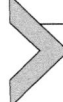

3. AZULENE DERIVATIVES WITH 5-MEMBERED RING HETEROCYCLES

3.1 Synthesis and Reactions of Thienylazulene Derivatives and Related Compounds

3.1.1 1-Thienylazulene Derivatives and Related Compounds

In 1999, Oda et al. have achieved the synthesis of 1-(2-thienyl)azulene **114** and 1,3-di(2-thienyl)azulene **115** by a similar method to that for the preparation of 1-(2-pyridyl)azulenes (Scheme 30) (99S1349). More recently, they have succeeded in the synthesis of derivative **115** by a Stille coupling reaction of 1,3-diiodoazulene (**44a**) with 2-(tri-*n*-butylstannyl)thiophene (Scheme 30) (07T10608). As a result of examining various palladium catalysts and additives, the catalytic system using Pd(PPh$_3$)$_4$, CsF, and CuI was found to give **115** in the best yield (98%).

Ho and coworkers found that photoirradiation (254–350 nm) of a *n*-hexane solution of azulene **1** in the presence of several iodoarenes leads to the formation of 1-arylazulenes including the compound **114** (Scheme 30) (01T715). The mechanism of this photo-assisted coupling reaction is believed to take place by the generation of aryl radicals by homolytic cleavage of the C–I bond of iodoarenes, followed by radical coupling with azulene (**1**) to give the 1-arylazulenes.

The 2-position (or 5-position) of the thiophene moiety in the thienylazulenes can be halogenated in several ways (Scheme 31) (07T10608).

Scheme 30 Synthesis of 1-(2-thienyl) and 1,3-di(2-thienyl)azulenes **114** and **115**.

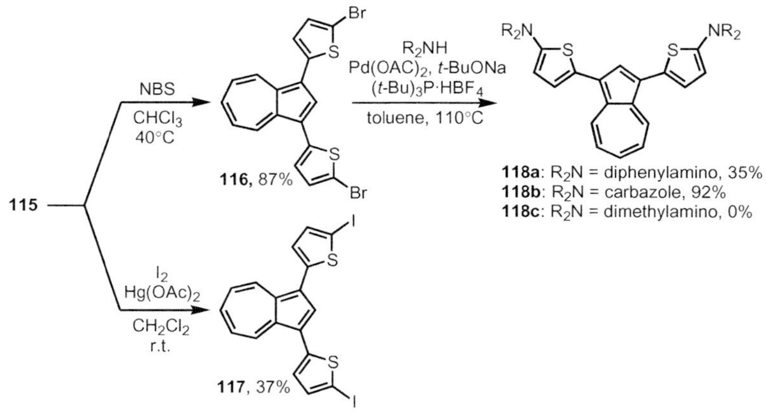

Scheme 31 Synthesis and Hartwig–Buchwald amination of **116** and **117**.

1,3-Di(2-thienyl)azulene (**115**) reacts with NBS to produce 1,3-bis(5-bromo-2-thienyl)azulene (**116**) in 87% yield. A similar reaction of **115** with NIS afforded 1,3-bis(5-iodo-2-thienyl)azulene (**117**) but only in trace amounts; however, compound **117** is generated in 37% yield by electrophilic substitution with I_2, using $Hg(OAc)_2$ as a catalyst. The Hartwig–Buchwald reaction of **116** with diphenylamine and carbazole yields the thienylazulenes **118a** and **118b** bearing arylamino substituents in 35% and 92% yields, respectively. Although amination of **116** with dimethylamine in a sealed tube was investigated under various reaction conditions, the coupling product **118c** could not be obtained.

Organic light-emitting (OLE) properties of **118a** and **118b** have been investigated in OLE devices composed of layered structures (07T10608). The study revealed that the highest occupied molecular orbital (HOMO) energy level of derivative **118a** lies between the indium–tin–oxide work function and the HOMO energy levels of commonly used hole-transporting materials. On the other hand, the HOMO energy level of **118b** is greater than the energy levels of the hole-transporting materials and is the same as the indium–tin–oxide work function.

1-Thienylazulene derivative **121** has been obtained in 88% yield by the reaction of di(1-azulenyl)ketone **119** with Lawesson's reagent (Scheme 32) (08JOC2256). Oxidation of **121** by 2,3-dichloro-5,6-dicyanobenzoquinone (DDQ) and subsequent treatment with HPF_6 produce the tropylium ion derivative **122$^+$** as a hexafluorophosphate in 67% yield. The neutral species **121** shows a weak absorption band in the visible region, derived from azulene moiety, but the UV/vis spectra of cation **122$^+$** show a strong absorption band

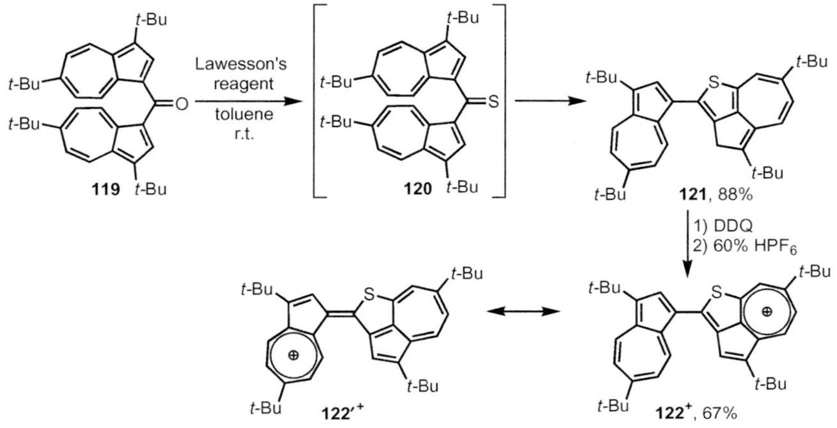

Scheme 32 Reaction of di(1-azulenyl)ketone **119** with Lawesson's reagent.

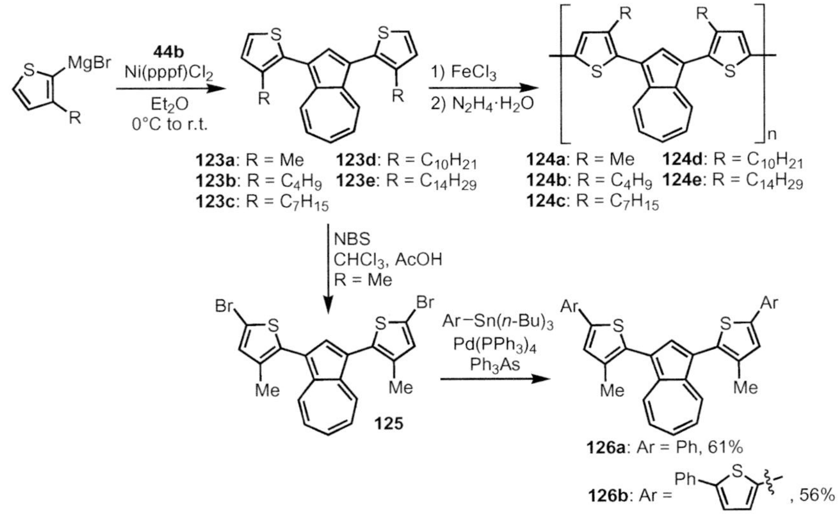

Scheme 33 Preparation of poly[1,3-bis(3-alkyl-2-thienyl)azulene]s **124a–124e** and 1,3-di(2-thienyl)azulene derivatives **125** and **126a,b**.

at λ_{max} 639 nm (log ε 4.56). The strong absorption band of **122$^+$** can be explained by a CT transition from the substituted 1-azulenyl group owing to the resonance structure described in Scheme 32.

Lai and Wang et al. have reported the synthesis and characterization of poly[1,3-bis(3-alkyl-2-thienyl)azulene]s prepared by Scholl-type polymerization of 1,3-bis(3-alkyl-2-thienyl)azulenes (Scheme 33) (03MM536, 04MM3222).

1,3-Bis(3-alkyl-2-thienyl)azulene derivatives **123a–123e** were prepared by Kumada–Tamao–Corriu coupling reactions of 1,3-dibromoazulene **44b** with the corresponding 2-thienylmagnesium reagent in the presence of Ni(dppp) Cl$_2$ catalyst (72JA4374, 72JCS(CC)144). FeCl$_3$-mediated Scholl-type polymerization of the derivatives **123a–123e**, followed by the treatment with hydrazine, yields the poly[1,3-bis(3-alkyl-2-thienyl)azulene]s **124a–124e**.

The average molecular weight (M_n) of these azulene polymers has been shown to be 16,000–41,000 by gel permeation chromatography analysis by using polystyrene standards. Although the UV/vis spectrum of the polymer **124c** in CHCl$_3$ shows a weak absorption band derived from the azulene ring in the visible region, **124c** in 30% trifluoroacetic acid/CHCl$_3$ reveals a strong absorption band at λ_{max} 694 nm due to the formation of an azulenium ion substructure.

The 2-thienyl moiety at the 5-position of compound **123a** can be brominated with NBS to give **125** in 86% yield. The brominated compound **125** when reacted with tri-n-butylphenyltin and tri-n-butyl(5-phenylthienyl)tin in the presence of tetrakis(triphenylphosphine)palladium(0) and triphenylarsine affords the coupling products **126a** and **126b** in 61% and 56% yields, respectively (Scheme 33) (12JMC10448). The UV/vis spectrum of compounds **126a** and **126b** in CHCl$_3$ shows a red shift of their absorption maxima along with the expansion of π-conjugated system. Protonation of compounds **126a** and **126b** with trifluoroacetic acid leads to noticeable changes of their spectra in both the visible and near-IR regions. For instance, compound **126b** shows a new band at λ_{max} 548 nm in 30% trifluoroacetic acid/CHCl$_3$, whereas the compound **126b** exhibits an absorption band in the protonated form at around λ_{max} 1700 nm.

Synthesis of 1,3-bis(2-benzo[b]thienyl)azulene (**127**) by the Stille coupling reaction of 1,3-dibromoazulene **44b** with 2-(trimethylstannyl)benzo[b]thiophene under microwave irradiation has been reported by Hawker et al. (Scheme 34) (14CS4483). Compound **127** shows a weak absorption band at λ_{max} 630 nm in the UV/vis spectrum due to the transition of the azulene ring. Compound **127** does not change the absorption wavelength

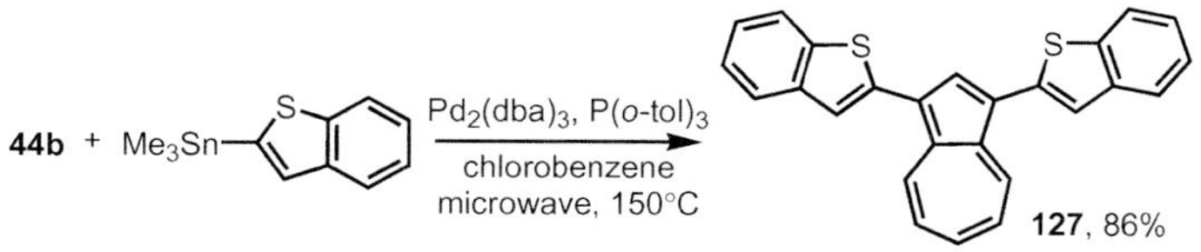

Scheme 34 Synthesis of 1,3-bis(2-benzo[b]thienyl)azulene **127** by Stille coupling reaction.

even with the addition of trifluoroacetic acid, although the similar compounds **126a** and **126b**, described earlier, exhibited significant spectral changes. Compound **127** shows a quasi-reversible oxidation wave on cyclic voltammetry (CV) derived from the formation of a stabilized radical cationic species.

To construct photochromic systems, Nica et al. reported the preparation of the dithienylethenes **130a** and **130b** with 1-thienylazulene substituents by Suzuki–Miyaura cross-coupling reactions of the bromoazulene derivatives **44b** and **129** with the thienylboronic acid ester **128** in 24% and 43% yields, respectively (Scheme 35) (15RSCA63282). Compound **130b** has been converted into **130c** in 20% yield by pyrrole-mediated deformylation in acetic acid (99BCJ2543).

These compounds **130a–130c** exhibit photochromism due to the ring-closing reaction of the dithienylethane moiety under visible light irradiation. When the CH_2Cl_2 solutions of the **130a–130c** are irradiated with λ 405 nm wavelength light, the photocyclization reaction proceeds to generate the ring-closed systems **131a–131c**, along with a color change of the solutions from greenish-yellow to bluish-purple due to extension of the π-conjugation.

3.1.2 2-Thienylazulene Derivatives and Related Compounds

2-Thienylazulene derivatives are usually prepared by palladium-catalyzed cross-coupling reactions using 2-haloazulene or 2-azulenyl-metal reagents

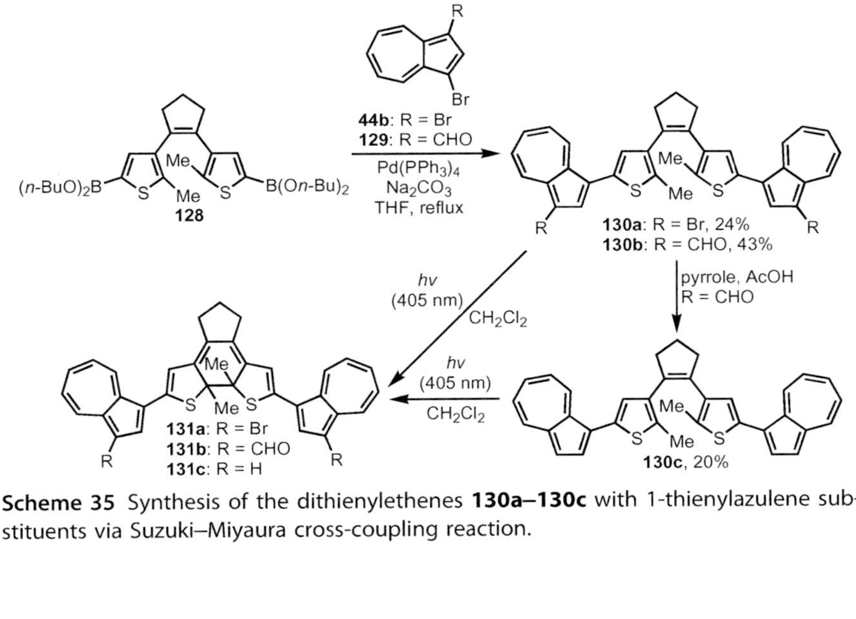

Scheme 35 Synthesis of the dithienylethenes **130a–130c** with 1-thienylazulene substituents via Suzuki–Miyaura cross-coupling reaction.

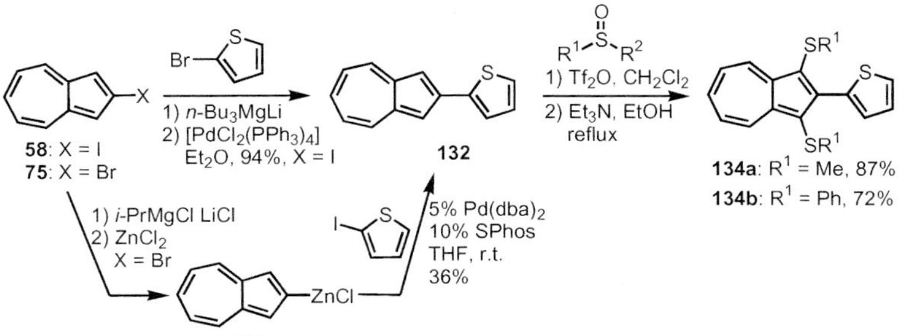

Scheme 36 Synthesis and sulfanylation of 2-thienylazulene (**132**).

as a precursor (Scheme 36). As in the preparation of 2-pyridylazulenes (Scheme 15), the parent 2-(2-thienyl)azulene (**132**) can be obtained by a cross-coupling between 2-iodoazulene **58** and 2-thienylmagnesium ate complex in the presence of a palladium catalyst (05H91). This method is also applicable to the preparation of 2-(3-thienyl)azulene. Synthesis of derivative **132** has also been achieved by a Negishi coupling reaction of the 2-azulenylzinc reagent **133**, which is prepared from 2-bromoazulene **75**, with 2-iodothiophene (15S2663).

1,3-Bis(methylthio)- and 1,3-bis(phenylthio)thienylazulenes **134a** and **134b** have been prepared in a two-step process via azulenyldisulfonium ions; the reaction of **132** with dimethyl sulfoxide (DMSO) and methyl phenyl sulfoxide in the presence of Tf$_2$O, followed by the treatment with Et$_3$N, affords the products **134a** and **134b** (Scheme 36) (09EJO4307). Electrochemical reduction of the thiophene derivative **132** shows a reversible reduction wave at a half-wave potential of -1.80 V (in volts vs Ag/AgNO$_3$) on CV due to the formation of a radical anionic species, although the reduction of **134a** and **134b** exhibits irreversible waves.

The preparation of 5,5′-di(2-azulenyl)-2,2′-bithiophene (**137**) and 2,5-di(2-azulenyl)-thieno[3,2-*b*]thiophene (**139**) via Suzuki–Miyaura cross-coupling reactions has been reported by Katagiri et al. (Scheme 37) (12OL2316). The cross-coupling reaction of 2-iodoazulene **58** with the 2-thienylboronate **135**, with a 5-chloro substituent, gives 2-(5-chloro-2-thienyl)azulene (**136**) in 73% yield. Compound **136** is converted into **137** in 60% yield by a homocoupling reaction using a Ni catalyst. 2,5-Di(2-azulenyl)thieno[3,2-*b*]thiophene (**139**) was also synthesized in 40% yield by a Suzuki–Miyaura cross-coupling reaction of 2-iodoazulene **58** with the diboronic acid ester **138** of thieno[3,2-*b*]thiophene. The herringbone packing in the crystal structures of **137** and **139** has been clarified by

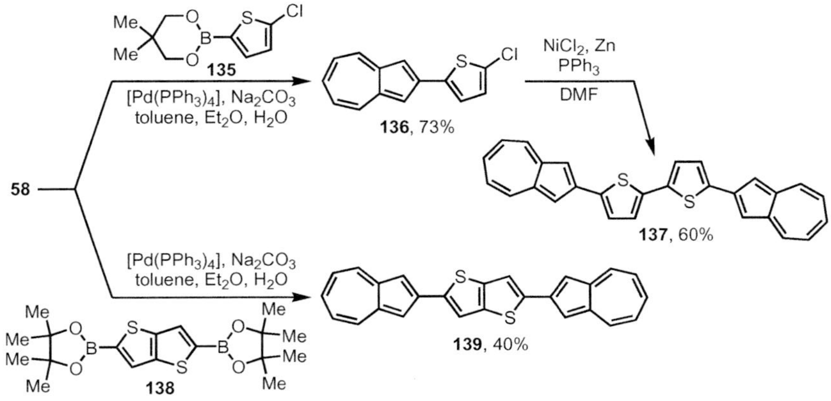

Scheme 37 Synthesis of 5,5′-di(2-azulenyl)-2,2′-bithiophene (**137**) and 2,5-di(2-azulenyl)-thieno[3,2-*b*]thiophene (**139**) by Suzuki–Miyaura coupling reaction.

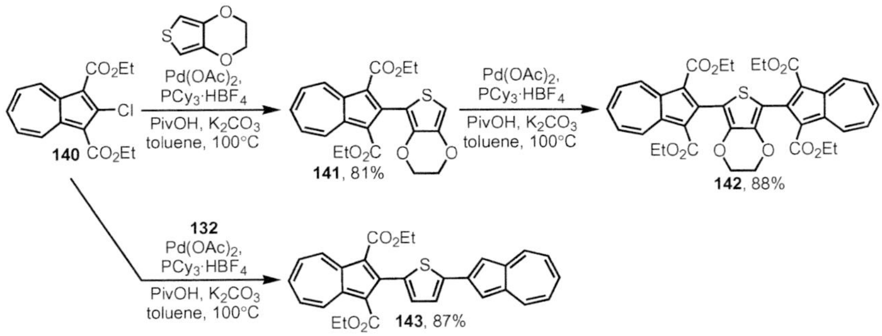

Scheme 38 Synthesis of 2-(2-thienyl)azulene derivatives **141–143** by palladium-catalyzed C–H bond activation reaction.

single-crystal X-ray analysis. Compounds **137** and **139** are aligned almost perpendicular to the substrate in the film form and show the characteristics of an OFET having a hole mobility.

As described earlier, in the most cases, azulene derivatives with heteroaryl substituents have been prepared by a cross-coupling reaction of either azulenyl metal reagents with aryl halides or haloazulenes with organometallic reagents. A more direct preparation of 2-thienylazulene derivatives without using an organometallic reagent has also been reported recently (Scheme 38) (15OBC10191). The 2-chloroazulene derivative **140**, which is easier to synthesize than 2-iodoazulene derivatives, reacts with 3,4-ethylenedioxythiophene (EDOT) in the presence of $Pd(OAc)_2$, $PCy_3 \cdot HBF_4$, pivalic acid (PivOH), and K_2CO_3 to give the 2-thienylazulene derivative **141** in 81% yield. The 2-thienylazulene derivatives **141** and **132** also react with **140** under similar conditions to afford the symmetric and unsymmetric 2,5-di(2-azulenyl)thiophene derivatives **142** (88%) and **143** (87%).

Cross-coupling by C–H bond activation can also be applied to the synthesis of 2-azulenyltetrathiafulvalenes (Scheme 39) (17JOC1657). In 2011, Yorimitsu and coworkers reported the preparation of tetrathiafulvalene (TTF) derivatives with multiple aryl substituents by the palladium-catalyzed direct arylation of TTF with aryl bromides (11CS2017). The 2-chloroazulene derivatives **57a** and **57b** react with TTF, under conditions similar to those reported by Yorimitsu et al., to generate the tetra(2-azulenyl)TTF derivatives **144a** (70%) and **144b** (57%), along with tri(2-azulenyl)TTF **145** (9%) in the case of the reaction of **57b**. Despite the lower solubility of compound **144a** in common organic solvents, due to the strong π–π stacking of the molecules, compound **144b** possesses considerable solubility, since the steric hindrance of the isopropyl groups at 5-position of the azulene rings effectively prevents the π–π stacking.

Although the parent TTF exhibits a two-stage reversible oxidation wave, the CV experiments on **144b** and **145** reveal only one quasi-reversible wave under the electrochemical oxidation conditions owing to the overlapping of the second wave with the following irreversible waves. This result indicates the instability of compounds **144b** and **145** toward electrochemical oxidation, probably due to the existence of the unsubstituted and reactive 1- and 3-positions on the azulene ring.

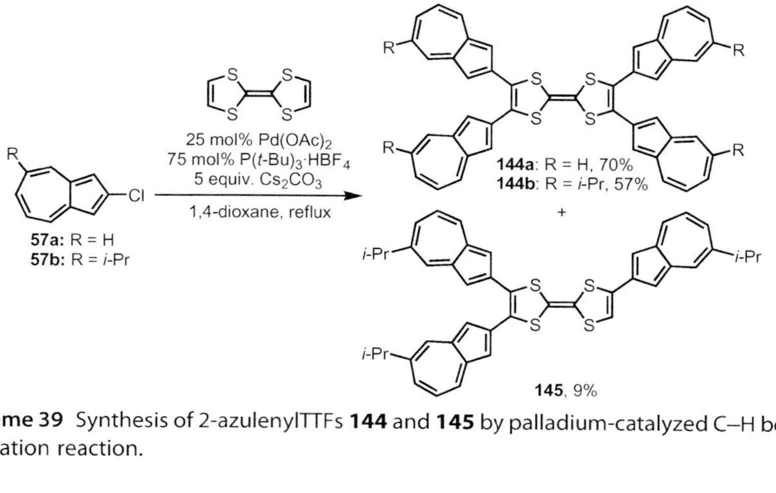

Scheme 39 Synthesis of 2-azulenylTTFs **144** and **145** by palladium-catalyzed C–H bond activation reaction.

When the electrochromic property was evaluated by the spectroelectrochemical measurements, compounds **144b** and **145** showed spectral changes under the electrochemical reduction conditions. Especially, **145** exhibits new absorption bands, which spread into the near-IR region, gradually developed during the electrochemical reduction owing to the generation of anionic species.

3.1.3 4-Thienyl- and 5-Thienylazulene Derivatives and Related Compounds

Syntheses of azulene derivatives having a thienyl group at the 4- and/or 5-positions by transition metal-catalyzed cross-coupling reactions have been reported. However, there are few examples of such derivatives, since the preparation of their precursors (4- and 5-haloazulenes) is substantially difficult.

In 2004, Danheiser and his coworkers achieved the synthesis of the 4-thienylazulene derivatives **148** and **149** by cross-coupling with the 4-iodoazulene derivative **147**, which is prepared by a rhodium-catalyzed ring-expansion reaction of diazoketone **146** and thienyl organometallic reagents (Scheme 40) (04JOC8652). The Stille coupling reaction of 5,5′-(trimethylstannyl)-2,2′-bithiophene with iodo derivative **147** gives 5,5′-di(4-azulenyl)-2,2′-bithiophene **148** in 73% yield. On the other hand, **147** reacts with 2-thienylzinc reagent in the catalytic system of $Pd_2(dba)_3$ and Ph_3As to produce the 4-thienylazulene **149** in 81% yield.

The synthesis of a series of 4,7-di(2-thienyl)azulene derivatives **151–153** by Stille coupling reactions of 4,7-dibromoazulene (**150**) with 2-thienyltin

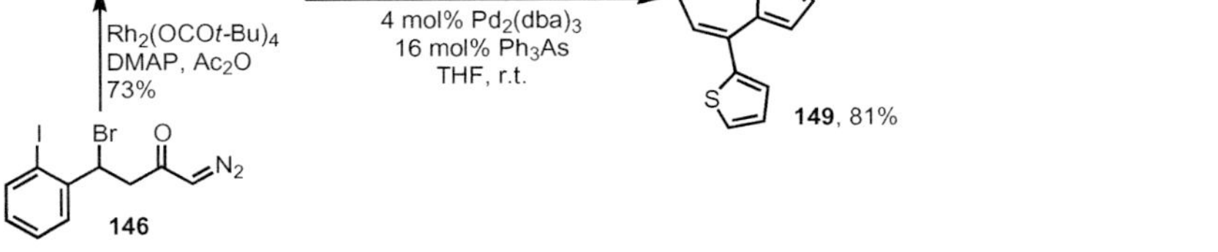

Scheme 40 Synthesis of 4-(2-thienyl)azulenes **148** and **149** by palladium-catalyzed cross-coupling reaction.

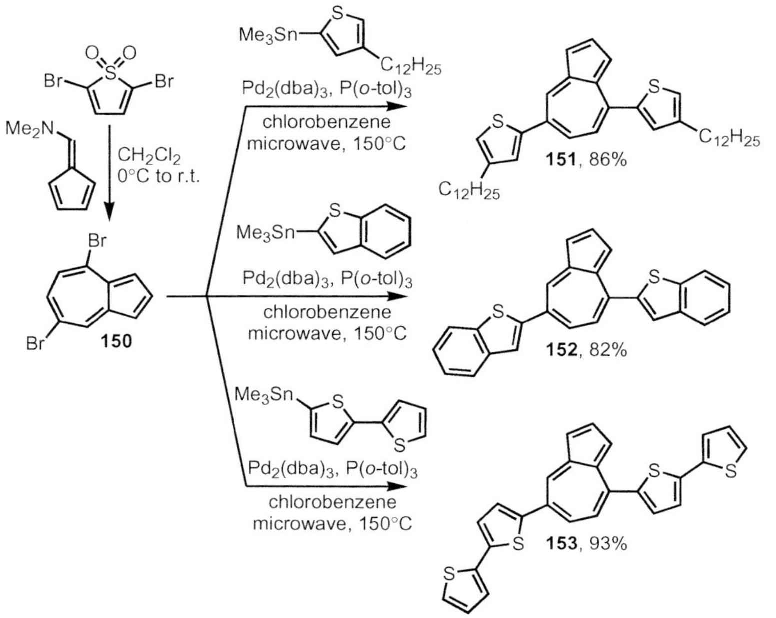

Scheme 41 Synthesis of 4,7-di(2-thienyl)azulene derivatives **151–153** by Stille coupling reaction.

reagents under microwave irradiation has been reported by Hawker et al. (Scheme 41) (11JA10046, 14CS4483). 4,7-Dibromoazulene (**150**) can be prepared by the [6 + 4] cycloaddition of 2,5-dibromothiophene 1,1-dioxide with 6-dimethylaminofulvene.

In UV/vis spectra, the 4,7-di(2-thienyl)azulene derivatives **151–153** in CH_2Cl_2 show a weak absorption band due to the $\pi-\pi^*$ transition of the azulene ring itself in the visible region. By the addition of CF_3CO_2H to the solutions, a strong absorption band is generated at around λ_{max} 500 nm owing to the formation of tropylium ion substructures by the protonation of the five-membered ring of azulene moiety. The protonated species of **151** reveals a strong fluorescence at λ_{FL} 573 nm, although the neutral species does not exhibit the fluorescent.

For the purpose of the construction of a new photochromic system, the 5,6-di(3-thienyl)azulene-based dithienylethene analogue **155** was prepared by Uchida and coworkers (Scheme 42) (12JOC3270). The synthesis of analogue **155** was achieved by the Suzuki–Miyaura cross-coupling reaction of 5,6-dibromoazulene (**154**) with the corresponding 3-thienylboronic acid ester in the presence of $Pd(PPh_3)_4$ as a catalyst.

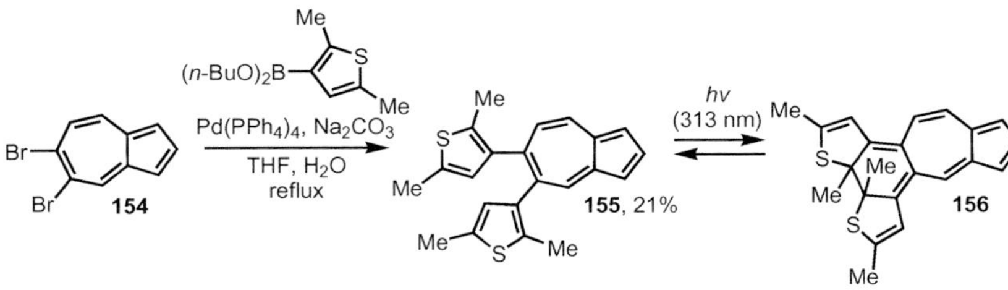

Scheme 42 Synthesis of 5,6-di(3-thienyl)azulene derivative **155** by Suzuki–Miyaura coupling reaction.

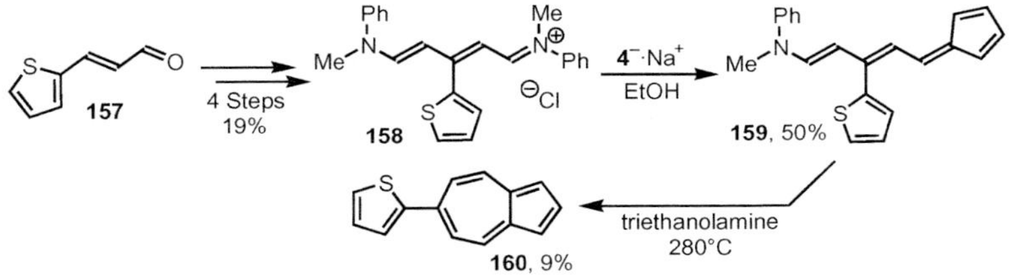

Scheme 43 Synthesis of 6-(2-thienyl)azulene (**160**) by Ziegler–Hafner's azulene synthesis.

The UV/vis spectrum of compound **155** in n-hexane shows a weak absorption band at λ_{max} 594 nm. When UV light (λ 313 nm) is irradiated on **155**, new absorption bands at λ_{max} 495 nm and at around λ_{max} 700 nm appear due to the formation of the ring-closed system **156**, along with the color change to brownish pink from the blue color of **155**.

3.1.4 6-Thienylazulene Derivatives and Related Compounds

Although 6-thienylazulene derivatives have formerly been prepared by the Ziegler–Hafner's method, these are currently synthesized from 6-haloazulenes by transition metal-catalyzed cross-coupling reactions in the most cases.

In 1974, Porshnev et al. reported the synthesis of 6-(2-thienyl)azulene (**160**) via Ziegler–Hafner's azulene synthesis starting from the thiophene-substituted conjugated ketone **157** (Scheme 43) (74KG1156). However, this method requires a multiple-step synthesis and the overall yield of **160** is 0.9% because of the low yield of the each step.

Dorofeenko and coworkers also reported the preparation of 6-(2-thienyl)azulenes **162–164** from the thiophene-substituted pyrylium salt **161** by a modified Ziegler–Hafner's azulene synthesis (Scheme 44) (80KG807). The sequential addition–cyclization reaction between **161** and cyclopentadienide

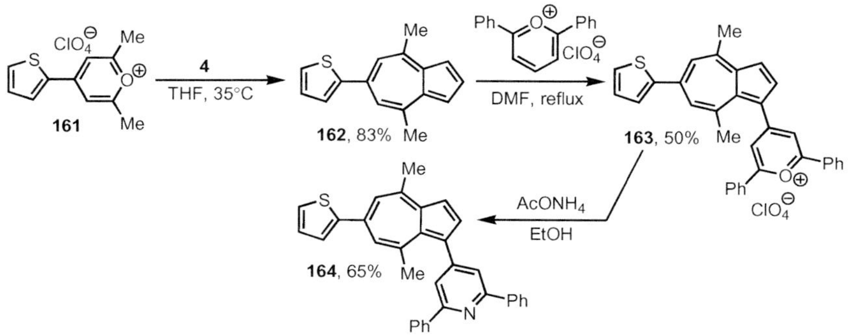

Scheme 44 Synthesis of 6-(2-thienyl)azulenes **162–164** by a modified Ziegler–Hafner's azulene synthesis.

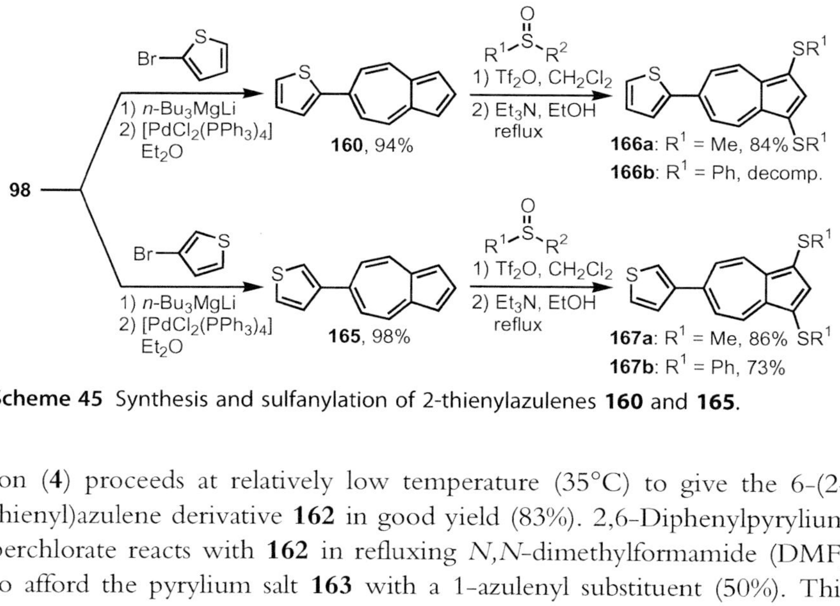

Scheme 45 Synthesis and sulfanylation of 2-thienylazulenes **160** and **165**.

ion (**4**) proceeds at relatively low temperature (35°C) to give the 6-(2-thienyl)azulene derivative **162** in good yield (83%). 2,6-Diphenylpyrylium perchlorate reacts with **162** in refluxing N,N-dimethylformamide (DMF) to afford the pyrylium salt **163** with a 1-azulenyl substituent (50%). This can be converted into the 1-(4-pyridyl)-6-(2-thienyl)azulene derivative **164** in 65% yield by the treatment with $AcONH_4$.

The 6-pyridyl- and quinolylazulene derivatives, 6-(2- and 3-thienyl) azulenes **160** and **165**, can be prepared by palladium-catalyzed cross-coupling of 6-bromoazulene (**98**) with the corresponding thienyl magnesium ate complexes (Scheme 45) (09EJO4307). The reactions of the 6-(thienyl)azulenes **160** and **165** with DMSO or methyl phenyl sulfoxide, followed by the treatment with Et_3N in the presence of Tf_2O, give the 1,3-bis(methylthio)- and 1,3-bis (phenylthio)-6-thienylazulenes **166a**, **167a**, and **167b**, except for **166b**.

6-Thienylazulenes exhibit a reversible reduction wave on CV and the first reduction potential of 1,3-bis(methylthio)- and 1,3-bis(phenylthio)-6-thienylazulenes **166a** ($-1.66\,V$), **167a** ($-1.57\,V$), and **167b** ($-1.54\,V$) shows lower values compared with those of **160** ($-1.72\,V$) and **165** ($-1.82\,V$). These results suggest that the methylthio and the phenylthio groups at the 1,3-positions of the azulene ring decrease the lowest unoccupied molecular orbital (LUMO) energy level. The CV experiments also reveal the reversible oxidation waves of **166a** ($+0.23\,V$), **167a** ($+0.23\,V$), and **167b** ($+0.47\,V$) due to the generation of a stabilized radical cationic species.

Like 2-(thienyl)azulene derivatives (Scheme 24), 6-(thienyl)-2-aminoazulene derivatives are synthesized by a palladium-catalyzed direct arylation (Scheme 46) (15OBC10191). The cross-coupling reaction of the 2-amino-6-bromoazulene derivative **93** with EDOT in the presence of Pd(OAc)$_2$, PCy$_3$·HBF$_4$, PivOH, and K$_2$CO$_3$ generates the 6-thienylazulene derivative **168** with a 2-amino function in 80% yield. Furthermore, the product **168** reacts with 2-chloroazulene **140** under similar conditions to afford the unsymmetrically substituted thiophene derivative **169** having both 2-azulenyl and 6-azulenyl groups in 82% yield. Bathochromic shift of the absorption band of **169** is observed in the UV/vis spectrum relative to that of **168**, owing to the effective π-conjugation expanded by the thiophene ring substituted by two azulene rings.

3.2 Synthesis and Reactions of Pyrrolylazulene Derivatives and Related Compounds

Eichen et al. reported that the Stille coupling reaction of 1,3-dibromoazulene **44b** with N-(*tert*-butoxycarbonyl)-2-(trimethylstannyl)pyrrole, followed by the thermal deprotection of the Boc moiety of product **170**, gives 1,3-di(2-pyrrolyl)azulene **171** (Scheme 47) (05EJO2207). Although the compound **171** shows blue color in DMSO solution, the addition of fluoride ion to this solution changes the color which has been characterized by spectral measurements. Furthermore, **171** shows weak emission in DMSO, but the

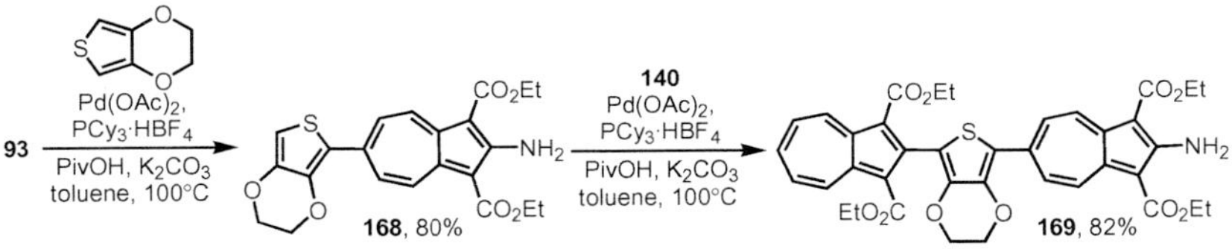

Scheme 46 Synthesis of 6-(2-thienyl)azulene derivatives **168** and **169** by palladium-catalyzed C–H bond activation reaction.

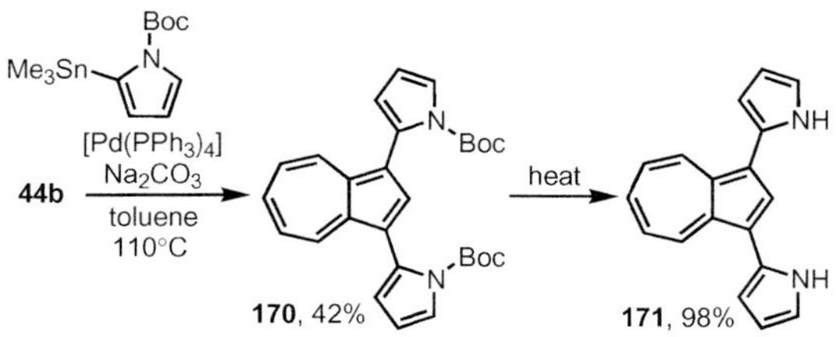

Scheme 47 Synthesis of 1,3-di(2-pyrrolyl)azulene **171** by Stille coupling reaction.

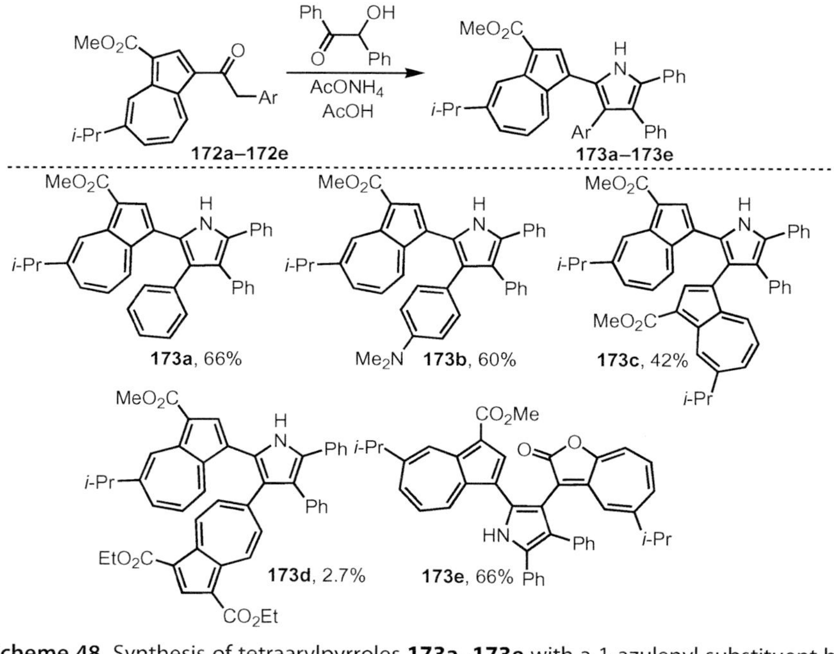

Scheme 48 Synthesis of tetraarylpyrroles **173a–173e** with a 1-azulenyl substituent by the cyclization reaction with benzoin and ammonium acetate.

addition of tetrabutylammonium fluoride to the solution increases the quantum yield of fluorescence by more than 10-fold.

As a synthetic method for 1-(2-pyrrolyl)azulene derivatives (except for the metal-catalyzed cross-coupling reaction), preparation via cyclization reactions has been reported in recent years. Preparation of a series of tetraarylpyrroles **173a–173e** with a 1-azulenyl substituent was achieved by the cyclization reaction of 1-azulenyl ketones **172a–172e**, bearing various aryl groups on the α-position, with benzoin in the presence of large excess of ammonium acetate (Scheme 48) (17H1870). In this reaction, the yield of the products is significantly affected by the aryl substituents at the α-position on **172a–172e**. The reaction of the ketones with an electron-donating aryl substituent affords the products in relatively good yields. On the other hand, strong electron-withdrawing groups on the ketones decrease the product yields. Similar 1-azulenyl ketone derivatives with a phenyl substituent on the 3-azulenyl group also exhibit similar reactivity.

Indole derivatives, which are pyrrole analogues, are found in numerous natural products and numerous biologically active pharmaceuticals. There are a few reports concerning the synthesis of indole derivatives containing an azulene ring (07H951). The indole derivatives **175a–175c** and **176a–176c** substituted with a 1-azulenyl group are obtained by Vilsmeier–Haack-type reaction of the azulene precursors **1**, **27**, and **174** with 2-indolinones in the presence of Tf$_2$O (Scheme 49) (12TL1493). The reaction of azulene **1** with 1-phenyl-2-indolinone gives the anticipated product, but in low yield (27%), since the steric effect of the *N*-phenyl moiety prevents the electrophilic reaction. 6-*tert*-Butylazulene (**27**) affords the disubstituted product **177** (13%) in addition to the major product **175b** (64%). Formation of **177** indicates that the *tert*-butyl group at the 6-position enhances the reactivity toward the electrophilic reaction.

For the construction of both linear and nonlinear optical materials, Gryko and coworkers reported the synthesis and optical properties of the pyrrolo[3,2-*b*]pyrroles **180a** and **180b** possessing two 5-azulenyl and 6-azulenyl substituents, respectively (Scheme 50) (17JMCC2620). The

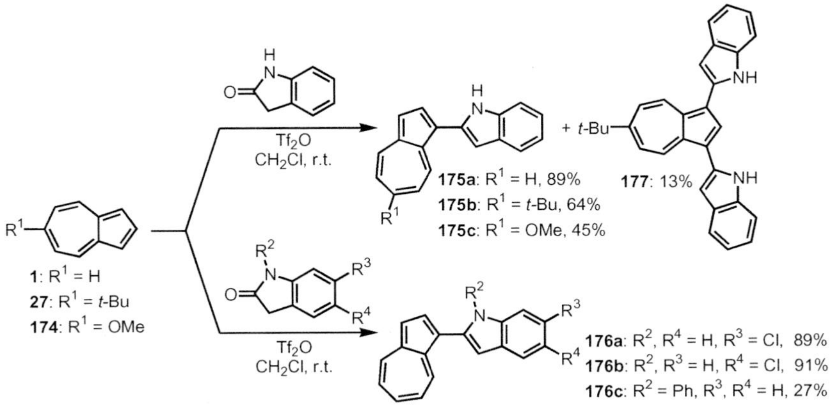

Scheme 49 Synthesis of indole derivatives **175a–175c** and **176a–176c** substituted with a 1-azulenyl group by Vilsmeier–Haack-type arylation reaction.

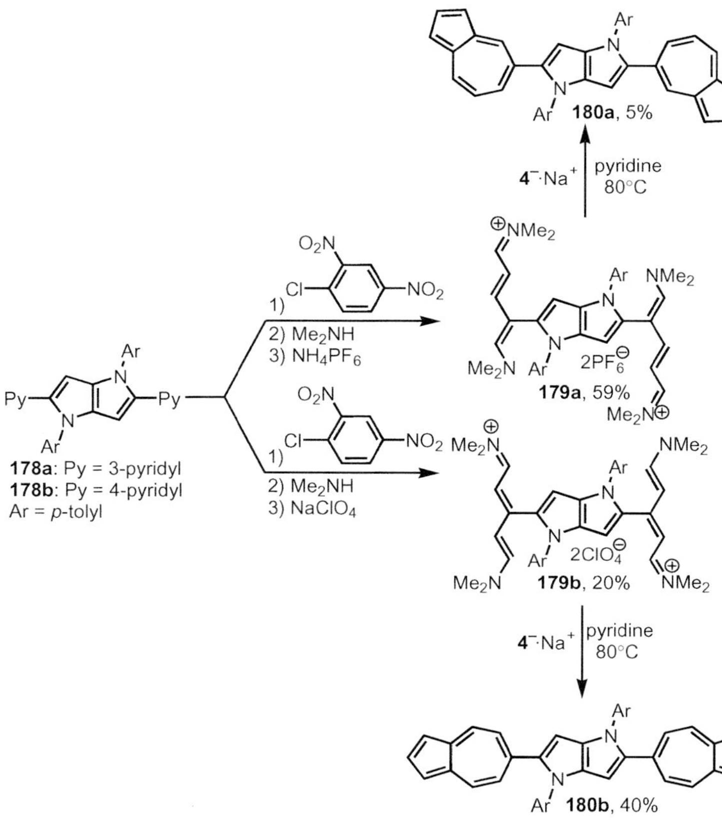

Scheme 50 Synthesis of pyrrolo[3,2-*b*]pyrroles **180a** and **180b** possessing two 5-azulenyl and 6-azulenyl substituents by Ziegler–Hafner's azulene synthesis.

preparation of compounds **180a** and **180b** was accomplished by the Ziegler–Hafner's method. The reaction of 2,5-di(3-pyridy)- and 2,5-di(4-pyridyl) pyrrolo[3,2-*b*]pyrroles **178a** and **178b** with 1-chloro-2,4-dinitrobenzene and subsequent treatment with dimethylamine give the corresponding Zincke salts **179a** and **179b** in 59% and 20% yields, respectively. These salts **179a** and **179b** react with cyclopentadienide ion (**4**$^-$) to afford **180a** (5%) and **180b** (40%). Although the product **180a** substituted by 5-azulenyl groups reveals weak absorption in the visible region (600–700 nm) in the UV/vis spectrum, **180b** substituted by 6-azulenyl groups exhibits a strong absorption band at λ_{max} 504 nm arising from ICT between the pyrrolo[3,2-*b*]pyrrole moiety and the two 6-azulenyl substituents.

3.3 Synthesis and Reactions of Furylazulene Derivatives and Related Compounds

The preparation of 1-(2-furyl)azulenes **181a**, **181b**, and **182** (Fig. 2) has been reported by Oda et al. using the same method as the corresponding 1-(2-thienyl)azulene derivative described in Section 3.1.1 (99S1349). The molecular structure of **182** has been clarified by a single-crystal X-ray structure analysis, in which the out-of-plane deformation of the azulene ring by the substitution of two furyl groups is observed (09H451).

Wang et al. demonstrated the synthesis of 3-(2-furyl)guaiazulene derivatives **184** by phosphine-mediated intermolecular cyclization of 1-(3-phenyl-2-cyanopropenoyl)guaiazulenes **183**, which are prepared by Knoevenagel condensation of the 1-cyanoacetylguaiazulene with aldehydes, with benzoyl chlorides (Scheme 51) (13CCL622). [Guaiazulene is 1,4-dimethyl-7-isopropylazulene and is a natural product.] In this reaction, the yield is slightly decreased when the substituent R^2 is an electron-withdrawing group, e.g., NO_2. In contrast, in the case of electron-donating R^2 substituents (e.g., methyl and methoxy groups), the reaction gives the desired (1-azulenyl)furans in excellent yields.

2-(1-Azulenyl)benzo[*b*]furans **187a–187c** can be obtained in one pot via Sonogashira–Hagihara reaction of 1-ethynylazulenes **185a–185c** with 2-iodophenol (Scheme 52) (16RSCA78303). The formation of **187a–187c**

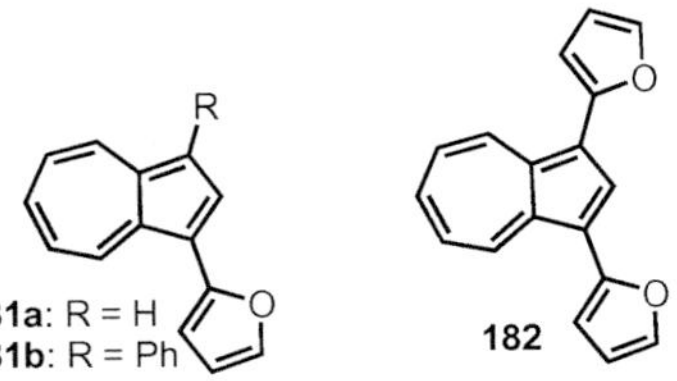

Fig. 2 Structure of 1-(2-furyl)azulenes **181a** and **181b**, and 1,3-di(2-furyl)azulene **182**.

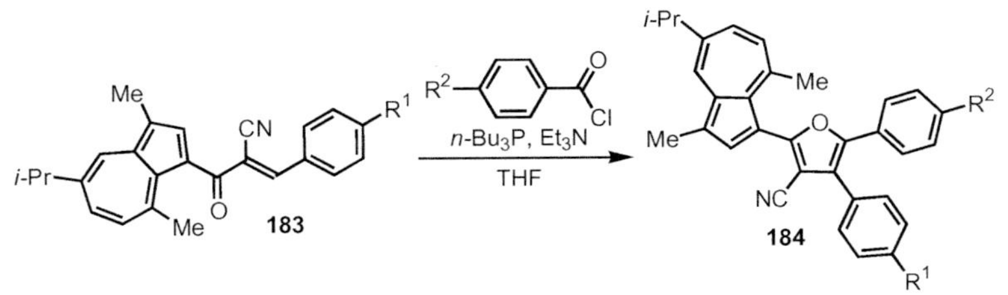

Scheme 51 Phosphine-mediated synthesis of 3-(2-furyl)guaiazulene derivatives **184**.

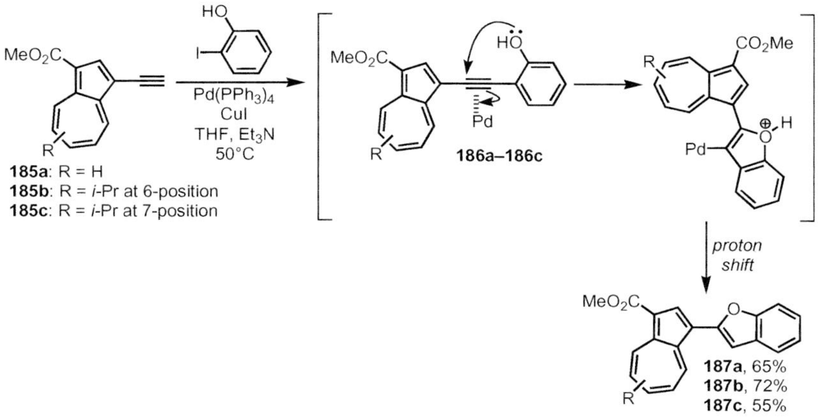

Scheme 52 Synthesis of 1-(2-benzofuryl)azulene **187a–187c**.

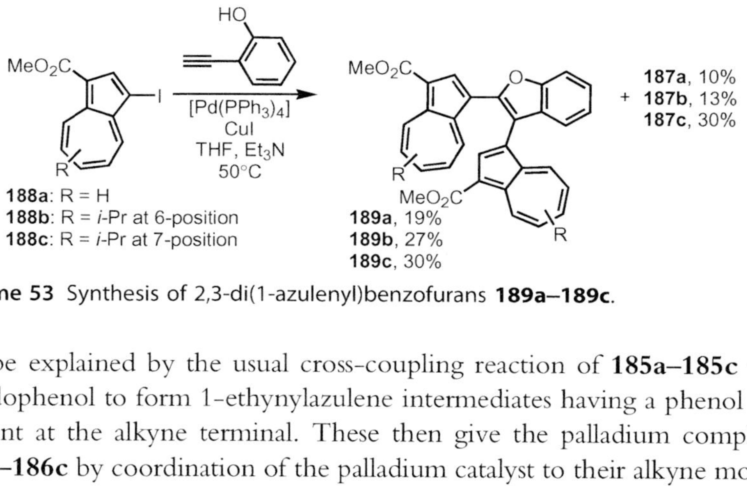

Scheme 53 Synthesis of 2,3-di(1-azulenyl)benzofurans **189a–189c**.

can be explained by the usual cross-coupling reaction of **185a–185c** with 2-iodophenol to form 1-ethynylazulene intermediates having a phenol substituent at the alkyne terminal. These then give the palladium complexes **186a–186c** by coordination of the palladium catalyst to their alkyne moiety, and then intramolecular nucleophilic attack of the phenol oxygen forms the benzofuran ring resulting in **187a–187c**.

Unlike the results described earlier, the reaction of the 1-iodoazulenes **188a–188c** with 2-ethynylphenol unexpectedly gives the 2,3-di(1-azulenyl)benzo[*b*]furans **189a–189c**, along with the presumed **187a–187c**, under Sonogashira–Hagihara conditions (Scheme 53) (18OBC480). Generation of **189a–189c** by the reaction of **188a–188c** with 2-ethynylphenol is in accordance with the reaction mechanism via an oxypalladation process proposed by Arcadi et al. (96JOC9280).

It is well known that the reaction of electron-rich alkynes with tetracyanoethylene via [2 + 2] cycloaddition–retroelectrocyclization affords tetracyanobutadiene derivatives in high yields (10CC1994, 11CRV2306, 17CEJ5126). On the other hand, the 3-(1-azulenyl)propargyl alcohols **190a** and **190b** with tetracyanoethylene produce the 2-aminofuran derivatives **191a** (93%) and **191b** (93%) in excellent yields (Scheme 54) (17CEJ5126). The 2-aminofuran derivative **191b** can be converted into the 6-aminopentafulvenes **192** and 6,6-diaminopentafulvenes **193** in moderate to excellent yields by treatment with an excess of primary and secondary amines.

In 2007, Wang et al. reported the preparation of the 2-(2-benzo[*b*]furyl)azulene derivatives **197a–197e** via annulation of 2-bromomethylazulene **196** with salicylaldehydes under both thermal and microwave irradiation conditions (Scheme 55) (07YCHHDX1305). Compared to the reaction under thermal conditions, the yield of the products is improved under microwave irradiation, despite the shorter reaction time (within 10 min).

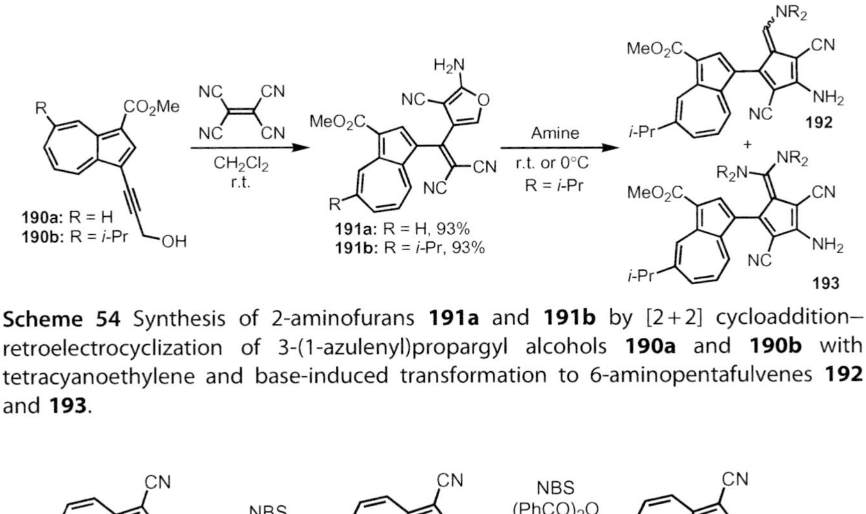

Scheme 54 Synthesis of 2-aminofurans **191a** and **191b** by [2+2] cycloaddition–retroelectrocyclization of 3-(1-azulenyl)propargyl alcohols **190a** and **190b** with tetracyanoethylene and base-induced transformation to 6-aminopentafulvenes **192** and **193**.

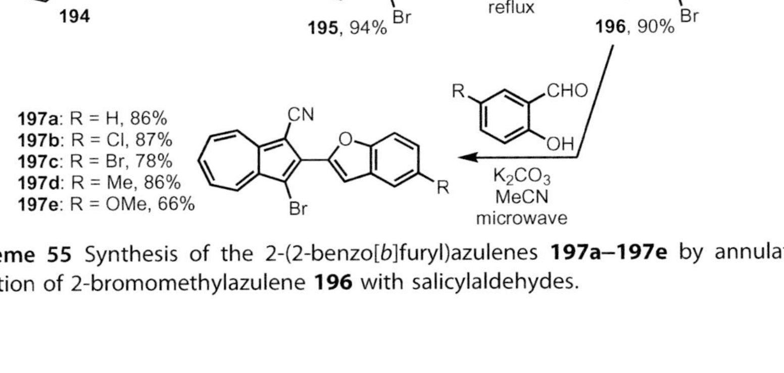

Scheme 55 Synthesis of the 2-(2-benzo[*b*]furyl)azulenes **197a–197e** by annulation reaction of 2-bromomethylazulene **196** with salicylaldehydes.

Synthesis and optical properties of π-extended 2-(2-furyl)azulene derivative **200**, as well as 1-(2-furyl)azulene **199**, with a diketopyrrolopyrrole core were reported by Hawker and his colleagues (Scheme 56) (14CS3753). The two regioisomers **199** and **200** were successfully obtained by microwave-assisted Suzuki–Miyaura cross-coupling reaction of the 1- and 2-azulenylboronic acid esters **30** and **62** (see Schemes 9 and 16) with 3,6-bis(5-bromo-2-furyl)pyrrolo[3,4-c]pyrrole-1,4-dione (**198**) in 69% and 84% yields, respectively (03EJO3663). The esters 2-azulenylboronic acid pinacol ester **30** and **62** can be prepared by Ir-catalyzed direct borylation of azulene (**1**).

In the UV/vis spectra in CH_2Cl_2, the furylazulene derivatives **199** and **200** exhibit strong absorption in the visible region between λ_{max} 550 and 700 nm, suggesting the narrowing of the HOMO–LUMO gap due to the extension of the π-system. In common with other azulene derivatives, **199** and **200** in CH_2Cl_2 solution show remarkable color and spectral changes by the addition of trifluoroacetic acid. Compounds **199** having two 1-azulenyl groups show a hypsochromic shift by about 25 nm, whereas a bathochromic shift of about 200 nm is observed in the isomers **200** substituted by two 2-azulenyl groups in their UV/vis spectra.

Cross-polarized, nonpolarized transmission optical microscopy and high-resolution specular X-ray diffraction also reveal the influence of introduction of the two azulene units into **199** and **200** on the planarity, molecular self-assembly, and crystallization during thin-film formation.

Hawker et al. have also reported the preparation of 4,7-di(2-furyl)azulene (**201**) from 4,7-dibromoazulene **150** (Scheme 57) (14CS4483), in a similar manner to that for the preparation of 4,7-di(2-thienyl)azulene derivatives **151–153** (Scheme 41).

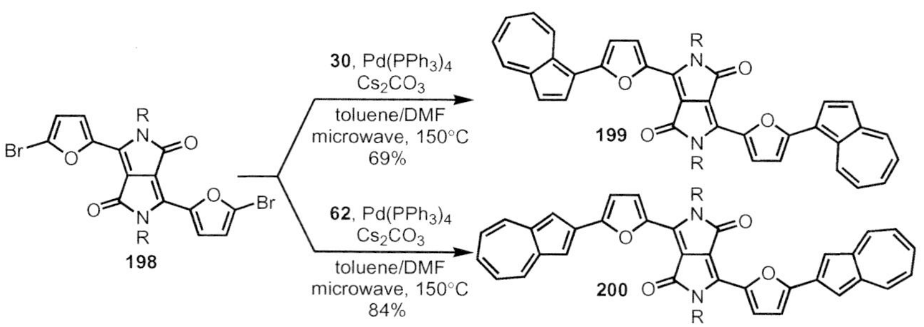

Scheme 56 Synthesis of 1- and 2-(2-furyl)azulenes **199** and **200** with a diketopyrrolopyrrole by Suzuki–Miyaura coupling reaction.

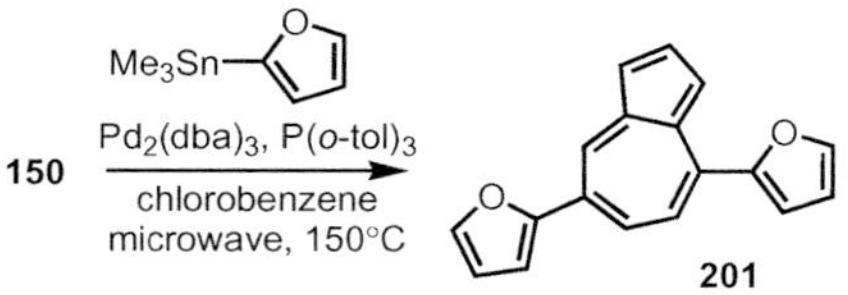

Scheme 57 Preparation of 4,7-di(2-furyl)azulene (**201**) by Stille coupling reaction.

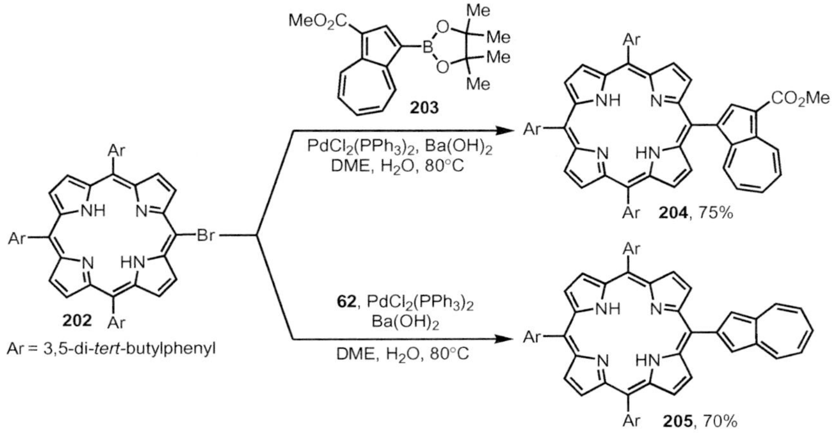

Scheme 58 Synthesis of 1- and 2-azulenyl porphyrins **204** and **205** by Suzuki–Miyaura coupling reaction.

3.4 Synthesis and Reactions of Porphyrin- and BODIPY-Substituted Azulene Derivatives

Osuka and Kurotobi have reported the preparation of 1-, 2-, 4-, and 6-azulenylporphyrins substituted at the *meso*-position (Schemes 58 and 59) (05OL1055). Synthesis of *meso*-1- and 2-azulenylporphyrins **204** and **205** is accomplished by Suzuki–Miyaura cross-coupling of the *meso*-bromoporphyrin derivative **202** with the 1- and 2-azulenylboronic acid esters **203** and **62** in 75% and 70% yields, respectively. In an alternative approach, the *meso*-5- and *meso*-6-azulenylporphyrins **207** and **208** have been prepared in 15% and 55% yields, respectively, by the Ziegler–Hafner method using the *meso*-(3- and 4-pyridyl)porphyrins **206a** and **206b**.

The photophysical properties of these azulenylporphyrins have been investigated as their Zn(II)-complexes, which can be obtained by the treatment with $Zn(OAc)_2$. The Soret bands of all Zn(II)-complexes of azulenylporphyrins show substantial red shifts and broadening in their UV/vis spectra, compared to those of tetrakis(3,5-di-*tert*-butylphenyl) Zn(II)-porphyrins. The order of the red shifts of the Soret bands is

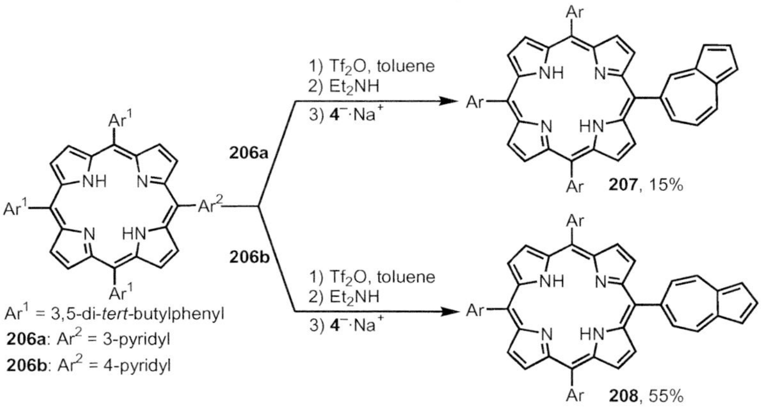

Scheme 59 Synthesis of 1- and 2-azulenyl porphyrins **207** and **208** by Ziegler–Hafner's azulene synthesis.

Zn(II)-**205** (λ_{max} 428 nm) = Zn(II)-**207** (λ_{max} 428 nm) > Zn(II)-**208** (λ_{max} 427 nm) > Zn(II)-**204** (λ_{max} 424 nm); the broadening for the spectrum of Zn(II)-**204** is the most noticeable in this series. The fluorescence spectra of Zn(II)-complexes of azulenylporphyrins measured by excitation at λ_{EX} 428 nm in CH$_2$Cl$_2$ show the emission nearly at the same wavelength (λ_{FL} 602 and 649 nm), although the fluorescence intensities and quantum yields are quite low compared to those of tetrakis(3,5-di-*tert*-butylphenyl) Zn(II)-porphyrins.

Synthesis of the azulene-fused porphyrin **213** (Ar = 2,4,6-*tert*-butylphenyl) has been achieved by oxidative ring closure at the 3-position of the azulene rings with the pyrrole moieties of the 5,10,15,20-tetra(4-azulenyl)porphyrin **212** (Scheme 60) (06AGE3944). 4-Formylazulene **211**, a precursor of **212**, is obtained by the reaction of 4-methylazulene **209** with *N*,*N*-dimethylformamide dimethylacetal (DMFDMA), followed by oxidative cleavage of the generated enamine by NaIO$_4$. Porphyrin **212** is prepared in 12% yield by using the aldehyde **211** under Lindsey conditions. The Scholl reaction of **212** with FeCl$_3$ leads to azulene-fused porphyrin **213** in 60% yield, but the addition of AgPF$_6$ does not cause any ring-closing reaction and the use of DDQ/Sc(OTf)$_3$ results in formation of a complex mixture (10CB1734). The azulene-fused porphyrin **213** shows intensely perturbed absorption bands at λ_{max} 684 and 1136 nm covering the both visible and near-IR regions up to 1200 nm.

Synthesis of a series of 1- and 2-azulenylBODIPYs **215–218** by Suzuki–Miyaura cross-coupling reactions was reported by Mack and Lu in 2016 (Scheme 61) (16RSCA32124). Mono-azulenylBODIPYs **215** and **216** are

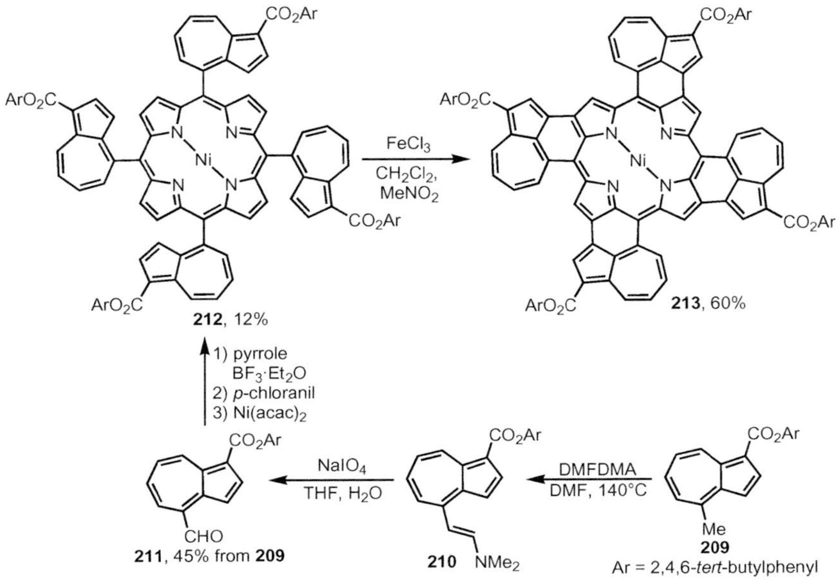

Scheme 60 Synthesis of the azulene-fused porphyrin **213**.

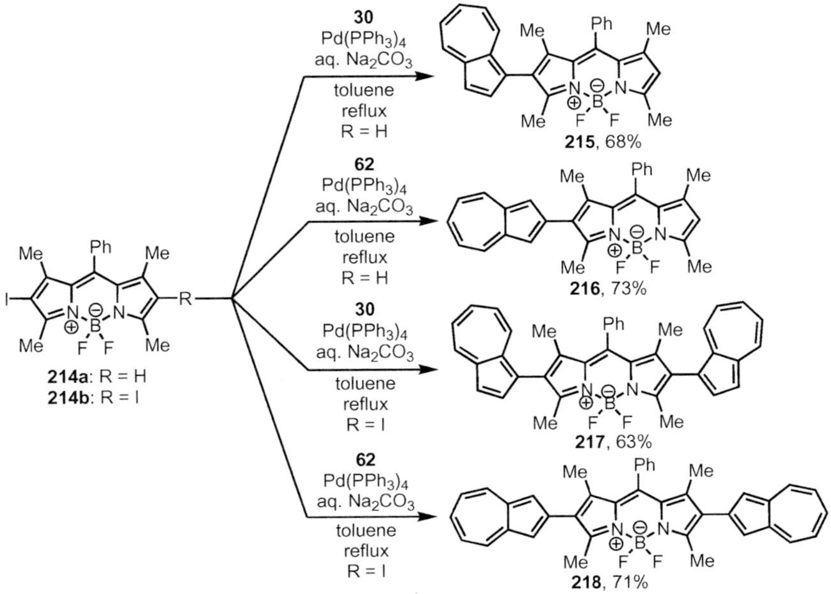

Scheme 61 Synthesis of 1- and 2-azulenylBODIPYs **215–218** by Suzuki–Miyaura coupling reaction.

available by the cross-coupling between 2-iodoBODIPY **214a** and the 1- and 2-azulenylboronic acid esters **30** and **62** in 68% and 73% yields, respectively, whereas the Suzuki–Miyaura cross-coupling reaction of 2,6-diiodoBODIPY **214b** affords bis-azulenylBODIPYs **217** and **218** in 63% and 71% yields, respectively. In their UV/vis spectra, azulenylBODIPYs show an absorption band in the 500–600 nm region, along with weaker and broader bands in the 350–450 nm region, which is attributable to the $S_0 \rightarrow S_2$ transition of the BODIPY moiety. The longest wavelength absorption bands of bis-azulenylBODIPYs **217** and **218** exhibit bathochromic shifts of about 30 nm compared with those of the mono-azulenylBODIPYs **215** and **216**.

4. AZULENE DERIVATIVES WITH AZOLES AND BENZOAZOLES

4.1 Synthesis and Reactions of 1-(Azolyl)- and 1-(Benzoazolyl)Azulenes and Related Compounds

In 2004, Imafuku and coworkers reported the preparation of 1-azulenylthiazole derivatives by cyclization of the α-bromoketone function of the 1-substituted azulenes **219** with thioamides (Scheme 62) (04JHC723). In this synthetic methodology, 1-(bromoacetyl)azulenes **219a** and **219b** react with thioamides in refluxing EtOH to generate the 1-(4-thiazolyl)azulenes **220a** and **220b** in good to excellent yields (72%–99%). The ester function of derivatives **220a** can be removed by heating in 100% phosphoric acid to yield the 1-(4-thiazolyl)azulene parent derivatives **221** (45%–96%).

Imafuku et al. also demonstrated the preparation of the 1-(3-isoxazolyl)- and 1-(3-pyrazolyl)azulene derivatives **223** and **224a–224f** via the 1-azulenylenaminone **222** (Scheme 63) (07H2237). Enaminone **222**, which is obtained by the reaction of the 1-acetylazulenes **19b** (see Scheme 7) with DMFDMA, reacts with hydroxylamine to afford 1-(3-isoxazolyl)azulene

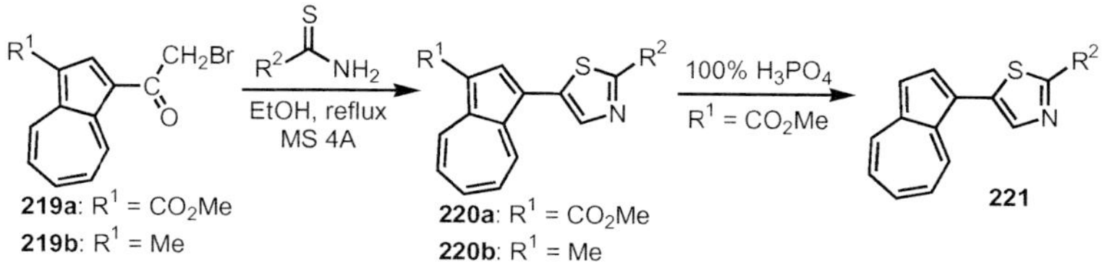

Scheme 62 Synthesis of 1-(4-thiazolyl)azulenes **220a** and **220b** and their parent derivatives **221**.

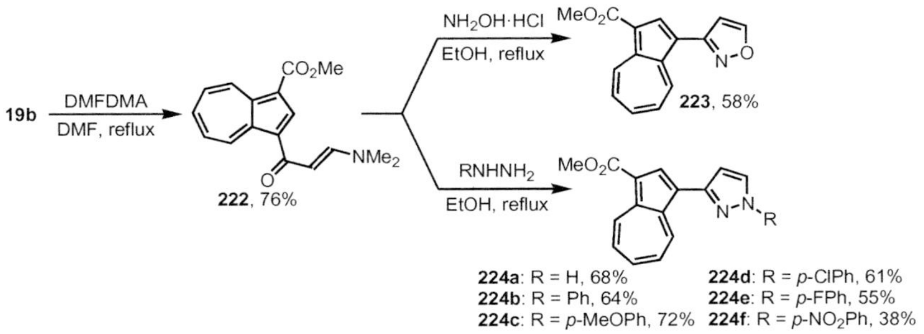

Scheme 63 Synthesis of 1-(3-pyrazolyl)azulenes **223** and **224a–224f**.

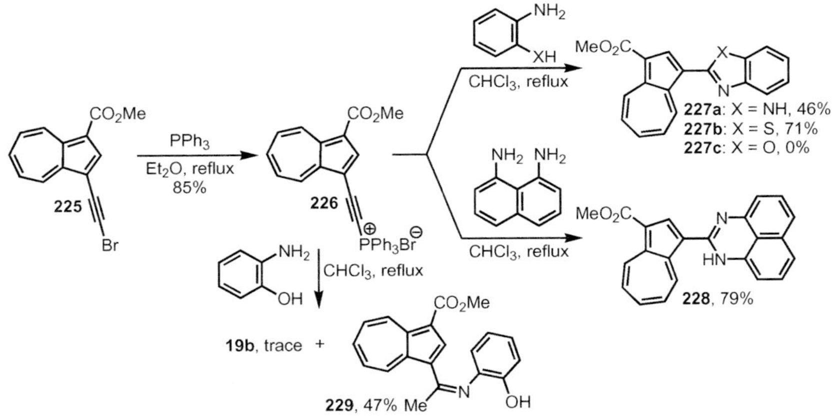

Scheme 64 Synthesis of 1-(2-benzoazolyl)azulenes **227** and 1-(1H-perimidin-2-yl) azulene **228**.

223 in 58% yield. Similarly, the reaction of **222** with hydrazine derivatives gives the 1-(3-pyrazolyl)azulenes **224a–224f** in moderate to good yields.

1-(2-Benzoazolyl)azulenes **227a** and **227b** and 1-(1H-perimidin-2-yl) azulene **228** have been obtained by cyclization of the phosphonium salt **226** with o-substituted aniline derivatives and 1,8-diaminonaphthalene, respectively (Scheme 64) (04H305). In the case of the reaction of **226** with o-aminophenol, imine **229** and a trace amount of **19b** were obtained instead of **227c**.

The preparation and chromogenic behavior toward Hg^{2+} ion of 1,3-di (azolyl)azulenes **230–232** have been investigated by Wakabayashi et al. (Scheme 65) (08H383). Synthesis of compounds **230–232** was achieved by Stille coupling reactions of 1,3-diiodoazulene **44a** with 2–3 equiv. of

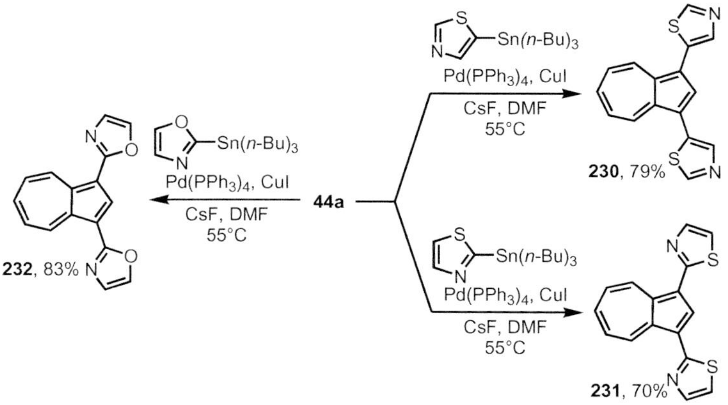

Scheme 65 Synthesis of 1,3-di(azolyl)azulenes **230–232** by Stille coupling reaction.

the appropriate (tri-*n*-butylstannyl)azole in the presence of Pd(PPh$_3$)$_4$, CsF, and CuI as a catalytic system.

Although the addition of Li$^+$, Na$^+$, Ca^{2+}, Zn^{2+}, and Cd^{2+} ions into the acetone solution of **230–232** causes negligible color changes, the addition of heavy metal ions (Hg^{2+}, Ag$^+$, Pb^{2+}, Cu^{2+}, and Cr^{3+}) leads to a color change from blue to red or reddish-brown colors. Especially, the absorption band of compound **231** in the visible region shows a considerable blue shift to λ_{max} 604 nm from λ_{max} 531 nm on the addition of 1 equiv. of Hg(ClO$_4$)$_2$, due to the formation of a Hg complex of **231**.

4.2 Synthesis and Reactions of Other Azolyl- and Benzoazolylazulenes and Related Compounds

In a manner similar to the synthesis of 1-(benzoazol-2-yl)- and 1-(1*H*-perimidin-2-yl)azulenes (Scheme 64), 2-(benzoazol-2-yl)azulenes **235a–235c** and 2-(1*H*-perimidin-2-yl)azulene **236** are available by the reaction of 2-(2-azulenyl)ethynyltriphenylphosphonium bromide **234** with 2-substituted anilines and 1,8-diaminonaphthalene (Scheme 66) (03H339). Although 2-(1-azulenyl)benzoxazole **227c** could not be obtained by a similar method (see Section 4.1), 2-(2-azulenyl)benzoxazole **235c** can be synthesized in low yield (11%).

The Tf$_2$O-mediated heteroarylation reaction is applicable to the synthesis of 5-azulenyl benzothiazoles (Scheme 67) (10EJO1059). In this case, the reaction of 1,3-*tert*-butylazulene **84** with 5 equiv. of benzothiazole in the presence of 1.5 equiv. of Tf$_2$O affords **237**, which is convertible to **238** in

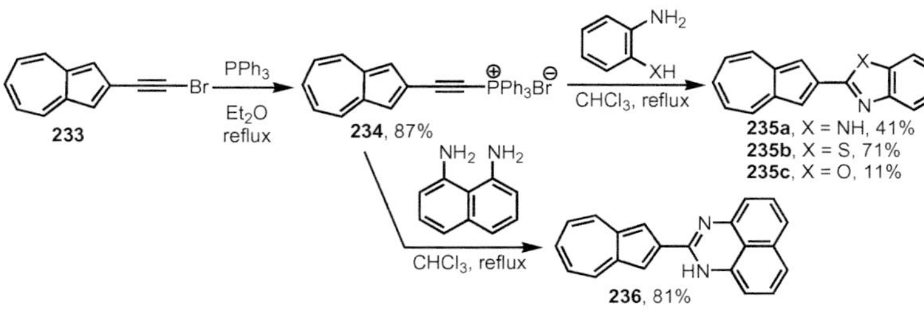

Scheme 66 Synthesis of 2-(benzoazol-2-yl)azulenes **235a–235c** and 2-(1H-perimidin-2-yl)azulene **236**.

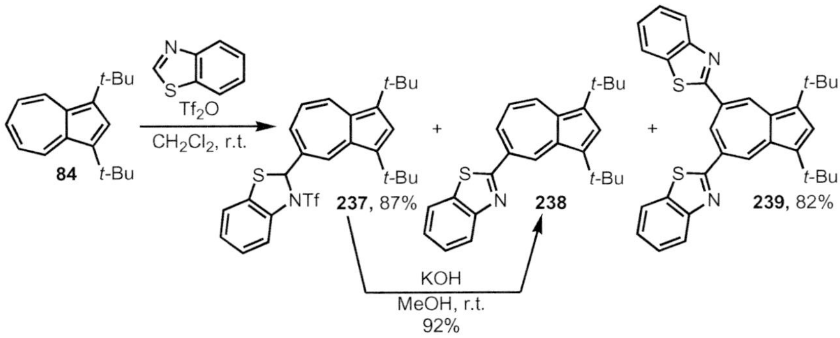

Scheme 67 Synthesis of 5-(2-benzolyl) and 5,7-di(2-benzothiazolyl)azulenes **238** and **239**.

92% yield by base-induced aromatization. When the reaction is carried out with a large excess of benzothiazole and Tf$_2$O, 5,7-bis(2-benzothiazolyl) azulene **239** is obtained in good yield (82%).

As for the synthesis of the 1- and 2-azulenyl derivatives, 6-(benzoazol-2-yl)- and 6-(1H-perimidin-2-yl)azulenes can also be synthesized by the reaction of ethynyltriphenylphosphonium bromide **242** with 2-substituted anilines and 1,8-diaminonaphthalene, respectively (Scheme 68) (11H1283). The 1- and 2-azulenylethynyl bromides **225** and **233** (Schemes 64 and 66) are prepared by Corey–Fuchs reactions, followed by an E2 elimination using 1,8-diazabicyclo[5.4.0]undec-7-ene because of the easy availability of the starting formylazulenes. However, 6-azulenylethynyl bromide **241** is prepared by the reaction of 6-ethynylazulene with NBS in the presence of AgNO$_3$ because of difficulties in the synthesis of the starting 6-formylazulene derivative. Ethynylphosphonium bromide **242**, which is generated by the reaction of **241** with triphenylphosphine, reacts with

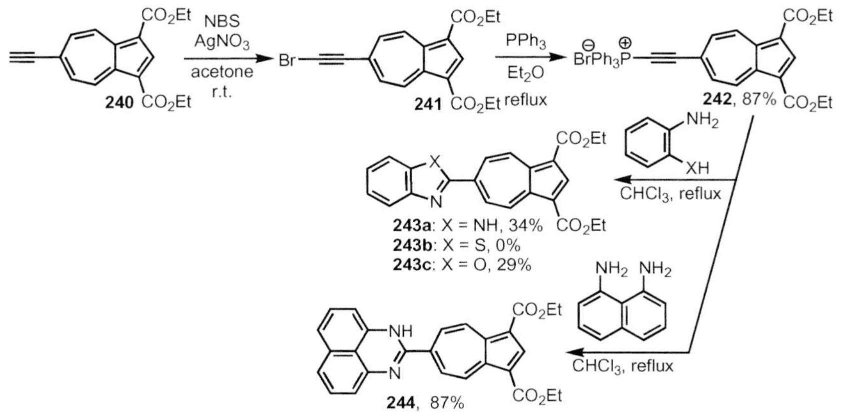

Scheme 68 Synthesis of 6-(benzoazol-2-yl)- and 6-(1H-perimidin-2-yl)azulenes **243** and **244**.

o-phenylenediamine, 2-aminophenol, and 1,8-diaminonaphthalene to yield the corresponding derivatives **243a**, **243c**, and **244** in 34%, 29%, and 87% yields, respectively; the reaction with o-aminobenzenethiol does not give the benzothiazole derivative **243b**.

5. CONCLUDING REMARKS

We have summarized the preparation, reactivity, and physical properties of heterocycle-substituted and hetero-fused azulene derivatives. The synthetic methods for these compounds are diverse, since the electronic properties of azulenes greatly depend on the substitution position.

1-Heteroarylazulenes have been prepared by various procedures, such as electrophilic substitution, cyclization, radical coupling, and transition metal-catalyzed cross-coupling. On the other hand, 2- and 6-heteroarylazulenes are mainly synthesized from the corresponding haloazulenes by cross-coupling reactions, since the reactivity toward electrophiles at these sites is low. There is only one example of heteroarylation at the 2-position via an electrophilic substitution reaction, but a strong electron-donating group, such as a dimethylamino group, is indispensable at the 6-position of the azulene ring in this case. The method of Ziegler–Hafner, in which a 4-substituted pyridine is used as a starting material, is also applicable to the synthesis of 6-heteroarylazulenes.

The methods described earlier are very effective for the preparation of novel heteroarylazulene derivatives and hetero-fused azulene derivatives which have potential to be organic electronics and pharmaceutical agents.

REFERENCES

37HCA224	P.A. Plattner and A.S. Pfau, *Helv. Chim. Acta*, **20**, 224 (1937).
53JACS4980	A.G. Anderson Jr, J.A. Nelson, and J.J. Tazuma, *J. Am. Chem. Soc.*, **75**, 4980–4989 (1953).
55AG301	K. Ziegler and K. Hafner, *Angew. Chem.*, **67**, 301 (1955).
62BCJ1990	T. Nozoe, S. Seto, and S. Matsumura, *Bull. Chem. Soc. Jpn.*, **35**, 1990–1998 (1962).
62LA40	K. Hafner and K.L. Moritz, *Liebigs Ann. Chem.*, **656**, 40–53 (1962).
71T3357	T. Nozoe, K. Takase, T. Nakazawa, and S. Fukuda, *Tetrahedron*, **27**, 3357–3368 (1971).
72JA4374	K. Tamao, K. Sumitani, and M. Kumada, *J. Am. Chem. Soc.*, **94**, 4374–4376 (1972).
72JCS(CC)144	R.J.P. Corriu and J.P. Masse, *J. Chem. Soc. Chem. Commun.*, 144 (1972).
74KG1156	Y.N. Porshnev, E.M. Tereshchenko, and V.B. Mochalin, *Chem. Heterocycl. Compd.*, **10**, 1156–1158 (1974).
76JA7095	L.C. Dunn, Y.-M. Chang, and K.N. Houk, *J. Am. Chem. Soc.*, **98**, 7095–7096 (1976).
76JOC1811	R.N. McDonald, J.M. Richmond, J.R. Curtis, H.E. Petty, and T.L. Hoskins, *J. Org. Chem.*, **41**, 1811–1821 (1976).
80KG807	G.N. Dorofeenko, A.V. Koblik, T.I. Polyakova, and L.A. Murad'yan, *Chem. Heterocycl. Compd.*, **16**, 807–810 (1980).
80S31	M. Hanke and C. Jutz, *Synthesis*, 31–32 (1980).
81JSOCJ1172	K. Takase and M. Yasunami, *J. Synth. Org. Chem. Jpn.*, **39**, 1172–1182 (1981).
88M1113	A. Messmer, G. Hajós, and G. Timári, *Monatsh. Chem.*, **119**, 1113–1119 (1988).
92H429	H. Wakabayashi, P.-W. Yang, C.-P. Wu, K. Shindo, S. Ishikawa, and T. Nozoe, *Heterocycles*, **34**, 429–434 (1992).
95JOC7508	T. Ishiyama, M. Murata, and N. Miyaura, *J. Org. Chem.*, **60**, 7508–7510 (1995).
96BCJ1645	T. Ueno, H. Toda, M. Yasunami, and M. Yoshifuji, *Bull. Chem. Soc. Jpn.*, **69**, 1645 (1996).
96JHC841	T. Mori, K. Imafuku, M.-Z. Piao, and K. Fujimori, *J. Heterocycl. Chem.*, **33**, 841–846 (1996).
96JOC9280	A. Arcadi, S. Cacchi, M.D. Rosario, G. Fabrizi, and F. Marinelli, *J. Org. Chem.*, **61**, 9280–9288 (1996).
99BCJ2543	S. Ito, N. Morita, and T. Asao, *Bull. Chem. Soc. Jpn.*, **72**, 2543–2548 (1999).
99S1349	M. Oda, T. Kajioka, K. Haramoto, R. Miyatake, and S. Kuroda, *Synthesis*, 1349–1353 (1999).
00JHC1019	D.-L. Wang and K. Imafuku, *J. Heterocycl. Chem.*, **37**, 1019–1031 (2000).
01T715	T.-I. Ho, C.-K. Ku, and R.S.H. Liu, *Tetrahedron Lett.*, **42**, 715–717 (2001).
02AGE3056	T. Ishiyama, J. Takagi, J.F. Hartwig, and N. Miyaura, *Angew. Chem. Int. Ed.*, **41**, 3056–3058 (2002).
02S1013	K. Kurotobi, H. Tabata, M. Miyauchi, T. Murafuji, and Y. Sugihara, *Synthesis*, 1013–1016 (2002).
03EJO3663	K. Kurotobi, M. Miyauchi, K. Takakura, T. Murafuji, and Y. Sugihara, *Eur. J. Org. Chem.*, 3663–3665 (2003).
03H339	S. Ito, S. Moriyama, M. Nakashima, M. Watanabe, T. Kubo, M. Yasunami, K. Fujimori, and N. Morita, *Heterocycles*, **61**, 339–348 (2003).

03MM536	F. Wang and Y.-H. Lai, *Macromolecules*, **36**, 536–538 (2003).
04EJOC899	T. Asao, S. Ito, and I. Murata, *Eur. J. Org. Chem.*, 899–928 (2004).
04H305	N. Morita, S. Moriyama, T. Shoji, M. Nakashima, M. Watanabe, S. Kikuchi, S. Ito, and K. Fujimori, *Heterocycles*, **64**, 305–316 (2004).
04JHC723	H. Takao, D.-L. Wang, S. Kikuchi, and K. Imafuku, *J. Heterocycl. Chem.*, **41**, 723–729 (2004).
04JOC8652	A.L. Crombie, J.L. Kane Jr, K.M. Shea, and R.L. Danheiser, *J. Org. Chem.*, **69**, 8652–8667 (2004).
04MM3222	F. Wang, Y.-H. Lai, and M.-Y. Han, *Macromolecules*, **37**, 3222–3230 (2004).
05EJO2207	H. Salman, Y. Abraham, S. Tal, S. Meltzman, M. Kapon, N. Tessler, S. Speiser, and Y. Eichen, *Eur. J. Org. Chem.*, 2207–2212 (2005).
05H91	T. Shoji, S. Kikuchi, S. Ito, and N. Morita, *Heterocycles*, **66**, 91–94 (2005).
05OL1055	K. Kurotobi and A. Osuka, *Org. Lett.*, **7**, 1055–1058 (2005).
06AGE3944	K. Kurotobi, K.S. Kim, S.B. Noh, D. Kim, and A. Osuka, *Angew. Chem. Int. Ed.*, **45**, 3944–3947 (2006).
07H1413	M. Oda, S. Kishi, N.C. Thanh, and S. Kuroda, *Heterocycles*, **71**, 1413–1416 (2007).
07H2237	D.-L. Wang, J.-J. Deng, J. Xu, and K. Imafuku, *Heterocycles*, **71**, 2237–2242 (2007).
07H951	M. Nishiura, I. Ueda, and K. Yamamura, *Heterocycles*, **74**, 951–960 (2007).
07JHC245	A.C. Razus, L. Birzan, C. Pavel, O. Lehadus, A. Corbu, F. Chiraleu, and C. Enache, *J. Heterocycl. Chem.*, **44**, 245–250 (2007).
07JOC744	S. Wakabayashi, Y. Kato, K. Mochizuki, R. Suzuki, M. Matsumoto, Y. Sugihara, and M. Shimizu, *J. Org. Chem.*, **72**, 744–749 (2007).
07T10608	M. Oda, N.C. Thanh, M. Ikai, H. Fujikawa, K. Nakajima, and S. Kuroda, *Tetrahedron*, **63**, 10608–10614 (2007).
07TL1099	T. Shoji, R. Yokoyama, S. Ito, M. Watanabe, K. Toyota, M. Yasunami, and N. Morita, *Tetrahedron Lett.*, **48**, 1099–1103 (2007).
07TL3009	T. Shoji, S. Ito, M. Watanabe, K. Toyota, M. Yasunami, and N. Morita, *Tetrahedron Lett.*, **48**, 3009–3012 (2007).
07YCHHDX1305	D.-L. Wang, J. Xu, S. Han, and Z. Gu, *Youji Huaxue*, **27**, 1305–1308 (2007).
08EJO5823	J. Higashi, T. Shoji, S. Ito, K. Toyota, M. Yasunami, and N. Morita, *Eur. J. Org. Chem.*, 5823–5831 (2008).
08H383	S. Wakabayashi, R. Uriu, T. Asakura, C. Akamatsu, and Y. Sugihara, *Heterocycles*, **75**, 383–390 (2008).
08JOC2256	S. Ito, T. Okujima, S. Kikuchi, T. Shoji, N. Morita, T. Asao, T. Ikoma, S. Tero-Kubota, J. Kawakami, and A. Tajiri, *J. Org. Chem.*, **73**, 2256–2263 (2008).
09EJO1554	T. Shoji, S. Ito, T. Okujima, J. Higashi, R. Yokoyama, K. Toyota, M. Yasunami, and N. Morita, *Eur. J. Org. Chem.*, 1554–1563 (2009).
09EJO4307	T. Shoji, S. Ito, K. Toyota, T. Iwamoto, M. Yasunami, and N. Morita, *Eur. J. Org. Chem.*, 4307–4315 (2009).
09H1079	N. Furuta, T. Mizutani, A. Ohta, and K. Fujimori, *Heterocycles*, **77**, 1079–1088 (2009).
09H451	A. Ohta, S. Kuroda, N.C. Thanh, K. Terasawa, K. Fujimori, K. Nakajima, and M. Oda, *Heterocycles*, **79**, 451–470 (2009).

10CB1734	R. Scholl and J. Mansfeld, *Ber. Dtsch. Chem. Ges.*, **43**, 1734–1746 (1910).
10CC1994	S.-I. Kato and F. Diederich, *Chem. Commun.*, **46**, 1994–2006 (2010).
10EJO1059	T. Shoji, S. Ito, K. Toyota, and N. Morita, *Eur. J. Org. Chem.*, 1059–1069 (2010).
10TL5127	T. Shoji, K. Okada, S. Ito, K. Toyota, and N. Morita, *Tetrahedron Lett.*, **51**, 5127–5130 (2010).
11CRV2306	T. Michinobu, *Chem. Soc. Rev.*, **40**, 2306–2316 (2011).
11EJO5311	T. Shoji, S. Ito, J. Higashi, and N. Morita, *Eur. J. Org. Chem.*, 5311–5322 (2011).
11H1283	N. Morita, S. Moriyama, K. Toyota, M. Watanabe, S. Kikuchi, S. Ito, and K. Fujimori, *Heterocycles*, **82**, 1283–1296 (2011).
11JA10046	E. Amir, R.J. Amir, L.M. Campos, and C.J. Hawker, *J. Am. Chem. Soc.*, **133**, 10046–10049 (2011).
11SL2279	S. Ito, T. Shoji, and N. Morita, *Synlett*, **16**, 2279–2298 (2011).
11CS2017	Y. Mitamura, H. Yorimitsu, K. Oshima, and A. Osuka, *Chem. Sci.*, **2**, 2017–2021 (2011).
12CL1644	T. Shoji, A. Yamamoto, Y. Inoue, E. Shimomura, S. Ito, J. Higashi, and N. Morita, *Chem. Lett.*, **41**, 1644–1646 (2012).
12H305	T. Shoji, J. Higashi, S. Ito, M. Oda, M. Yasunami, and N. Morita, *Heterocycles*, **86**, 305–315 (2012).
12H35	T. Shoji, Y. Inoue, S. Ito, T. Okujima, and N. Morita, *Heterocycles*, **85**, 35–41 (2012).
12JOC3270	J.-I. Kitai, T. Kobayashi, W. Uchida, M. Hatakeyama, S. Yokojima, S. Nakamura, and K. Uchida, *J. Org. Chem.*, **77**, 3270–3276 (2012).
12OL2316	Y. Yamaguchi, Y. Maruya, H. Katagiri, K.-I. Nakayama, and Y. Ohba, *Org. Lett.*, **14**, 2316–2319 (2012).
12TL1493	T. Shoji, Y. Inoue, and S. Ito, *Tetrahedron Lett.*, **53**, 1493–1496 (2012).
12JMC10448	F. Wang, T.T. Lin, C. He, H. Chi, T. Tanga, and Y.-H. Lai, *J. Mater. Chem.*, **22**, 10448–10451 (2012).
13CL638	T. Shoji, A. Yamamoto, E. Shimomura, M. Maruyama, S. Ito, T. Okujima, K. Toyota, and N. Morita, *Chem. Lett.*, **42**, 638–640 (2013).
13JOC12513	T. Shoji, M. Maruyama, E. Shimomura, A. Maruyama, S. Ito, T. Okujima, K. Toyota, and N. Morita, *J. Org. Chem.*, **78**, 12513–12524 (2013).
13CCL622	D.-L. Wang, Z. Dong, J. Xu, and D. Li, *Chin. Chem. Lett.*, **24**, 622–624 (2013).
14BCJ141	T. Shoji, A. Maruyama, M. Maruyama, S. Ito, T. Okujima, J. Higashi, K. Toyota, and N. Morita, *Bull. Chem. Soc. Jpn.*, **87**, 141–154 (2014).
14H2588	T. Shoji, A. Maruyama, S. Ito, T. Okujima, M. Yasunami, J. Higashi, and N. Morita, *Heterocycles*, **89**, 2588–2603 (2014).
14T2796	S. Ito, S. Yamazaki, S. Kudo, R. Sekiguchi, J. Kawakami, M. Takahashi, T. Matsuhashi, K. Toyota, and N. Morita, *Tetrahedron*, **70**, 2796–2803 (2014).
14CS3753	M. Murai, S.-Y. Ku, N.D. Treat, M.J. Robb, M.L. Chabinyc, and C.J. Hawker, *Chem. Sci.*, **5**, 3753–3760 (2014).
14CS4483	E. Amir, M. Murai, R.J. Amir, J.S. Cowart Jr, M.L. Chabinyc, and C.J. Hawker, *Chem. Sci.*, **5**, 4483–4489 (2014).
15IC16	A.E. Ion, S. Nica, A.M. Madalan, S. Shova, J. Vallejo, M. Julve, F. Lloret, and M. Andruh, *Inorg. Chem.*, **54**, 16–18 (2015).

15OBC10191	T. Shoji, A. Maruyama, T. Araki, S. Ito, and T. Okujima, *Org. Biomol. Chem.*, **13**, 10191–10197 (2015).
15S2663	J. Dubovik and A. Bredihhin, *Synthesis*, **47**, 2663–2669 (2015).
15S538	J. Dubovik and A. Bredihhin, *Synthesis*, **47**, 538–548 (2015).
15RSCA63282	E.A. Dragu, A.E. Ion, S. Shova, D. Bala, C. Mihailciuc, M. Voicescu, S. Ionescu, and S. Nica, *RSC Adv.*, **5**, 63282–63286 (2015).
16AGE2564	P. Cowper, Y. Jin, M.D. Turton, G. Kociok-Kçhn, and S.E. Lewis, *Angew. Chem. Int. Ed.*, **55**, 2564–2568 (2016).
16TL4514	J.-I. Yamaguchi and S. Sugiyama, *Tetrahedron Lett.*, **57**, 4514–4518 (2016).
16BJOC1812	A.E. Ion, L. Cristian, M. Voicescu, M. Bangesh, A.M. Madalan, D. Bala, C. Mihailciuc, and S. Nica, *Beilstein J. Org. Chem.*, **12**, 1812–1825 (2016).
16RSCA32124	L. Gai, J. Chen, Y. Zhao, J. Mack, H. Lu, and Z. Shen, *RSC Adv.*, **6**, 32124–32129 (2016).
16RSCA78303	T. Shoji, M. Tanaka, T. Araki, S. Takagaki, R. Sekiguchi, and S. Ito, *RSC Adv.*, **6**, 78303–78306 (2016).
17CEJ5126	T. Shoji, D. Nagai, M. Tanaka, T. Araki, A. Ohta, R. Sekiguchi, S. Ito, S. Mori, and T. Okujima, *Chem. Eur. J.*, **23**, 5126–5136 (2017).
17H1870	T. Shoji, S. Takagaki, M. Tanaka, T. Araki, S. Sugiyama, R. Sekiguchi, A. Ohta, S. Ito, and T. Okujima, *Heterocycles*, **94**, 1870–1883 (2017).
17JOC1657	T. Shoji, T. Araki, S. Sugiyama, A. Ohta, R. Sekiguchi, S. Ito, T. Okujima, and K. Toyota, *J. Org. Chem.*, **82**, 1657–1665 (2017).
17T2488	L. Birzan, M. Cristea, C.C. Draghici, V. Tecuceanu, A. Hanganu, E.-M. Ungureanu, and A.C. Razus, *Tetrahedron*, **73**, 2488–2500 (2017).
17JMCC2620	Y.M. Poronik, L.M. Mazur, M. Samoć, D. Jacquemin, and D.T. Gryko, *J. Mater. Chem. C*, **5**, 2620–2628 (2017).
18OBC480	T. Shoji, M. Tanaka, S. Takagaki, K. Miura, A. Ohta, R. Sekiguchi, S. Ito, S. Mori, and T. Okujima, *Org. Biomol. Chem.*, **16**, 480–489 (2018).

FURTHER READING

03S30	K. Kurotobi, H. Tabata, M. Miyauchi, R.A.F.M. Mustafizur, K. Migita, T. Murafuji, Y. Sugihara, H. Shimoyama, and K. Fujimori, *Synthesis*, **12**, 30–34 (2003).

CHAPTER TWO

Recent Developments in the Chemistry of Pyrazoles

Andrew W. Brown[1]
Sygnature Discovery Ltd, Biocity, Nottingham, United Kingdom
[1]Corresponding author: e-mail address: a.brown@sygnaturediscovery.com

Contents

1.	Introduction	56
2.	Synthesis	57
	2.1 Condensation Reactions	57
	2.2 Cycloaddition Reactions	66
	2.3 Metal Catalysis and Metal-Promoted Synthesis	86
3.	Properties and Applications	93
	3.1 Biologically Active Pyrazoles	93
	3.2 Pyrazoles in Coordination Chemistry	96
	3.3 Miscellaneous Applications of Pyrazoles	99
4.	Concluding Remarks	101
References		102
Further Reading		107

Abstract

The pyrazole motif has been found to be amenable to extensive, wide-ranging applications. A long-standing history of success has been achieved in the pharmaceutical field. More recent exploits have seen exciting developments in the areas of metal–organic frameworks, liquid crystals, and electroluminescence to name but a few. Such diverse applications have resulted in a demand for access to evermore complex examples of this diazole, which has driven synthetic chemists to develop a multitude of novel methods. In particular, modern methods have strived to overcome limitations in regioselectivity that are associated with classical methods. They have also sought to install significant molecular complexity within a single transformation to avoid lengthy functionalization sequences. As highlighted in this review, covering advances in the synthesis and application of pyrazoles in the past two decades, significant progress has been made in the use of metal catalysis, cycloaddition reactions, and multicomponent reactions for the construction of 5-membered heterocycles.

Keywords: Condensation, Cycloaddition, Heterocycle, Metal catalysis, Multicomponent reaction, Pyrazole, Regioselectivity, Synthesis

1. INTRODUCTION

Pyrazoles are aromatic, 5-membered diazoles which exhibit a diverse array of chemical and biological properties, but they also have a rich and interesting history. Indeed, the very first pyrazole discovered, phenazone **1** (Scheme 1), saw commercial success as a marketed pharmaceutical. The compound was found to exhibit antipyretic properties, and patent rights were sold to Farbwerke in 1884 (1884BDCG2032, 1896BKW1061). However, it was only identified as a pyrazole shortly before entering the marketplace. Phenazone had been unintentionally prepared when Knorr attempted to synthesize a tetrahydroquinoline via the reaction of phenylhydrazine and acetoacetic ester (1883BDCG2597). Instead, he isolated a pyrazolone which, after subsequent N-methylation, furnished phenazone. Such a successful and intriguing beginning has no doubt added to the interest surrounding this heterocyclic motif. The reactions of 1,3-dicarbonyl compounds soon proved to be applicable to the synthesis of other heterocycles (1884BDCG2756, 1884BDCG2863) and have been utilized in the preparation of countless compounds with numerous applications over the following decades. A particular limitation of this method with regard to pyrazole synthesis is the control of regiochemistry of N-substituted pyrazoles. The method relies on either regioselective N-functionalization of N-unsubstituted pyrazoles or regioselective condensation of unsymmetrical hydrazines with 1,3-dicarbonyl equivalents. These strategies can often lead to regioisomeric mixtures, especially when highly complex pyrazoles are required (1996CHEC(3)1, 2008CHEC(4)1).

In the past two decades, a number of methods have been developed to address limitations in the regioselective synthesis of substituted pyrazoles. Substantial progress has also been made in the one-pot construction of complex examples from simple building blocks. Furthermore, transition metal catalysis has opened the door to new, exciting methods of construction from previously unused building blocks. The array of new methods available in

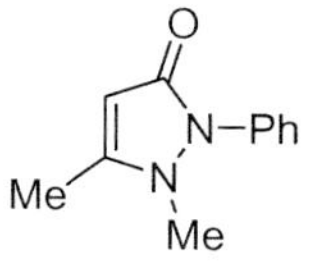

1
Phenazone

Scheme 1 Structure of phenazone.

the arsenal of the synthetic chemist has provided access to unusual structures which in turn have diversified the areas in which the pyrazole core has been employed. This review is intended to highlight some of the many advancements in both the synthesis and application of pyrazoles from the past two decades.

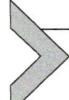

2. SYNTHESIS
2.1 Condensation Reactions

At the turn of the millennium, El Kaim and coworker reported the formation of pyrazoles from N-trichloroacylhydrazones (2000SL353). This interesting substrate class has unusual reactivity associated with it (1999JCS(CC)1893). Upon treatment with sodium carbonate in the presence of β-ketoesters, trisubstituted pyrazoles are formed in moderate to good yields (Scheme 2). Aliphatic hydrazones are successful substrates when an aprotic, polar solvent (NMP) is used. The proposed mechanism proceeds via deprotonation of hydrazone **2** to form **3**, which spontaneously undergoes chlorine transfer and then protonation by solvent to form intermediate **4**. The β-ketoester then reacts with **4** to form cyclization precursor **5**. Cyclization to **6** and subsequent amide bond cleavage affords the pyrazole product **7**.

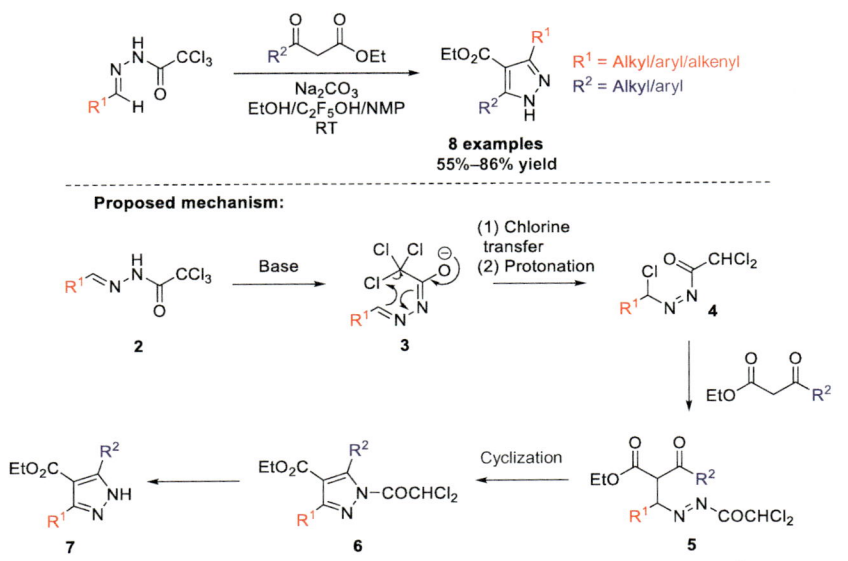

Scheme 2 Reaction of trichloroacetylhydrazones with β-ketoesters. *L. El Kaim and S. Lacroix, Synlett,* **3**, *353 (2000).*

Continuing their interest in hydrazones as pyrazole precursors, Kaim et al. reported the preparation of aminopyrazoles using isocyanides (2000SL489). Treatment of α-halohydrazones with sodium carbonate and isocyanides furnished a small variety of pyrazoles in respectable yields. The reaction is believed to proceed via an azoalkene intermediate. Two limitations of the reaction are the availability of α-halohydrazones and that aliphatic hydrazones are not effective substrates. The authors addressed the availability of starting materials by forming an azoalkene precursor from the Mannich reaction of hydrazones with aldehydes. This allowed them to utilize the chemistry for the synthesis of a number of analogues of the insecticide fipronil® **8** (Scheme 3) (2002TL8319). The authors suggested that the reaction proceeds via [4 + 1] cycloaddition, but mechanistic studies were not undertaken.

A number of years later, Zhang et al. disclosed an oxidative C–H activation route to azoalkenes which they applied to the synthesis of triazoles (2013AGE13324). Subsequently, Wang et al. reported a metal-free variant of the reaction using an iodine/*tert*-butyl peroxybenzoate (TBPB) oxidant system (2014OL5108). Inspired by these successes, Wang et al. then merged this system with that developed by El Kaim et al. to synthesize aminopyrazoles (2015OL1521). Optimization revealed iodine/*tert*-butyl hydrogen peroxide to be the optimal oxidant system. Aromatic *N*-sulfonylhydrazones reacted with a range of alkyl- and aryl isocyanides (Scheme 4). The reaction is not compatible with aliphatic hydrazones. Interestingly, *N*-bochydrazones, which had been successful for El Kaim, were not suitable substrates in this reaction. Impressively, the reaction was extended to a one-pot synthesis from aldehydes, albeit in lower yield. The proposed mechanism involves iodination

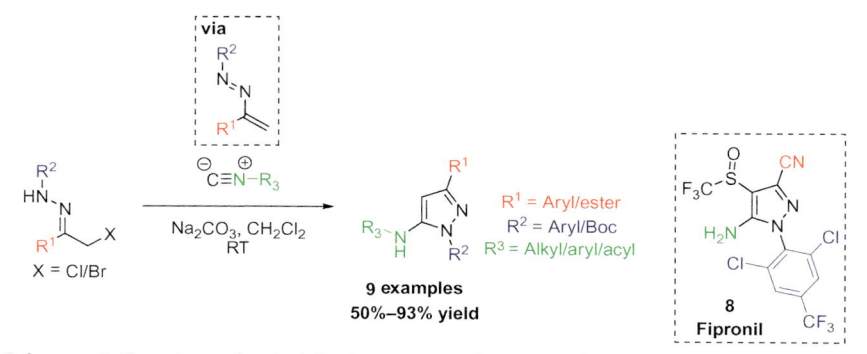

Scheme 3 Reaction of α-halohydrazones and isocyanides. J. E. Ancel, L. El Kaim, A. Gadras, L. Grimaud, and N.K. Jana, Tet. Lett., **43**, 8319 (2002).

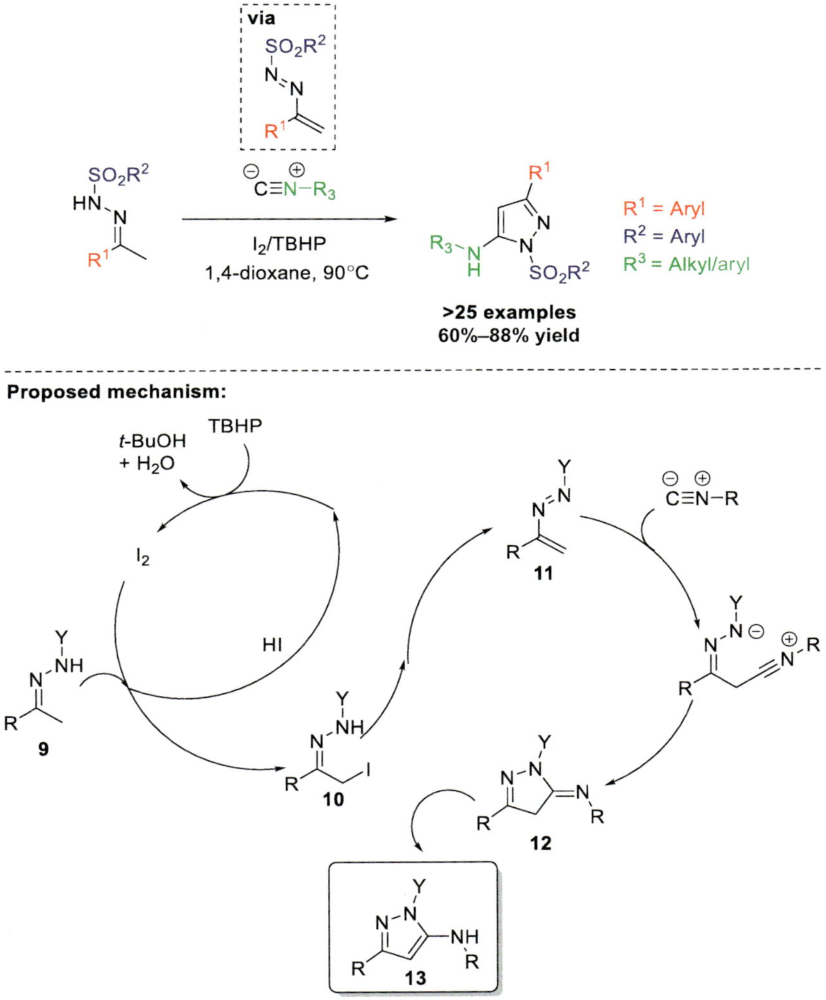

Scheme 4 Iodine–TBHP-promoted reaction of hydrazones with isocyanides. G.C. Senadi, W.-P. Hu, T.-Y. Lu, A.M. Garkhedkar, J.K. Vandavasi, and J.-J. Wang, Org. Lett., **17**, 1521 (2015).

of the hydrazones **9** to form **10** and HI. The hydrogen iodide is then oxidized to iodine to generate further **10**. The α-halohydrazones **10** then generate an azoalkene **11** which can react with an isocyanide to form a pyrazoline **12**, which after a [1,3]-*H* shift forms the target pyrazole **13**.

Kim et al. reported the synthesis of pyrazoles from the reaction of Bayliss–Hillman adducts with hydrazines (2003TL6737). The reaction is particularly noteworthy due to the fact that single isomers are isolated, as necessitated by the proposed mechanism (Scheme 5). An interesting array of pyrazoles has been successfully prepared, typically in high yields. Perhaps the most interesting example is the tetrasubstituted pyrazole **14** bearing four different carbon substituents. This clearly demonstrates the value of the method. A limitation of the method lies with the availability of Bayliss–Hillman adducts and the limitations these place on substitution patterns, particularly at the C4 position of the pyrazole.

An interesting addition to condensation methodologies was reported by Gerstenberger et al. who reported a one-pot procedure for the preparation of *N*-arylpyrazoles (2009OL2097). A number of arylhydrazines were formed in situ from the reaction of aryllithiums with di-*tert*-butylazodicarboxylate (Scheme 6). The arylhydrazines generated were then reacted with a variety of 3-carbon donors for regiocontrolled synthesis of the desired *N*-arylpyrazoles. A wide variety of aryl groups were successfully transformed, including three pyridyl examples (e.g., **19**). A clear limitation of the reaction is the formation of the aryllithium in the first step, which limits functional group tolerance on the aromatic ring. However, this limitation is partly mitigated by the versatility of arylhydrazines in pyrazole-forming reactions. Therefore, the methodology successfully broadens the range of *N*-arylpyrazoles available to synthetic chemists.

Ila et al. have developed a strategy toward complementary *N*-arylpyrazole substitution patterns based on monothio-1,3-diketones (2013JOC4960). High regioselectivity is obtained due to the stark difference in reactivity between the carbonyl and thiocarbonyl groups. The reactions are typically high yielding, and complete regioselectivity has been reported. In all

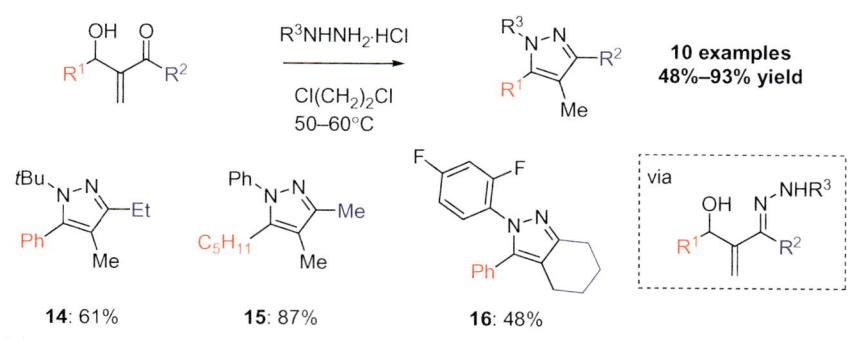

Scheme 5 Synthesis of pyrazoles from Bayliss–Hillman adducts. K.Y. Lee, J.M. Kim, and J.M. Kim, Tet. Lett., **44**, 6737 (2003).

Scheme 6 One-pot synthesis of N-arylpyrazoles from in situ-generated aryllithiums. B.S. Gerstenberger, M.R. Rauckhorst, and J.T. Starr, Org. Lett., **11**, 2097 (2009).

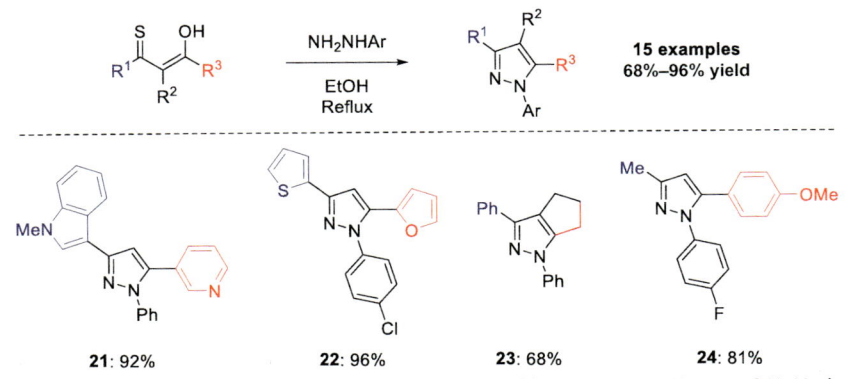

Scheme 7 Pyrazole synthesis from monothio-1,3-diketones. S.V. Kumar, S.K. Yadav, B. Raghava, B. Saraiah, H. Ila, K.S. Rangappa, and A. Hazra, J. Org. Chem., **78**, 4960 (2013).

examples the thioketone substituent is incorporated at the pyrazole C3 position (e.g., **21–24**) (Scheme 7). There are limited examples that include C4 substitution, such as bicyclic structure **23**.

In the same report, complementary regioselectivity is achieved by S-methylation of the monothio-1,3-diketones and running the reaction under basic conditions (Scheme 8) (2013JOC4960). Impressively, all examples exhibit reversed and complete regioselectivity with thioenolether

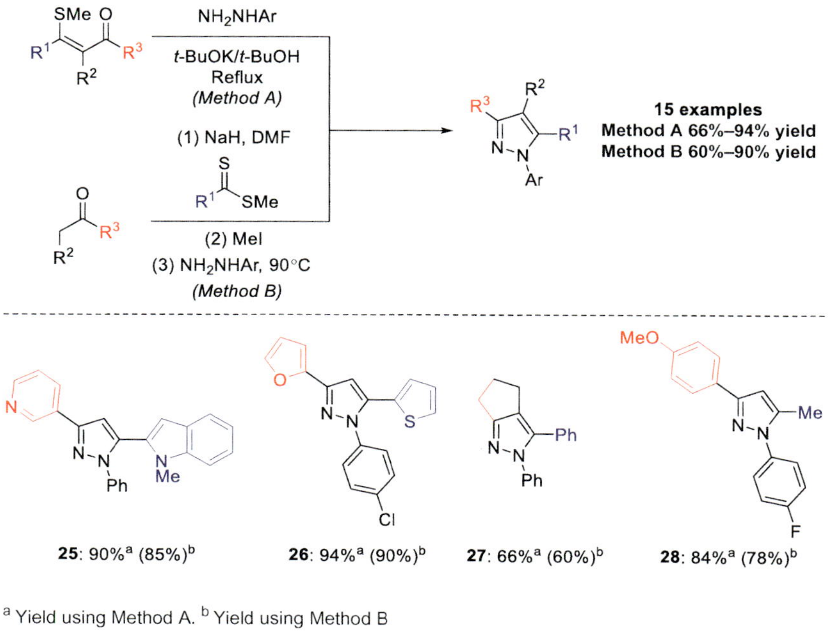

Scheme 8 Pyrazole synthesis from 3-methylthio-2-propenones. S.V. Kumar, S.K. Yadav, B. Raghava, B. Saraiah, H. Ila, K.S. Rangappa, and A. Hazra, J. Org. Chem., **78**, 4960 (2013).

substituent incorporation at the pyrazole C5 position, e.g., **25–28**. Furthermore, the procedure is amenable to adaptation to a one-pot, three-component reaction from ketones. Overall, the procedures developed are quite versatile and allow the construction of an array of interesting pyrazoles.

An intriguing alternative to the use of hydrazine condensations was reported by Nair et al. (2006OL2213). These authors reported the use of the Huisgen zwitterion, generated from azodicarboxylates **29** and triphenylphosphine and best known for its utility in the Mitsunobu reaction (1981S1), for the preparation of pyrazoles and pyrazolines from allenic esters. Isolated examples of use of the Huisgen zwitterion for the synthesis of pyrazoles had been reported previously by Cookson (1963JCS6062) and by Huisgen (1969AGE513). However, Nair's work generalized this strategy. The scope of the reaction is mostly limited by the availability of allenic esters, but a variety of alkyl and aryl substituents are tolerated (Scheme 9). Only a limited number of azodicarboxylates have been used in the reactions, but importantly diisopropyl azodicarboxylate is successful, and has an improved safety profile over the methyl and ethyl variants. The mechanism

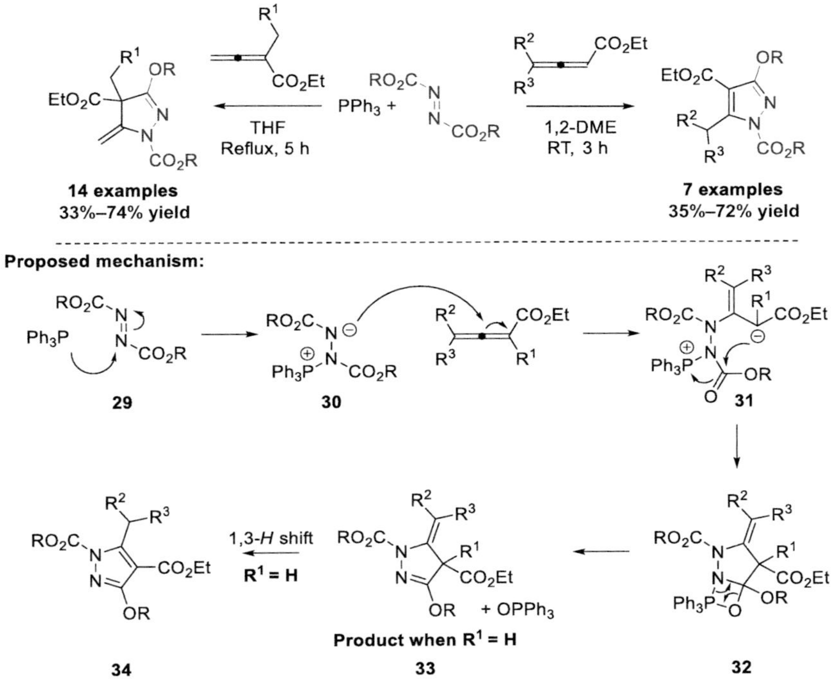

Scheme 9 Formation of pyrazoles and pyrazolines from allenic esters and azodicarboxylates. V. Nair, A.T. Biju, K. Mohanan, and E. Suresh, Org. Lett., **8**, 2213 (2006).

of the reaction is thought to proceed via formation of zwitterions **30**, which then undergo 1,4-conjugate addition with the allenic ester to form **31**. Intermediates **31** can then cyclize to form the pyrazoline core **32**, which can liberate triphenylphosphine oxide to form the products **33**. When R^1 is H then the pyrazolines **33** can undergo a 1,3-H shift to form the pyrazoles **34**.

2.1.1 Ynones in Pyrazole Synthesis

An interesting alternative class of electrophilic condensation partners to 1,3-dicarbonyl compounds is ynones. These compounds are advantageous due to the significant differentiation between their two electrophilic sites, thus providing an innate opportunity for regioselectivity in condensations with unsymmetrical hydrazines. An early example of the employment of this strategy was reported by Miller and Reiser (1993JHC755). Harrity et al. successfully extended this strategy to allow the use of borylated ynones and subsequently allow the synthesis of borylated pyrazoles (2012OL5354). The use of a stable boron moiety, namely, potassium trifluoroborates, was found to be fundamental to the success of the strategy. Condensation of ynone trifluoroborates with hydrazine proceeds smoothly and in high yield with aryl-substituted ynones, e.g., **36**. Lower yields are observed with alkyl-substituted ynones, e.g., **37**. When substituted hydrazines are employed, regioselectivity and the major product isolated are dependent on the nature of the hydrazine substituent. *N*-Alkyl hydrazines typically favor product **35a** (Scheme 10, e.g., **38**, **39**), whereas *N*-aryl hydrazines favor product **35b**, e.g., **40**. This observation is consistent with a mechanism involving initial conjugate addition of the most nucleophilic nitrogen, followed by condensation of the second nitrogen onto the carbonyl. In a follow-up paper, including examples **41**–**44**, an unusual reversal in expected regioselectivity was observed in the case of nitrile **43**, perhaps alluding to complexities and future scope extension (2017JOC1688). However, more complex ynone trifluoroborates are valid reaction partners affording interesting, late-stage intermediates such as piperidine **44**. The reaction has been extended to the formation of thiophene trifluoroborates (2018OL198).

A practical method in which ynones are generated as intermediates in a one-pot synthesis of pyrazoles (**45a**,**b**) (Scheme 11) from terminal alkynes has been reported by Togo et al. (2014JOC2049). The majority of the examples reported used hydrazine, and therefore, there were no issues around regioselectivity (Scheme 11, **46** and **47**). As observed in previous

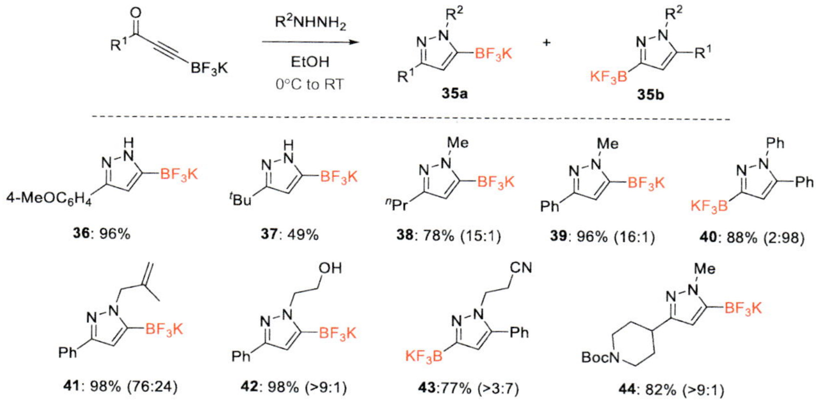

Ratio in parenthesis indicates ratio of regioisomers as determined by ^{1}H NMR

Scheme 10 Synthesis of pyrazole trifluoroborates from ynone trifluoroborates. *J.D. Kirkham, S.J. Edeson, S. Stokes, J.P.A. Harrity,* Org. Lett., ***14****, 5354 (2012).*

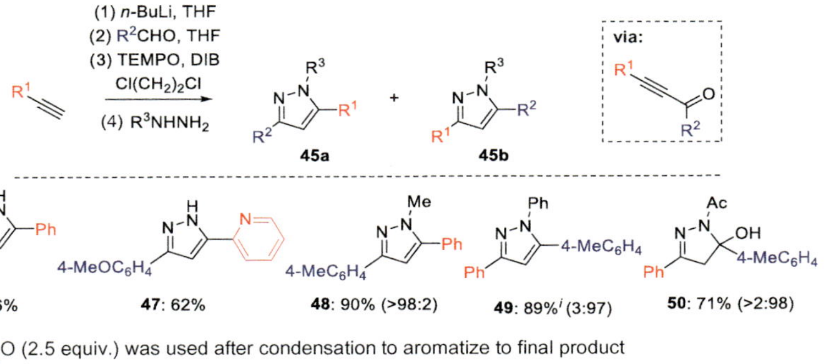

Scheme 11 The one-pot synthesis of pyrazoles from alkynes and hydrazines. R. Harigae, K. Moriyama, and H. Togo, J. Org. Chem., **79**, 2049 (2014).

reports, when substituted hydrazines were utilized the major regioisomer was dependent on the nature of the hydrazine substituent, e.g., **48** vs **49**. Interestingly, aromatization is not necessarily spontaneous; the formation of **49** required dehydration using tosic acid and **50** was isolated as the hydroxypyrazoline. A further feature of this procedure is that hydrazines can be replaced with hydroxylamine to enable the regioselective synthesis of isoxazoles.

Mori et al. described a palladium-catalyzed four-componant coupling of alkynes, carbon monoxide, hydrazines, and aryl iodides for the synthesis of both pyrazoles and isoxazoles (2005OL4487). The reaction presumably proceeds via the ynone, but mechanistic studies were not undertaken. Although an impressive transformation, the scope of the reaction is quite narrow. Electron–neutral aromatics and alkyl groups were shown to be tolerated on the alkyne (Scheme 12). Only hydrazine and methylhydrazine have been shown to be suitable substrates; phenylhydrazine failed to provide any product. A small selection of electron–neutral and electron-rich aryl iodides were shown to be compatible. In the case of methylhydrazine, the reaction was completely regioselective.

In 2017 Wang, Zhao, and coworkers also used palladium catalysis to generate ynones in situ from the oxidative Sonogashira-carbonylation of arylhydrazines (2017OL3466). A second equivalent of arylhydrazine is then able to condense onto the ynone and generate the pyrazole. A number of 1,3,5-trisubstituted pyrazoles, e.g., **51–54**, have been successfully prepared (Scheme 13). Aliphatic hydrazines and arylhydrazines with strong electron-withdrawing groups are not effective substrates. Similarly, electron-withdrawing groups are not tolerated on the alkyne partner, but aliphatic groups are well tolerated, e.g., **53**.

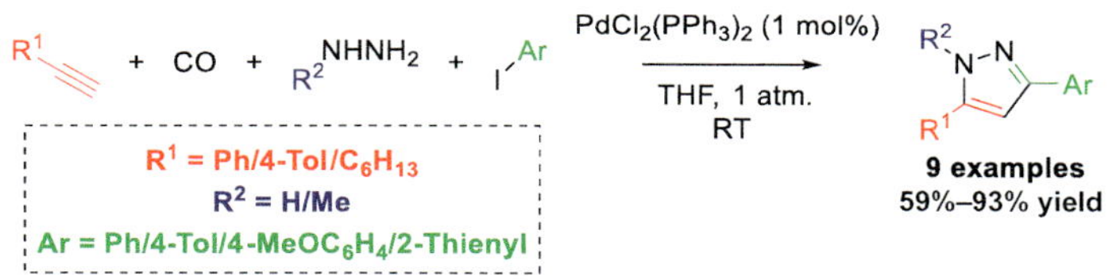

Scheme 12 4-Component coupling for the synthesis of pyrazoles. M.S.M. Ahmed, K. Kobayashi, and A. Mori, Org. Lett., **7**, 4487 (2005).

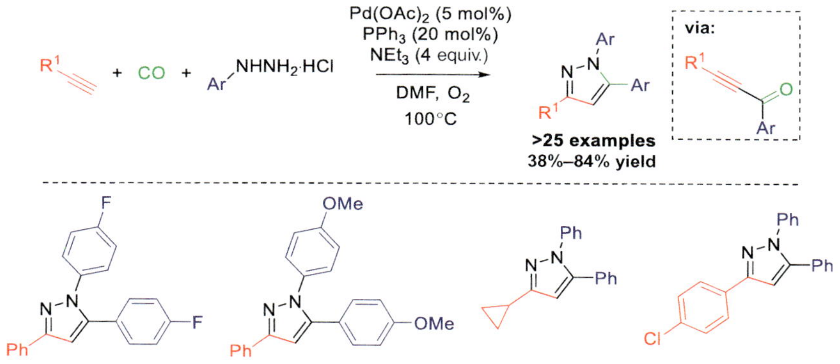

Scheme 13 Pd-catalyzed synthesis of pyrazoles from alkynes and arylhydrazines. Y. Tu, Z. Zhang, T. Wang, J. Ke, and J. Zhao, Org. Lett., **19**, 3466 (2017).

2.2 Cycloaddition Reactions

Cycloadditions represent a powerful method for the construction of heterocycles. Typically, they allow installation of significant functionality in a single step. These reactions are atom economical and have been a popular alternative to classic condensation routes to pyrazoles for a number of years.

2.2.1 Diazo Compounds in Pyrazole Synthesis

The past 20 years have seen rising interest in the use of diazo compounds for the construction of pyrazoles. In part, this has been due to advances in the preparation of the versatile diazo functional group (2001AGE1430, 2008AGE2359, 2015CRV9981). A particularly interesting example of the use of diazo compounds in pyrazole synthesis was reported by Namboothiri et al. (2007OL1125). The authors utilized the Bestmann–Ohira reagent (1989SC561, 1996SL521) with nitroalkenes for the construction of phosphoryl-substituted pyrazoles. Incredibly, the reactions were complete

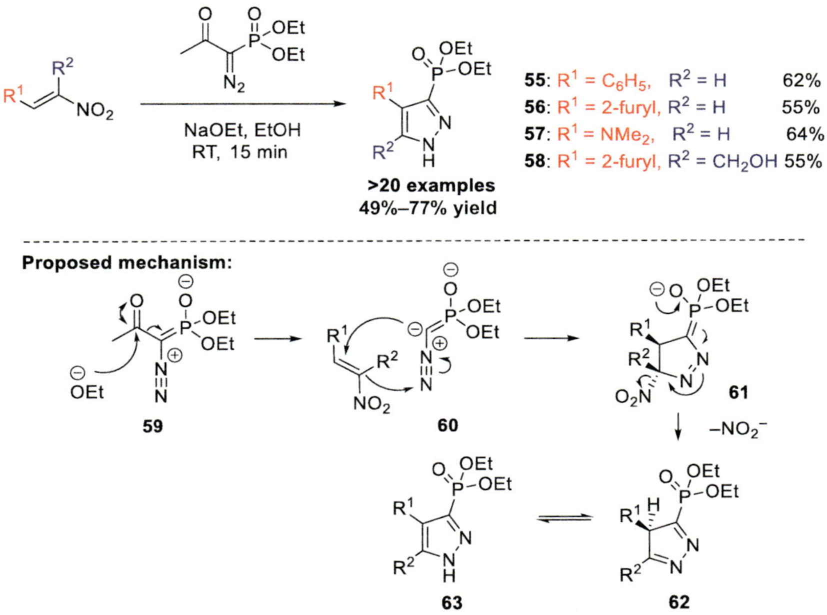

Scheme 14 Bestmann–Ohira reagent for the construction of phosphoryl-pyrazoles. R. Muruganantham, S.M. Mobin, and I.N.N. Namboothiri, Org. Lett., **9**, 1125 (2007).

in as little as 15 min at room temperature. A variety of phosphoryl-pyrazoles have been successfully prepared, e.g., **55**, including heteroaryl examples, such as **56** and **58** and aminopyrazole **57** (Scheme 14). The products were isolated as single isomers, with bond formation between the nitro group-bearing carbon and the diazo group nitrogen. This strategy has been successfully extended to trisubstituted nitroalkenes and NMR studies undertaken (2010JOC2197). Examples such as furan **58** were prepared by reacting the Bestmann–Ohira reagent with Morita–Bayliss–Hillman adducts. Overall, the strategy represents an efficient method for the preparation of phosphoryl-pyrazoles. The reaction represents a formal 1,3-dipolar cycloaddition and the authors propose that the reaction proceeds via cycloaddition. 1,3-Dipolar intermediates **60** are generated from the Bestmann–Ohira reagent **59**. The dipoles **60** then undergo [3 + 2] cycloaddition with the nitroalkene to form the adducts **61**. Loss of the nitro group results in initial products **62**, which tautomerize to the pyrazoles **63**. Alternatively, the reaction may proceed via 1,4-conjugate addition followed by cyclization. Indeed, the literature contains conflicting mechanistic proposals for these reactions. For clarity, the reactions have been listed under a single section in this review.

Namboothiri et al. expanded the substrate scope from nitroalkenes to chalcones with similar success to afford carbonyl-substituted, phosphoryl-pyrazoles (2011JOC4764). A similar strategy has been utilized by Mohanan et al. with α,β-unsaturated aldehydes to afford vinyl-substituted, phosphoryl-pyrazoles (Scheme 15) (2015OBC1492). The scope mostly encompasses aryl-substituted, α,β-unsaturated aldehydes, e.g., **64**–**66**, but did contain a single alkyl example, **67**. The proposed mechanism involves the formation of the key intermediates **68**, which react with a further equivalent of the Bestmann–Ohira reagent to form the alkynes **69**. Two subsequent 1,3-hydride shifts via the intermediates **70** furnish the pyrazoles **71**.

Ynones have also been shown to be effective substrates for the preparation of phosphoryl-pyrazoles, as demonstrated by Rastogi et al. (2014T5214). The reactions are extremely high yielding, and furnish carbonyl-substituted, phosphoryl-pyrazoles as single isomers, e.g., **72** and **73** (Scheme 16). Silyl-substituted ynones afforded exclusively proto-desilated products, e.g., **74** and **75**, providing a useful method for the preparation of less densely substituted phosphoryl-pyrazoles. An interesting observation is that the major tautomer observed in the solid state (X-ray crystal structure) and in solution (^{1}H and ^{31}P NMR) has the nitrogen adjacent to the phosphoryl group protonated. This is contrary to the observations of Namboothiri et al. in their early work (Scheme 14) (2010JOC2197).

The early success of Namboothiri et al. with the use of nitroalkenes inspired Smietana et al. to investigate whether simplified substrates can be

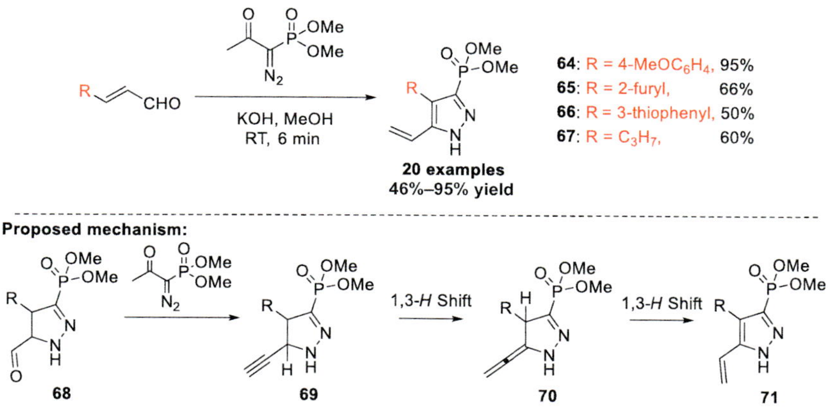

Scheme 15 Bestmann–Ohira reagent for the construction of vinyl-substituted phosphoryl-pyrazoles. S. Ahmed, A.K. Gupta, R. Kant, and K. Mohanan, Org. Biomol. Chem., **13**, 1492 (2015).

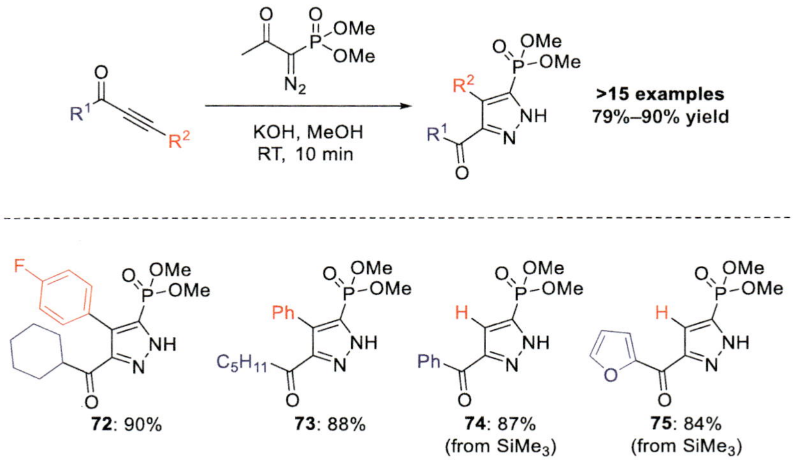

Scheme 16 Reaction of ynones with the Bestmann–Ohira reagent. *M.M.D. Pramanik, R. Kant, and N. Rastogi, Tetrahedron,* **70**, 5214 (2014).

employed in the reaction. Impressively, they discovered that a three-component coupling between aldehydes, malononitrile, and the Bestmann–Ohira reagent is possible (2010AGE3196). The scope is broad and the reactions high yielding. Various electron-withdrawing groups can be incorporated, as well as aryl, alkenyl, and alkyl substituents, e.g., **78–81** (Scheme 17). The authors propose that the Knoevenagel condensation-type product **76** is formed in situ, which then undergoes 1,4-addition with dipolar reagent **77**, followed by cyclization. Again, it is not known if the mechanism proceeds in this manner or by 1,3-dipolar cycloaddition. The same authors demonstrated that malononitriles can be replaced by ketones in a three-component reaction to facilitate a Claisen–Schmidt reaction followed by pyrazole formation with the Bestmann–Ohira reagent (2011EJO3184). A similar strategy was employed by Zhang and Yu et al. in which they reacted vinyl azides with in situ-generated diazo compounds (2013OL5967). The product pyrazoles are liberated by loss of azide.

In a similar manner, Kamal and Maurya et al. showed that the nitriles used by Smietana can be replaced with 1,3-dicarbonyl compounds (2015JOC4325). An advantage of this strategy is that toxic cyanide is not released during the reaction. The reaction is quite general and typically high yielding. Electron-rich and electron-poor aromatic aldehydes are successful, e.g., **84**, **85**, **86,** and **89** (Scheme 18) as well as aliphatic aldehydes, e.g., **87**. 1,3-Dicarbonyls bearing aliphatic, e.g., **82** and **83**

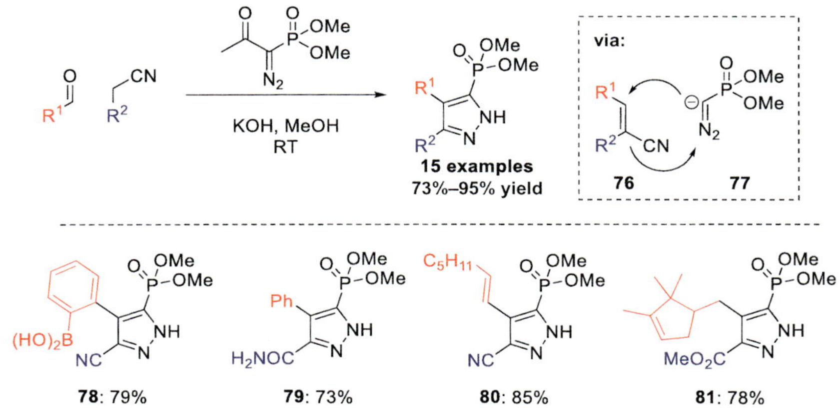

Scheme 17 3-Component reaction of aldehydes, malononitriles, and the Bestmann–Ohira reagent for the construction of phosphoryl-pyrazoles. *K. Mohanan, A.R. Martin, L. Toupet, M. Smietana, and J.-J. Vasseur*, Angew. Chem. Int. Ed., **49**, 3196 (2010).

and aromatic groups, e.g., **84** are tolerated in addition to esters, e.g., **88**. Substitution of the diazo compounds was slightly less varied—esters, e.g., **84**, **85** and aromatic groups, e.g., **86**, **87** are tolerated, but aliphatic groups are incompatible. The proposed mechanism begins with the Knoevenagel condensation of the aldehyde with the 1,3-dicarbonyl compound to form **90**, which undergo cycloaddition with the diazo compound to form the pyrazolines **91**. Intermediates **91** are in tautomeric equilibrium with **92** and attack of water on one of the carbonyls forms the hydrates **93**. The hydrates can then eliminate acetic acid to form pyrazolines **94**, which after subsequent oxidation forms the target pyrazoles **95**.

Namboothiri et al. have also shown that the α-diazo-β-ketoester substrates can react with chalcones to form pyrazoles (2016TL3146). The authors reported that only single pyrazole regioisomers are formed in this reaction (Scheme 19). However, the yields are typically low and the types of pyrazole available from this reaction are quite limited.

After success in the preparation of phosphoryl-pyrazoles, Namboothiri and coworker turned their attention to the preparation of sulfonylpyrazoles. In a similar manner to their use of the Bestmann–Ohira reagent, diazosulfones and nitroalkenes proved to be compatible reaction partners (2011OL4016). A wide variety of groups can be incorporated on the nitroalkene substrate, e.g., **96–98** (Scheme 20), and disubstituted nitroalkenes are successful reactants, e.g., **99** and **100**. A major limitation in the chemistry is in the sulfonyl groups exemplified. In this position, only

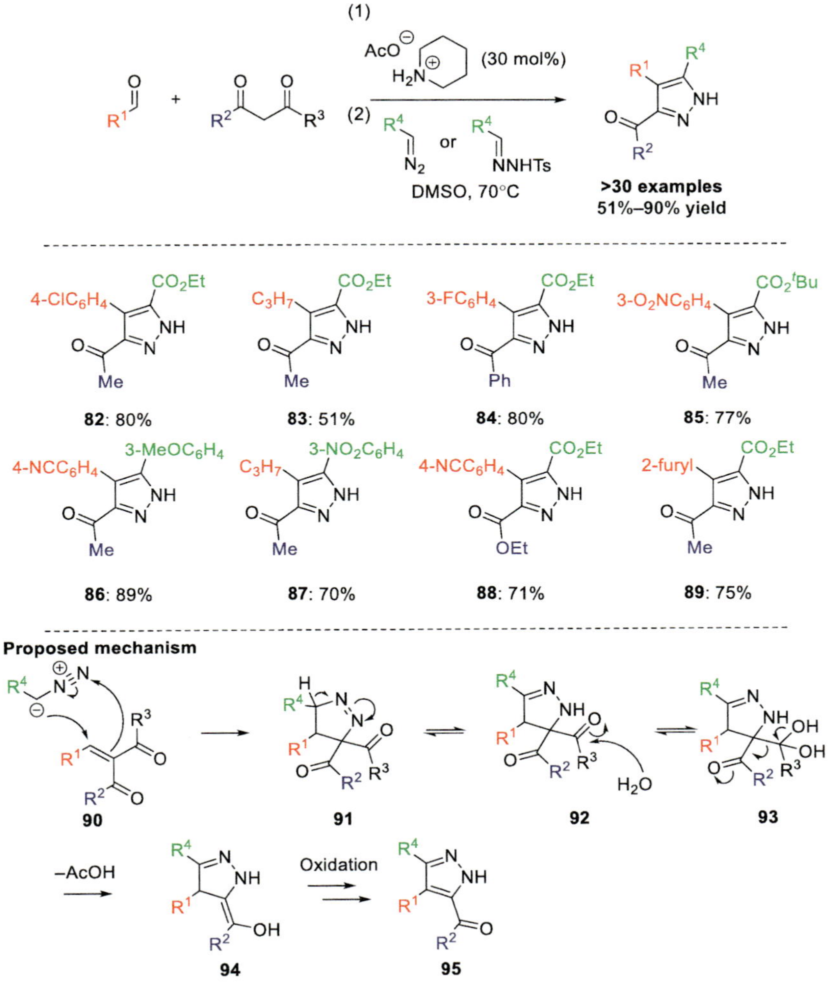

Scheme 18 Combinatorial approach to 3,4,5-trisubstituted pyrazoles. A. Kamal, K.N.V. Sastry, D. Chandrasekhar, G.S. Mani, P.R. Adiyala, J.B. Nanubolu, K.K. Singarapu, and R.A. Maurya, J. Org. Chem., **80**, 4325 (2015).

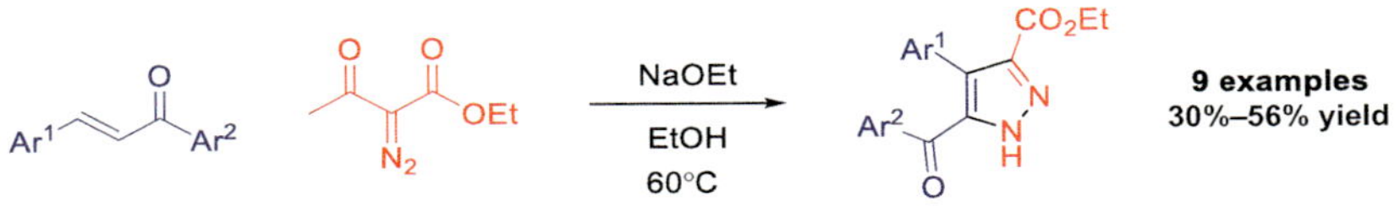

Scheme 19 Synthesis of pyrazoles from α-diazo-β-ketoesters. D. Nair, P. Pavashe, S. Katiyar, and I.N.N. Namboothiri, Tet. Lett., **57**, 3146 (2016).

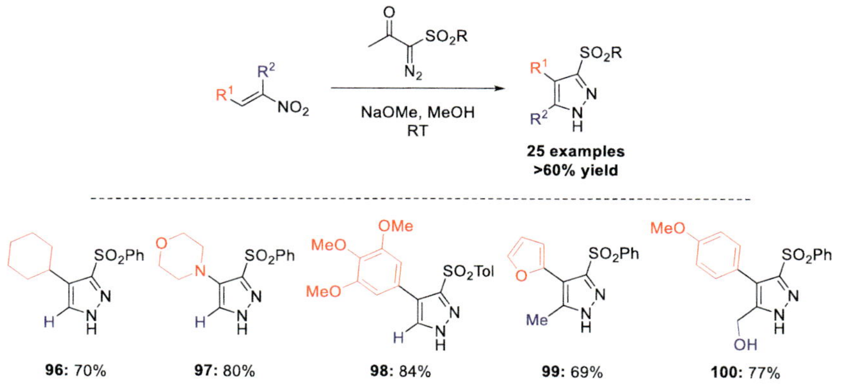

Scheme 20 Reaction of diazosulfones for the construction of sulfonylpyrazoles. R. Kumar and I.N.N. Namboothiri, Org. Lett., **13**, 4016 (2011).

phenyl and tolyl have been employed, possibly due to difficulties in preparation of the diazosulfonyl starting materials.

Aside from the synthesis of sulfonylpyrazoles, the utility of the chemistry has been demonstrated in the total synthesis of withasomnine **106** (2011OL4016). Withasomnine is a pyrazole-based alkaloid isolated from *Withania Somnifera,* which has been used in Indian herbal medicine (1966T2895). The synthesis began with reaction of the nitroalkene **101** and the diazosulfone **102** to form pyrazole **103** in very good yield (Scheme 21). Reduction of the ester **103** with lithium aluminium hydride proceeded smoothly to afford alcohol **104**. Appel reaction followed by stirring in base afforded bicyclic sulfone **105** in excellent yield over two steps. The sulfonyl moiety was then reductively cleaved using sodium amalgam to furnish withasomnine **106** in very good yield.

An undesirable feature of the above method is the reduction of the sulfone **105** with sodium amalgam. The same authors improved the method by changing the diazosulfone for trimethylsilyldiazomethane (2013JOC3482). Unlike the sulfone, the trimethylsilyl group spontaneously eliminates under the reaction conditions to leave the 3 position unsubstituted. This allowed the synthesis of the three known pyrazole-based withasomnine alkaloids and a number of unnatural analogues.

A limitation of the diazo compound/alkene cycloaddition approach was addressed by Shao and Wan et al. (2016OBC8486). Typically, the chemistry required a preinstalled leaving group (e.g., NO_2) on the alkene partner. Shao, Wan, and coworkers found that catalytic tetrabutyl ammonium iodide (TBAI) and *tert*-butyl hydroperoxide (TBHP) as primary oxidant resulted in

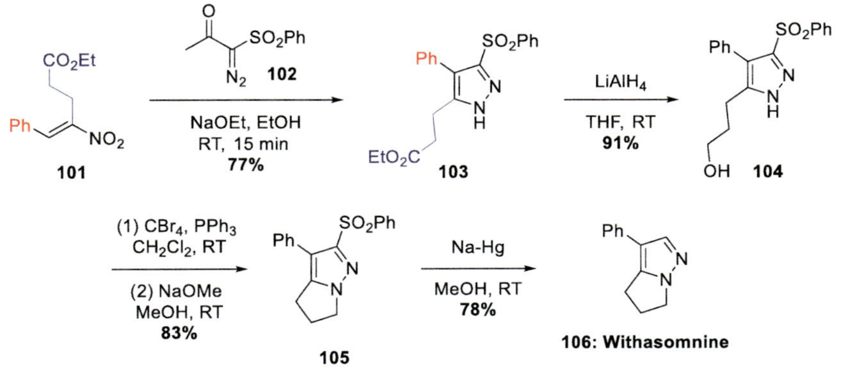

Scheme 21 Total synthesis of withasomnine 106 from the reaction of nitroalkene 101 and diazosulfone 102. R. Kumar and I.N.N. Namboothiri, Org. Lett., **13**, 4016 (2011).

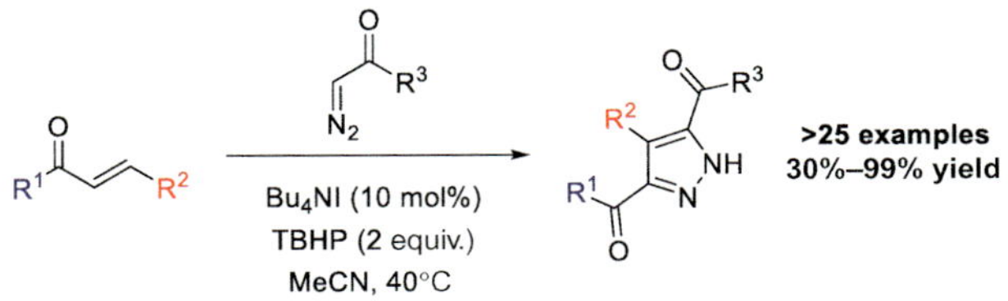

Scheme 22 Reaction of diazo compounds with alkenes. Y. Shao, J. Tong, Y. Zhao, H. Zheng, L, Ma, M. Ma, and X. Wan, Org. Biomol. Chem., **14**, 8486 (2016).

the formation of pyrazoles rather than pyrazolines from the reactions of simple enones (Scheme 22). The scope of this reaction is quite limited as both substrates required a carbonyl group. Furthermore, the only unsymmetrical alkenes employed were terminal alkenes, which underwent highly regioselective cycloaddition. Disubstituted alkenes are effective reaction partners, but typically required slightly elevated temperatures.

A different approach to pyrazole synthesis using diazo acetates has been reported by Frantz and coworkers (2011JOC5915). They utilized Pd-catalyzed crosscoupling of enol triflates to form vinyl diazomethanes which then undergo electrocyclic ring closure to form pyrazoles. This pericyclic pyrazole formation has been known for the best part of a century (1935JA2656, 1935JCS286). However, the method has largely been ignored as a general synthetic strategy for pyrazole synthesis. This is likely due to the difficulty in preparing the required vinyl diazomethanes. The functional group tolerance at the C4 position is impressive (Scheme 23). A range of protected amines are well tolerated, e.g., **109**, **112** and a boronic ester, e.g., **113** also survives the reaction and furnishes the product in good yield. An

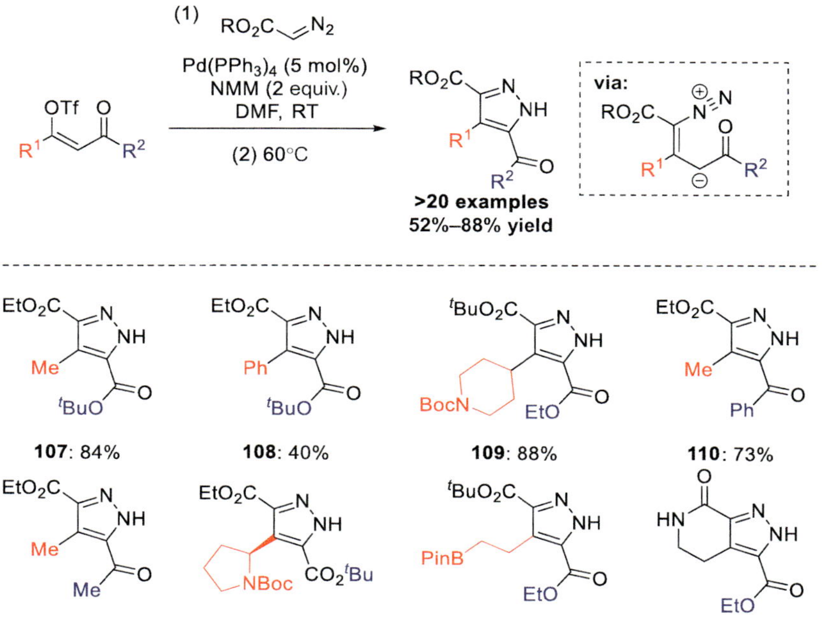

107: 84% **108**: 40% **109**: 88% **110**: 73%

111: 81% **112**: 53% **113**: 68% **114**: 62%

Scheme 23 Synthesis of pyrazoles by tandem crosscoupling/electrocyclization of diazoacetates and enol triflates. *D.J. Babinski, H.R. Aguilar, R. Still, and D.E. Frantz,* J. Org. Chem., ***76**, 5915 (2011).*

interesting finding was that the *Z*-triflates gave a cleaner reaction profile than the corresponding *E*-triflates. It was determined that this arose from the initial crosscoupling step. A clear limitation of the reaction lies in the available C3 and C5 substitutions, which are forced to be carbonyl derived, e.g., **107–114**.

A particularly interesting finding was made when the authors attempted to form fully substituted pyrazoles via the Alphen–Hüttel rearrangement shown in Scheme 24 (1943RTC485, 1960CB1425, 1960CB1433). They found instead that an unprecedented [1,3]-sigmatropic shift for 3*H*-pyrazoles occurred to form the 2-phenylethyl derivative **116**, presumably through intermediate **115** (Scheme 24).

In spite of the success of diazo compounds in [3 + 2] cycloadditions, their reactivity with alkynes is often found to be quite challenging. Early examples of intramolecular systems had some success (1995TL3087, 2000T4139), but the first intermolecular variant was discovered by Li and coworker (2004JCS(CC)394). They found that diazocarbonyl compounds can form

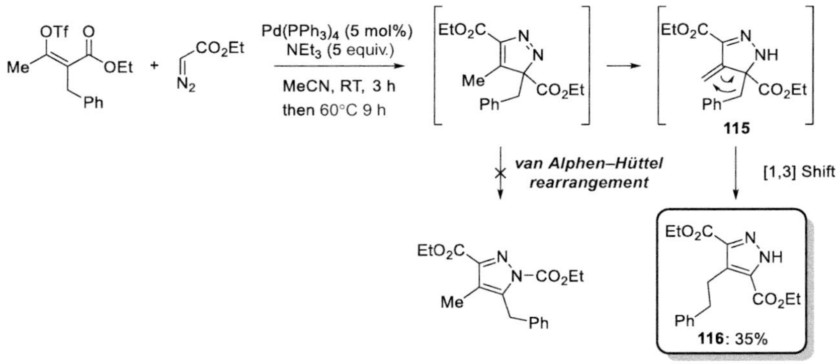

Scheme 24 [1,3]-Sigmatropic benzyl shift of 3H-pyrazoles. D.J. Babinski, H.R. Aguilar, R. Still, and D.E. Frantz, J. Org. Chem., **76**, 5915 (2011).

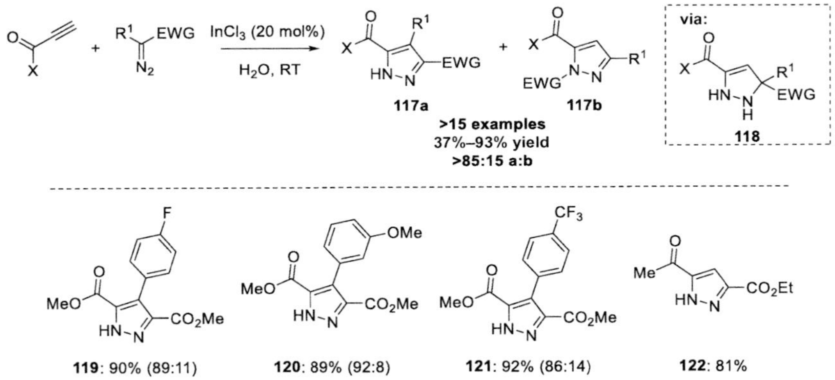

Scheme 25 Indium-catalyzed cycloaddition of alkynes with diazo compounds. N. Jiang and C.-J. Li, J. Chem. Soc. Chem. Commun., 394 (2004).

pyrazoles via [3 + 2] cycloaddition with alkynes in the presence of a suitable promoter. Impressively, the reactions are found to proceed using catalytic indium(iii) chloride in water at ambient temperature (Scheme 25). These reactions proceed in moderate–high regioselectivity and yield. The mechanism involves [3 + 2] cycloaddition to form intermediate **118**, with the final product (**117a** or **117b**) determined by the migration of either R^1 or the electron-withdrawing group (EWG). However, notable limitations in substrate scope are apparent. In particular, carbonyl groups are required on both the alkyne and diazo compounds for reactivity in the cycloadditions (**119**–**122**). No reactivity has been observed when phenylacetylene is subjected to the standard reaction conditions.

Nearly a decade later, Gong and Wang et al. reported the cycloaddition of diazoacetates with in situ-generated enamines under organocatalytic conditions as a metal-free alternative to Li's method (2013CEJ7555). The formation of the enamine raises the highest occupied molecular orbital energy in the inverse electron demand cycloaddition, thus significantly increasing the reactivity of carbonyl compounds in these reactions (2000AGE3702). This results in an impressively broad substrate scope and overall excellent yields (Scheme 26). Furthermore, the inherent electronic bias of the enamines results in single regioisomers being isolated from the reactions. It was found that 10 mol% of either pyrrolidine or diethylamine in dimethyl

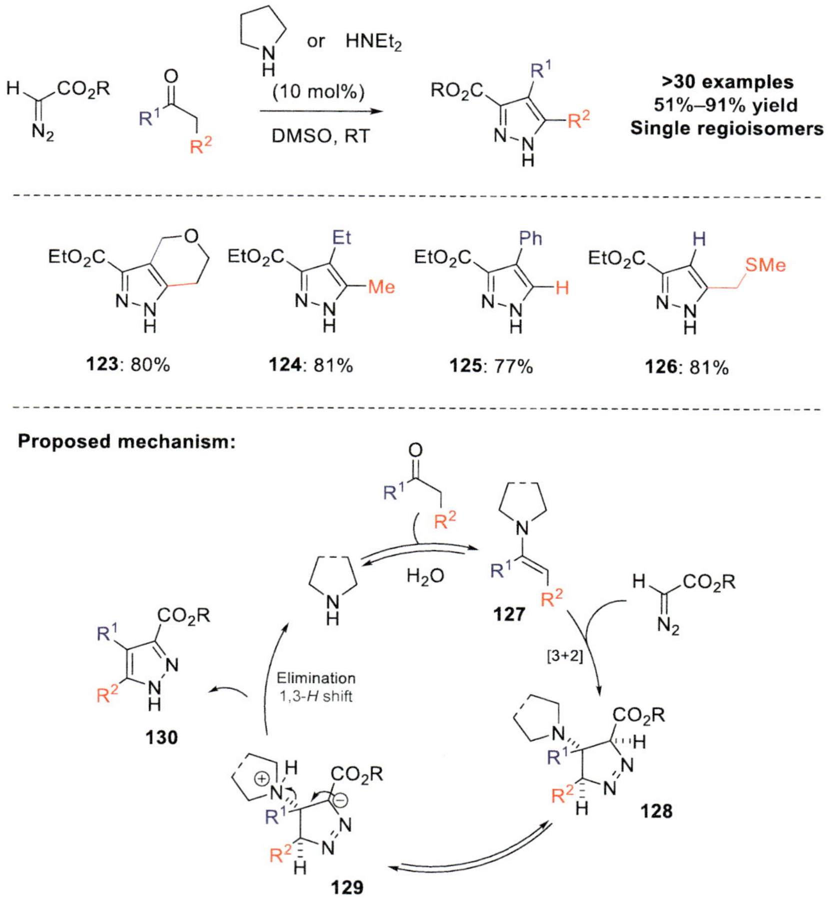

Scheme 26 Reaction of diazoacetates with in situ-generated enamines. L. Wang, J. Huang, X. Gong, and J. Wang, Chem. Eur. J., *19*, 7555 (2013).

sulfoxide (DMSO) at room temperature is sufficient for excellent reactivity, e.g., **123–126**. The major limitation in substrate scope arises from the diazoacetates. The proposed mechanism proceeds via the formation of an enamine **127**, which can then undergo [3+2] inverse electron demand cycloaddition with the diazoacetate to from a pyrazoline intermediate **128**. Proton transfer to generate intermediates **129** then allows for elimination of the catalysts and 1,3-hydride shift to generate the final products **130**.

2.2.2 Tosylhydrazones as Pyrazole Precursors

An interesting area of pyrazole synthesis involves the use of tosylhydrazones. The advantage of tosylhydrazones is starting material availability. In theory, tosylhydrazide can be condensed onto any aldehyde to prepare the azosulfone starting materials. Meng and Wang et al. developed a Lewis base-catalyzed [3+2] cycloaddition of allylic carbonates with arylazosulfones (2015OL872). Allylic carbonates have been known as effective [3+2] cycloaddition partners for a number of decades (1984TL5183, 1998JOC5031). The reactions proceed under incredibly mild conditions, at room temperature using catalytic tributylphosphine (Scheme 27). These reactions are typically high yielding, and a modest selection of substituents has been exemplified on the aromatic ring, e.g., **131** and **132**. Unfortunately, electron-deficient arenes are not viable substrates, e.g., **133**. When the allylic carbonate has a substituent, the reactions proceed well and form single regioisomers, e.g., **134** and **135**, presumably due to the stark difference in electronic character between the two nitrogens.

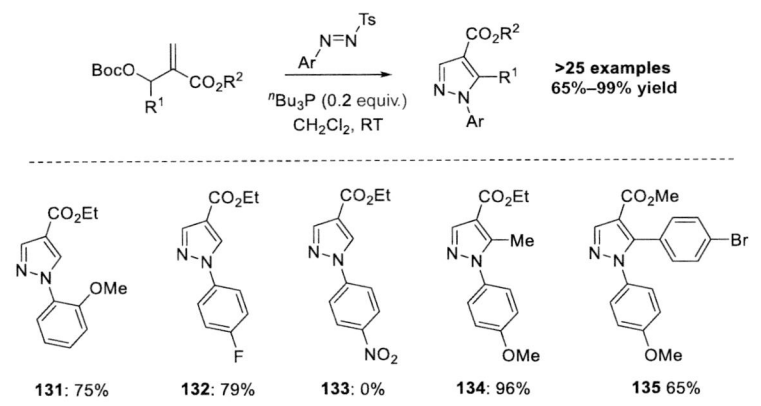

Scheme 27 Phosphine-catalyzed [3+2] cycloaddition of allylic carbonates with arylazasulfones. Q. Zhang, L.-G. Meng, K. Wang, and L. Wang, Org. Lett., **17**, 872 (2015).

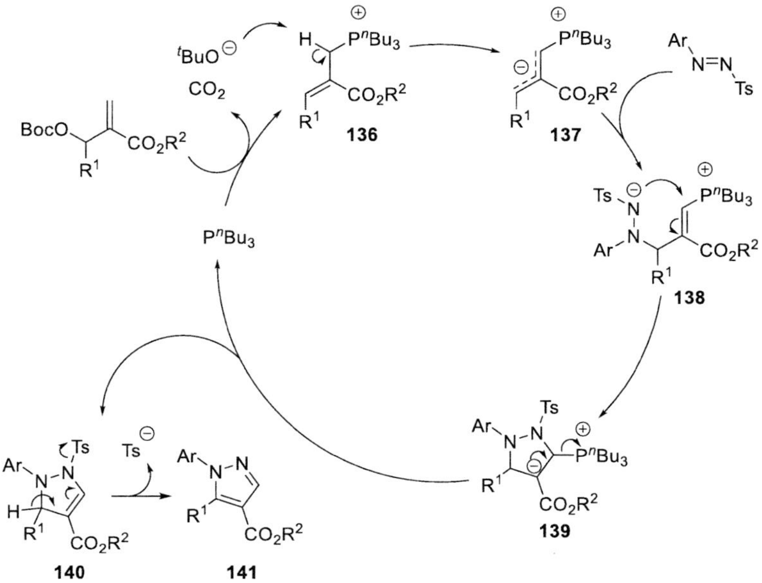

Scheme 28 Proposed mechanism of reaction of allylic carbonates with arylazasulfones. Q. Zhang, L.-G. Meng, K. Wang, and L. Wang, Org. Lett., **17**, 872 (2015).

The mechanism proposed by the authors involves a formal [3 + 2] cycloaddition of an in situ–generated ylide **137** (Scheme 28). Ylides **137** are generated from an S_N2' reaction involving tributylphosphine and the allylic carbonate to form **136**, carbon dioxide and *tert*-butoxide. **136** then deprotonates to form ylide **137** which reacts with the arylazasulfone to form **138**. Subsequent cyclization to **139** and elimination reforms the catalyst. The pyrazolines **140** then undergo elimination to form the pyrazoles **141**.

Aggarwal et al. reported the regioselective synthesis of disubstituted pyrazoles from aldehydes and alkynes, e.g., **143–146**, via the in situ formation of the hydrazone (2003JOC5381). A selection of aromatic aldehydes was successfully converted to the hydrazone and underwent highly regioselective cycloaddition with isomer **142a** as the favored product (Scheme 29). Monosubstituted pyrazoles can also be obtained by using *N*-vinylimidazole as the reaction partner. The reported reactions were limited to aromatic aldehydes and proceeded in moderate yields. However, the ready availability of aldehydes and alkynes is certainly a strength of the method.

Over a decade later, Tang et al. extended the methodology to include *N*-substitution. However, rather than using a one-pot procedure it was necessary to isolate the tosylhydrazone (2014OL576). The reaction suffers from

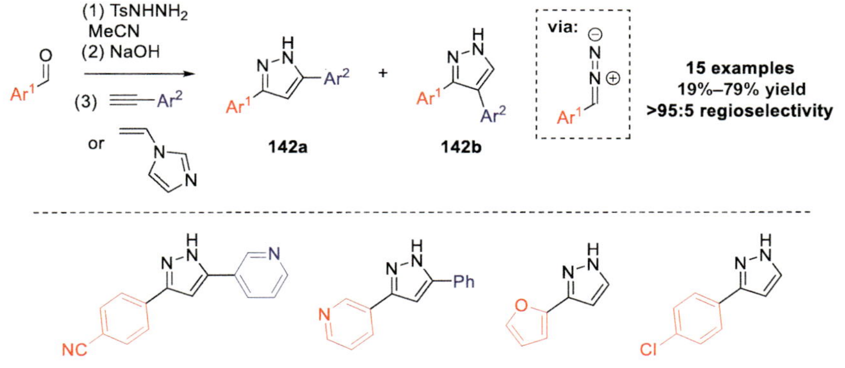

143: 19% **144**: 33% **145**: 33% **146**: 71%

Scheme 29 Reaction of tosylhydrazones with alkynes. *V.K. Aggarwal, J. de Vincente, and R.V. Bonnert,* J. Org. Chem., ***68**, 5381 (2003).*

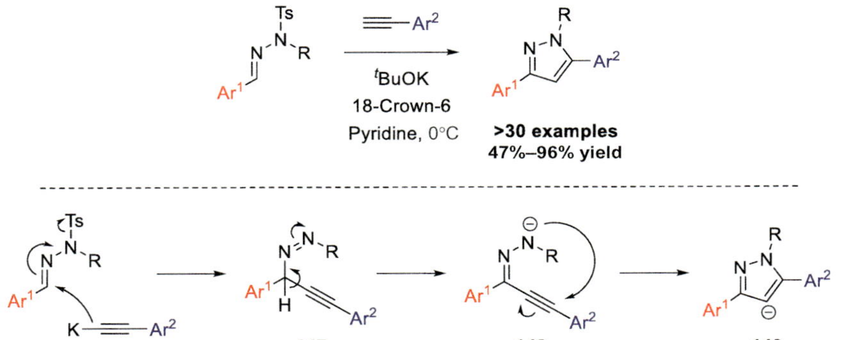

147 **148** **149**

Scheme 30 Synthesis of N-substituted pyrazoles from tosylhydrazones and alkynes. *Y. Kong, M. Tang, and Y. Wang,* Org. Lett., ***16**, 576 (2014).*

similar limitations to the Aggarwal method; only aromatic tosylhydrazones and aryl alkynes are tolerated, and electron-deficient aryl groups have not been reported. Furthermore, improved yields were only obtained when highly toxic 18-crown-6 was used. Nevertheless, a large number of trisubstituted pyrazole examples have been reported in moderate to high yield (Scheme 30). The proposed mechanism in this modified procedure is not a cycloaddition, but a nucleophilic addition of an 18-crown-6-activated alkynyl potassium to the tosylhydrazone to form **147**. The intermediate is then deprotonated forming **148** and undergoes cyclization to form **149**, which forms the target pyrazole upon protonation.

Maity, Manna, and coworker reported some isolated examples of alternative reaction partners for cycloadditions with tosylhydrazones

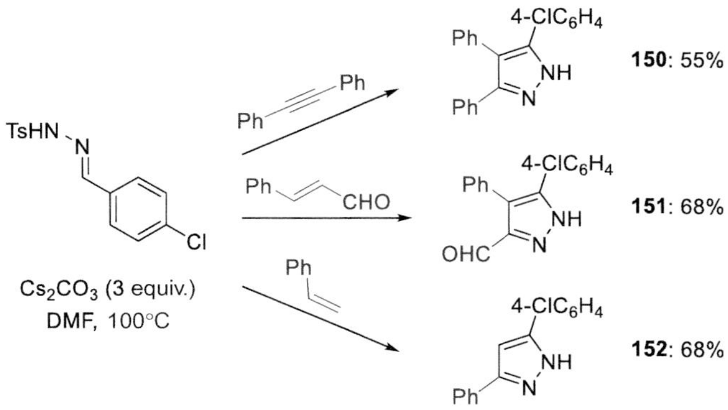

Scheme 31 Alternative cycloaddition partners for N-tosylhydrazones. S. Panda, P. Maity, and D. Manna, Org. Lett., **19**, 1534 (2017).

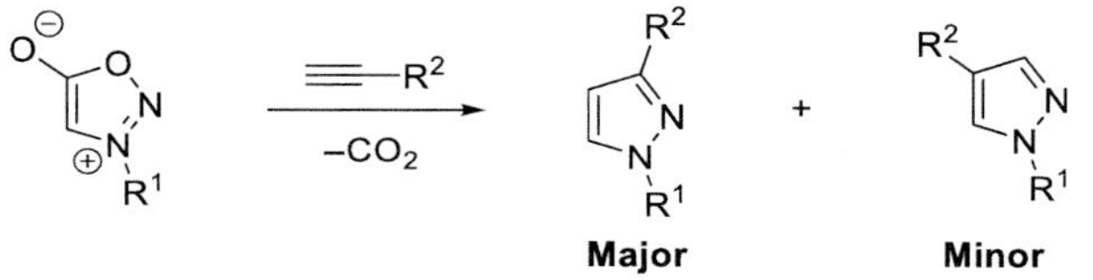

Scheme 32 [4+2] Cycloaddition of sydnones with alkynes.

(Scheme 31) (2017OL1534). Reaction with an internal alkyne furnished the corresponding trisubstituted pyrazole **150** in respectable yield. Impressively, reaction with an α,β-unsaturated aldehyde successfully provided a single regioisomer of trisubstituted pyrazole **151**. Finally, reaction with styrene furnished disubstituted pyrazole **152** in good yield. These results highlight the versatility of N-tosylhydrazone cycloadditions.

2.2.3 Sydnones in Pyrazole Synthesis

Interesting cycloaddition precursors for the formation of pyrazoles are the sydnones. These intriguing mesoionic compounds are bench stable and readily undergo [4+2] cycloaddition–retrocycloaddition with alkynes to liberate carbon dioxide and form pyrazoles in typically high regioselectivity (Scheme 32) (2010T553). The [4+2] cycloaddition of sydnones with alkynes was first reported by Huisgen in 1962 employing a range of monosubstituted alkynes (1962AGE48). The major regioisomer obtained in most cases was the 1,3-disubstituted pyrazole rather than the 1,4-isomer. More recent developments have sought to apply the methodology to more complex sydnones and alkynes and study effects on regioselectivity. Sydnones themselves are prepared by the N-nitrosation and subsequent cyclodehydration of amino acids (1950JCS1542). Sydnones are readily functionalized at the C4 position either by electrophilic aromatic substitution or by metalation–substitution chemistries (1996SC1441, 1986BCJ483, 2009S650, 2009TL3942, 2015JOC2467).

An important addition to sydnone cycloaddition methodology was reported by Harrity et al. They showed that alkynylboronates could undergo highly regioselective cycloaddition with sydnones to furnish synthetically valuable pyrazole boronic esters (2007AGE8656). Cycloadditions of C4-unsubstituted sydnones with alkynylboronates are successful in moderate to high regiocontrol (Scheme 33). Extensive studies have been undertaken on the origins of regioselectivity, and it was found that the regioselectivity obtained was dependent on the alkyne substitution (2009JA7762). Cycloaddition with the terminal alkynylboronate gave the pyrazole-3-boronate as the major product (**155**) with a regioselectivity of 1:7. Internal alkynylboronates gave inverse regioselectivity with pyrazole-4-boronates favored. The selectivity is also more variable—ranging from 98:2 for trimethylsilyl- and aryl-substituted alkynes, e.g., **154** to 5:2 in the case of *n*-butyl substitution, e.g., **153**. The selectivity of these cycloadditions is thought to be dominated by steric effects, with the most sterically demanding substituent being incorporated at the pyrazole C3 position.

Further studies highlighted the fact that consistently higher regioselectivities are observed with C4-substituted sydnones, e.g., **156** and **157** (Scheme 34) (2009OBC4052, 2017T3160). The C4 substituent is thought to further increase the steric differentiation of the alkynes' approach. The regiochemical outcome was again determined by the substitution of the alkyne.

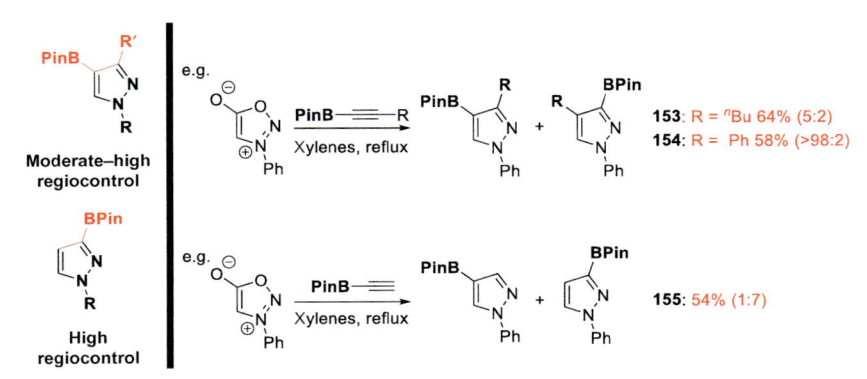

Scheme 33 Cycloaddition of alkynylboronates with C4-unsubstituted sydnones. D.L. Browne and J.P.A. Harrity, Tetrahedron, **66**, 553 (2010).

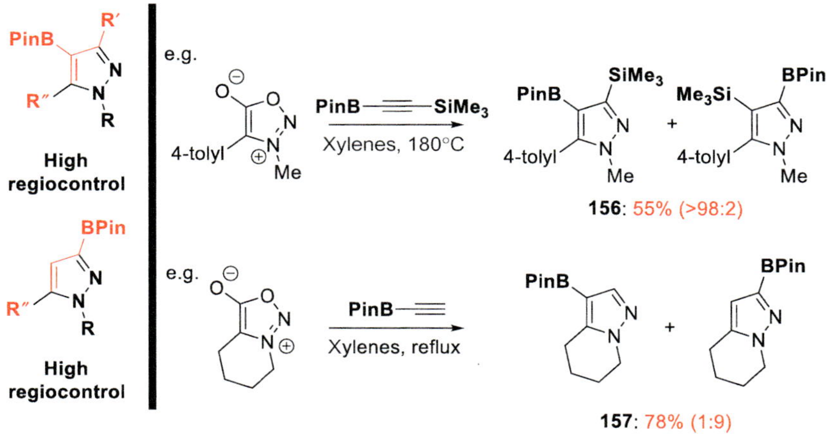

Scheme 34 Cycloaddition of alkynylboronates with C4-substituted sydnones. D.L. Browne, J.P.A. Harrity, Tetrahedron, **66**, 553 (2010).

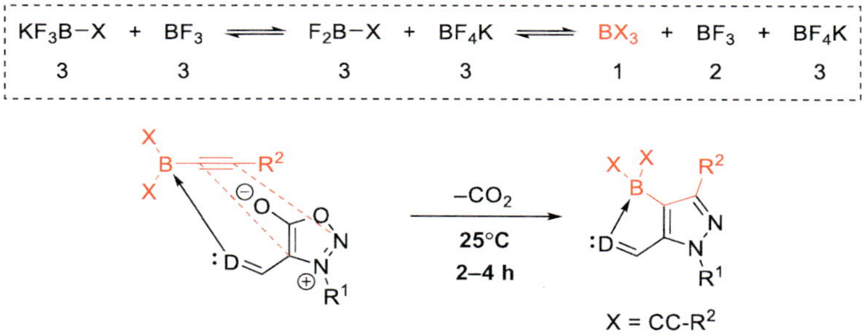

Scheme 35 Directed cycloaddition of sydnones with alkynylboranes.

A limitation of these chemistries is the high temperatures required to facilitate cycloaddition. Harrity et al. addressed this through use of Lewis acid/base-directed cycloadditions of alkynylboronates (2017CEJ5228). Lewis acidic trisalkynylboranes can be generated in situ from alkynyl trifluoroborates and boron trifluoride. The Lewis acidic intermediates then coordinated to a Lewis basic directing group on the sydnone, facilitating cycloaddition (Scheme 35). This strategy allows sydnone cycloadditions to take place at ambient temperature in only a few hours. Furthermore, the reactions are regiospecific, resulting in exclusive incorporation of the boron moiety at the pyrazole C4 position regardless of steric effects. Unusually, the products of these reactions were found to be dialkynylboranes ($X = CC-R^2$) and not the expected difluoroboranes ($X = F$) (Scheme 35).

These reactions are typically high yielding and tolerant of both N-alkyl- and N-arylsydnones. A major limitation was found in the directing groups tolerated for the reaction. Pyridyl directing groups were highly successful, e.g., **158**, **159**, but amides, e.g., **161** and 5-membered heterocycles failed to initiate reaction (Scheme 36). The only nonpyridine-based example of a successful directing group was an oxazine, e.g., **160**. The unusual dialkynylborane moiety was shown to successfully undergo typical organoborane reactions such as Suzuki–Miyaura coupling.

Further recent developments have seen the application of "click chemistry" to cycloaddition reactions of sydnones. Both stoichiometric and catalytic copper have been used to improve yields, lower temperature and reaction times, and even alter regioselectivity. Seminal work by Taran et al. showed that a range of terminal alkynes can undergo cycloaddition with sydnones at 60°C in the presence of catalytic copper (2013AGE12056). Phenanthroline and sodium ascorbate are integral to the catalytic nature of the reaction and the low reaction temperatures achieved. Astoundingly, the native regioselectivity is inverted under these conditions. The 1,4-substituted pyrazole products are furnished as single regioisomers in typically high yield, e.g., **162–164** (Scheme 37). Substrate scope in the study was limited to N-arylsydnones and terminal alkynes. C4 substitution is also not tolerated. The chemistry has since been extended to tolerate bromine substitution at the sydnone C4 position, allowing access to 1,4,5-trisubstituted pyrazoles

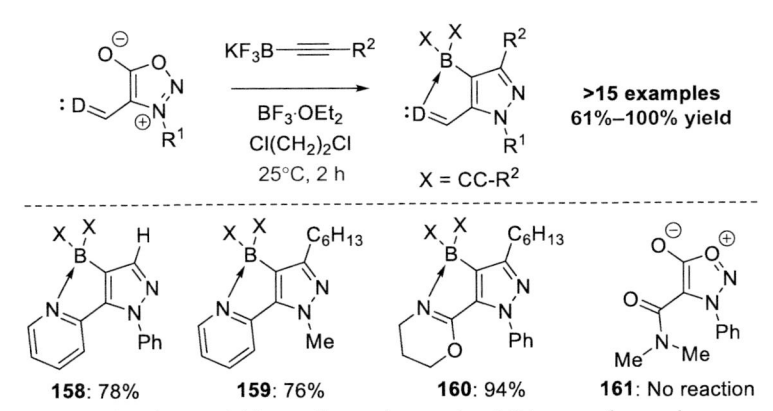

Scheme 36 Lewis acid/base-directed cycloaddition of sydnones with alkynyltrifluoroborates. *A.W. Brown, J. Comas-Barceló, and J.P.A. Harrity, Chem. Eur. J.,* **23**, *5228 (2017).*

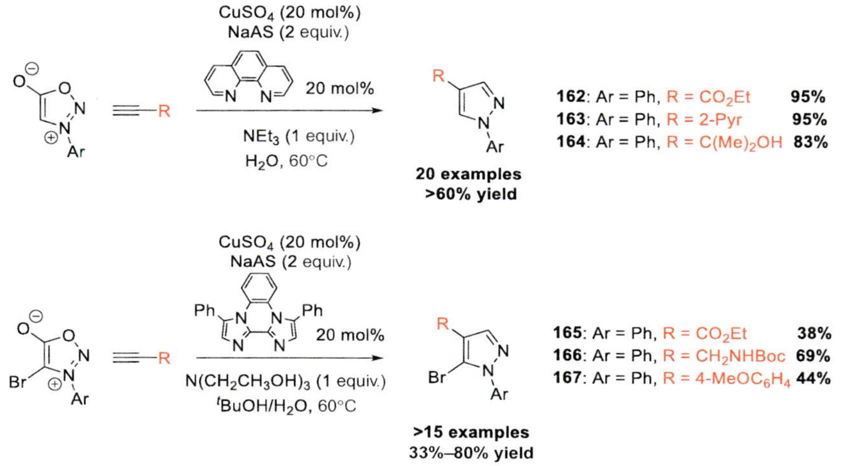

Scheme 37 Cu-catalyzed sydnone alkyne cycloaddition (CuSAC). S. Kolodych, E. Rasolofonjatovo, M. Chaumontet, M.-C. Nevers, C. Créminon, and F. Taran, Angew. Chem. Int. Ed. Engl., **52**, 12056 (2013).

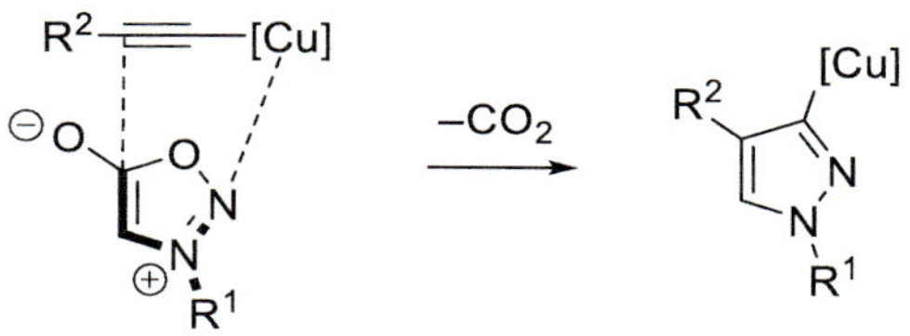

Scheme 38 Proposed mechanism of CuSAC reaction.

(2015OL362). It was found that simply changing the ligand to a bis-imidazole-based system expanded the scope with respect to the sydnone partner, e.g., **165–167**. The reactions proceeded in reduced yields compared to the unsubstituted sydnone variant, but the 5-bromopyrazole products can be further elaborated by crosscoupling. More recent work by the same authors has shown 4-fluorosydnones to be particularly reactive cycloaddition addition partners (2016AGE12073). Both copper-promoted and strain-promoted variants of the reaction proceed within minutes to afford a range of 5-fluoropyrazoles.

The proposed mechanism of the copper(I)-catalyzed sydnone alkyne cycloaddition (CuSAC) reaction involves the formation of the copper acetylide which undergoes cycloaddition with the sydnone (Scheme 38). It is believed that the copper has an interaction with the sydnone N2 position in addition to acting as the most sterically demanding group. Such an

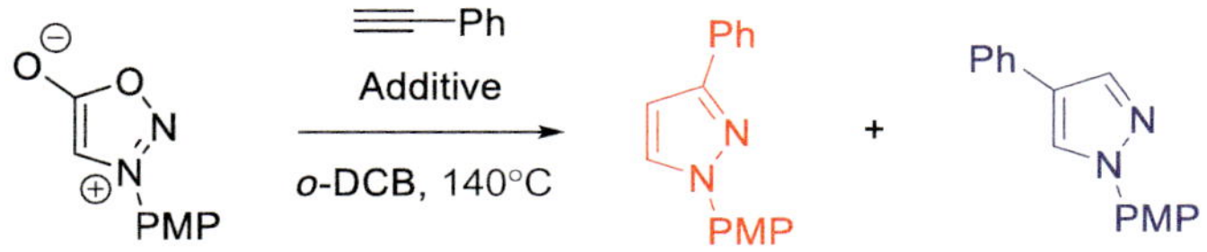

168a **168b**

Additive:
$Cu(OTf)_2$: 57% (>95:5), 1 h
$Cu(OAc)_2 \cdot H_2O$: 53% (<5:95), 5 h

Scheme 39 Effect of copper salt on cycloaddition regioselectivity. *J. Comas-Barceló, R.S. Foster, B. Fiser, E. Gomez-Bengoa, and J.P.A. Harrity,* Chem. Eur. J., ***21**, 3257 (2015).*

interaction is consistent with both the lower temperatures required for the cycloaddition as well as the inversion of native regioselectivity.

Interestingly, Harrity et al. found that the regioselectivity of the copper-promoted reaction can be altered by simply changing the copper salt used (Scheme 39) (2015CEJ3257). Copper triflate was found to accelerate the cycloaddition to yield 1,3-substituted pyrazoles (**168a**) as the major products. Copper acetate was found to accelerate the reaction in addition to inverting regioselectivity, preferentially giving 1,4-substituted pyrazoles (**168b**). Theoretical and practical mechanistic studies revealed copper triflate accelerated reaction by Lewis acidic activation of the sydnone. Conversely, copper acetate catalysis proceeded via the in situ formation of the appropriate copper acetylide in a similar manner to that proposed by Taran et al.

An alternative to metal-promoted cycloaddition reactions is to use inherently more reactive alkynes. Chin et al. and Taran et al. have shown reactions between sydnones and cyclooctynes to be particularly fast (2014JCS(CS)1742, 2014JCS(CC)9376). The high reactivity of cyclooctynes has been known for over 50 years (1961CB3260), but their reactivity with sydnones has led to application in biorthogonal reactions. These cycloadditions are successful in a matter of minutes under aqueous conditions at ambient temperature. Taran et al. subsequently showed the 4-fluorosydnones are even more reactive in these biorthogonal reactions, rendering them "ultrafast," which is a requirement for in vivo applications (2016AGE12073). In 2017 Taran and coworkers demonstrated that sydnone imines are of particular interest in this area and showed their use as cleavable linkers in proteins (2017AGE15612). The advantage of sydnone imines is that the urea by-product from the cycloaddition can also contain a biological tag (e.g., biotin), giving a biological readout as the reaction progresses (Scheme 40).

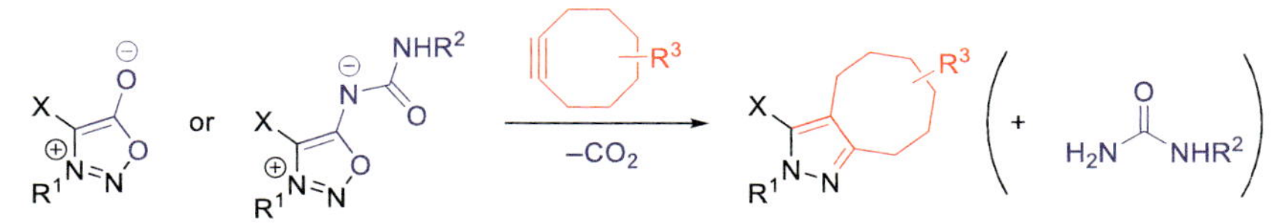

Scheme 40 Strain-promoted cycloaddition of sydnones and sydnone imines with cyclooctanes. S. Bernard, D. Audisio, M. Riomet, S. Bregant, A. Sallustrau, L. Plougastel, E. Decuypere, S. Gabillet, R.A. Kumar, J. Elyian, M.N. Trinh, O. Koniev, A. Wagner, S. Kolodych, and F. Taran, Angew. Chem. Int. Ed. Engl., **5**, 15612 (2017).

2.3 Metal Catalysis and Metal-Promoted Synthesis

Aside from advances in the classical methods of condensation and cycloaddition chemistry, innovative strides in the synthesis of pyrazoles have been taken using metal catalysis. A range of metals have successfully promoted the construction of pyrazoles, and no doubt further innovations will be found over the coming years. Furthermore, by having access to alternative mechanisms, a number of these metal-catalyzed reactions are regiospecific, thus overcoming a limitation of many classical syntheses. One such method using copper catalysis was reported by Buchwald et al. (2006AGE7079). Iodoenynes were found to undergo C–N crosscoupling followed by hydroamidation to afford nitrogen-containing heterocycles. Both pyrrole and pyrazole examples were reported, the latter requiring treatment with trifluoroacetic acid after hydroamination to furnish the final products. These reactions proceed in typically good to excellent yield and mono-, di-, and trisubstituted pyrazoles can be prepared, e.g., **169**–**171** (Scheme 41). Mechanistic studies revealed that the reactions proceed first by C–N bond formation to generate **172**, followed by intramolecular hydroamination to form **173**. Interestingly, the intramolecular hydroamination is promoted by both base and copper. A limitation of the reaction lies in the availability of the iodoenyne substrates and enforced inclusion of a CH_2 linker to the R^3 substituent in the products.

A number of years later, Zhan et al. used closely related propargylhydrazones as substrates to form tetrasubstituted pyrazoles under platinum catalysis (2014OL5940). In spite of the relatively specialized substrates, the range of substituents that can be incorporated at each position of the pyrazole is quite diverse, e.g., **174**–**176** (Scheme 42). The major limitation in scope resides with the N-substituent. The reaction failed with electron-withdrawing groups in this position, and the yield was also significantly decreased with alkyl groups, e.g., **177**. The transformation itself is

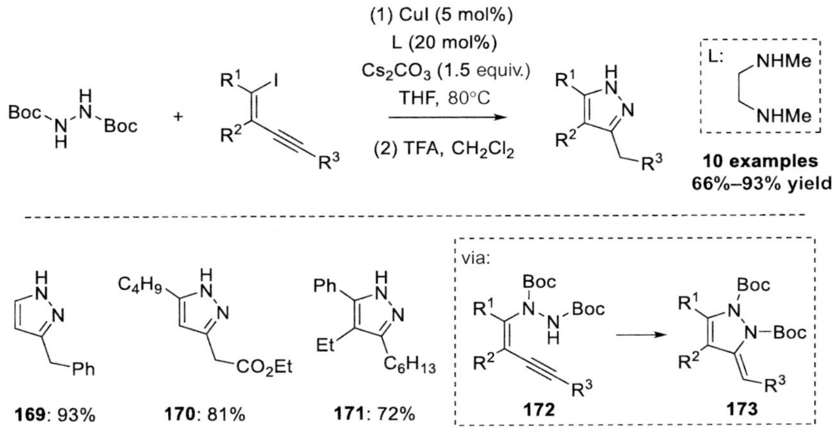

Scheme 41 Cu-catalyzed synthesis of pyrazoles from iodoenynes. R. Martín, M. Rodríguez Rivero, and S.L. Buchwald, Angew. Chem. Int. Ed. Engl., **45**, 7079 (2006).

interesting. Platinum coordinates to the triple bond forming **178** which facilitates attack by the proximal hydrazone to form vinyl platinum **179**. Following cyclization, **179** opens to form diazoallene **180**. Activation of the allene by platinum enables cyclization by the nitrogen to form the pyrazole skeleton **181**. Proton transfer provides the pyrazole products **182** and regenerates the platinum catalyst.

Rao et al. reported the use of ruthenium catalysis for the oxidative cyclization of N-arylhydrazones to form pyrazoles (2012OL5030). A large number of N-arylhydrazones were prepared from α,β-unsaturated aldehydes and N-arylhydrazines and successfully cyclized to provide tri- and tetrasubstituted pyrazoles in high yield (Scheme 43). Although limited to N-aryl substituents, a wide range of substituents on the aryl group are tolerated, and also around the other positions of the hydrazone, e.g., **183–186**. The utility of this method is exemplified by the preparation of the antiinflammatory agent celecoxib, **184**, in 77% yield.

Panda and coworker successfully demonstrated that iron catalysis can be used to form pyrazoles from diarylhydrazones and vicinal diols (2012JMC3644). A variety of 1,3-di- and 1,3,5-trisubstituted pyrazoles have been prepared (Scheme 44). The reaction proceeds via iron/peroxide-mediated oxidation of the vicinal diol to give an α-hydroxycarbonyl reagent which can then condense with the diarylhydrazone and form the pyrazole by subsequent cyclization. A range of C_3-aryl groups are well tolerated on the hydrazone, e.g., **187–190**, but electron-withdrawing groups on the N-aryl

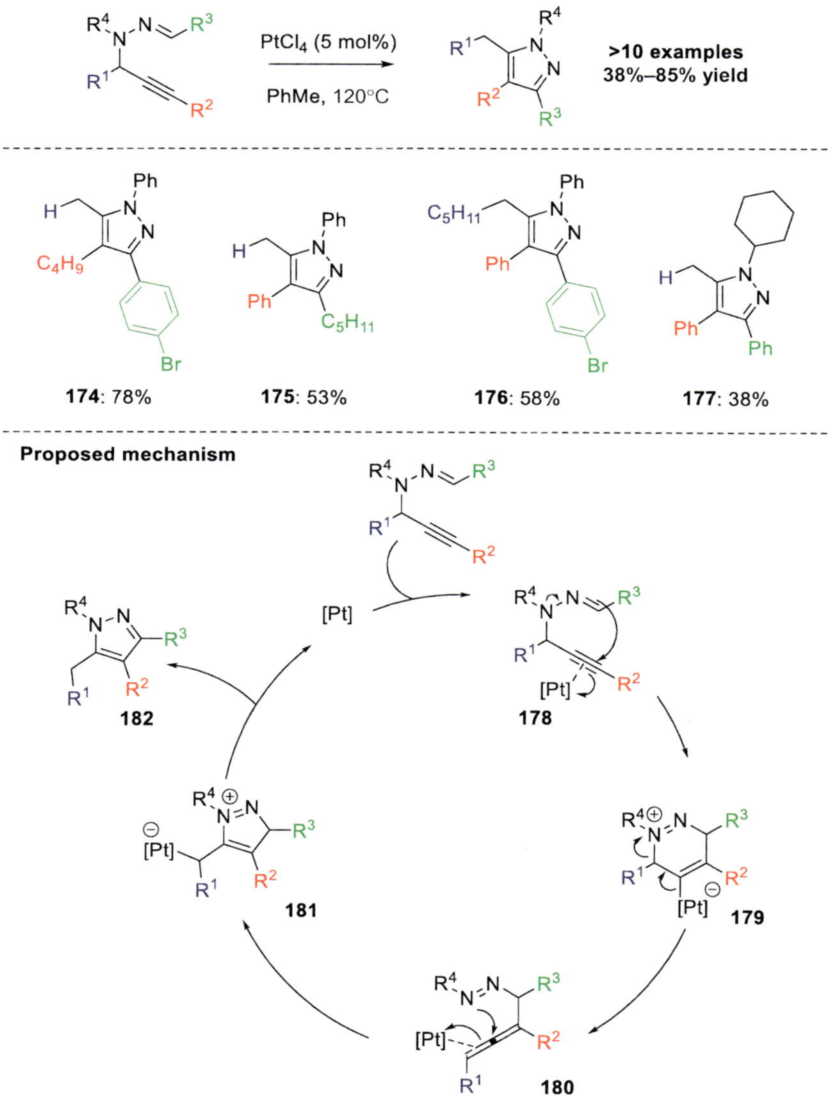

Scheme 42 Pt-catalyzed [3,3] sigmatropic rearrangement/cyclization of propargylhydrazones. J.-J. Wen, H.-T. Tang, K. Xiong, Z.-C. Ding, and Z.-P. Zhan, Org. Lett., **16**, 5940 (2014).

ring retarded reactivity. Furthermore, only a 5-methyl substituent has been exemplified in trisubstituted examples, e.g., **189** and **190**. Therefore, although the reaction represents an interesting advance, further investigation is needed to expand the scope.

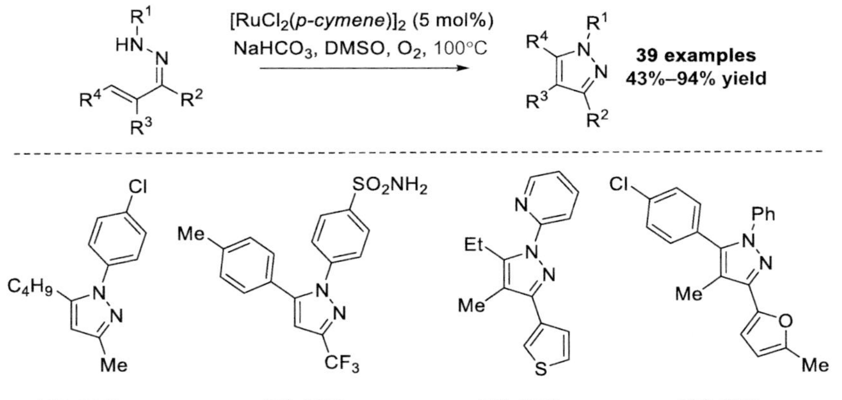

183: 72% **184**: 77% **Celecoxib** **185**: 63% **186**: 81%

Scheme 43 Ru-catalyzed oxidative cyclization of *N*-arylhydrazones. *J. Hu, S. Chen, Y. Sun, J. Yang, and Y. Rao,* Org. Lett., ***14****, 5030 (2012).*

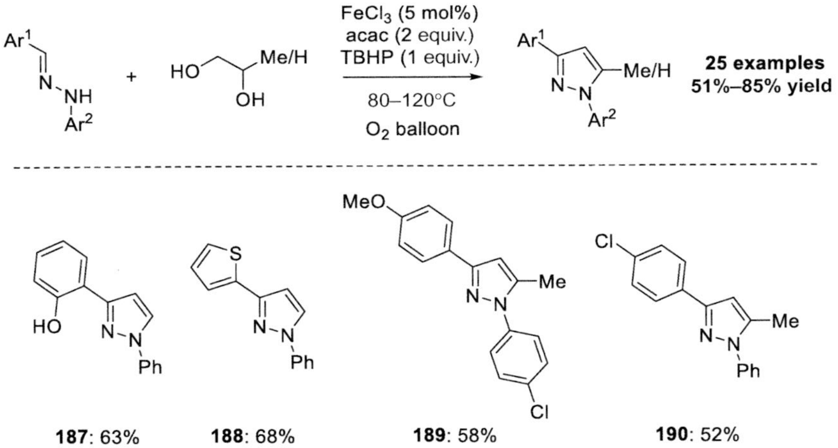

187: 63% **188**: 68% **189**: 58% **190**: 52%

Scheme 44 Fe-catalyzed synthesis of pyrazoles from diarylhydrazones and vicinal diols. *N. Panda and A.K. Jena,* J. Org. Chem., ***77****, 9401 (2012).*

Schmitt et al. have shown that 1,3-diols and hydrazines can form pyrazoles under ruthenium catalysis (2015OL1405). Ruthenium-mediated hydrogen transfer with crotononitrile afforded 1,4-disubstituted pyrazoles in moderate yield (Scheme 45). The ready availability of hydrazines and 1,3-diols allowed for the rapid construction of pyrazoles bearing interesting motifs, e.g., **191**. 1,3-Diols bearing heteroaryl motifs, e.g., **192**, and protected ketones, e.g., **193**, were successful substrates, albeit in diminished yields. The authors also showed that if β-hydroxyketones are used in place

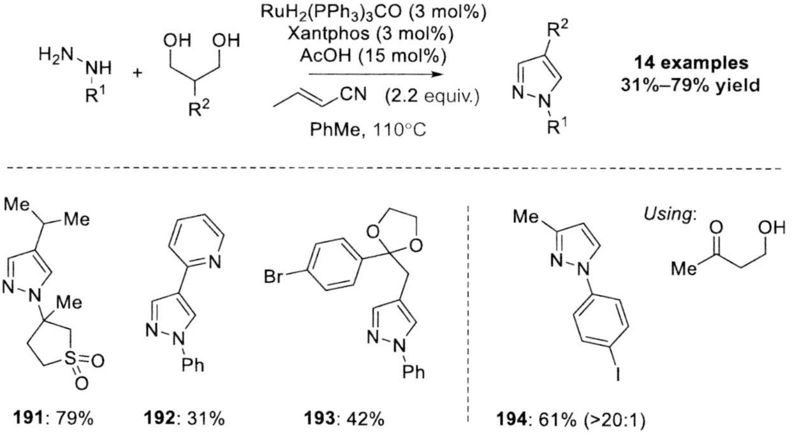

Scheme 45 Ru-catalyzed hydrogen transfer for the synthesis of pyrazoles. D.C. Schmitt, A.P. Taylor, A.C. Flick, and R.E. Kyne Jr., Org. Lett., **17**, 1405 (2015).

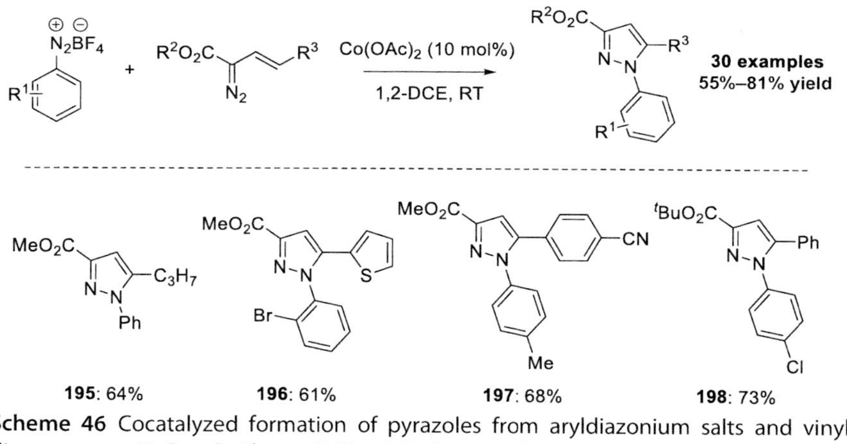

Scheme 46 Cocatalyzed formation of pyrazoles from aryldiazonium salts and vinyl diazoacetates. H. Guo, D. Zhang, C. Zhu, J. Li, G. Xu, and J. Sun, Org. Lett., **16**, 3110 (2014).

of 1,3-diols, then 1,3-disubstituted pyrazoles can be formed with excellent regioselectivity, e.g., **194**.

Sun and coworkers developed a strategy for the preparation of *N*-arylpyrazoles from vinyl diazoacetates and aryldiazonium salts (2014OL3110). Cobalt catalysis effectively promotes the transformation at ambient temperature to afford 1,3,5-trisubstituted pyrazoles, e.g., **195–198**, in moderate to very good yields (Scheme 46). The method affords single regioisomers of the products and is tolerant of electron–neutral and electron-poor aryldiazonium salts. A range of vinyl diazoacetates were tolerated; however, their commercial availability is a limitation of the method.

Also within group 9, rhodium, like cobalt, has been used as a catalyst in pyrazole synthesis. Liu et al. found that rhodium promotes an addition–cyclization cascade of hydrazines to electron-deficient alkynes (2014OL3476). The scope of this reaction is quite limited due to the requirement for electron-deficient alkynes. Conversely, electron-deficient *N*-aryl groups have a negative effect on yield, e.g., **199** (Scheme 47), but aryl and alkyl groups are tolerated on the carbonyl unit, e.g., **200–202**. The proposed mechanism involves deprotonation of the hydrazide by the active rhodium catalyst to generate **203**. This then coordinates to the alkyne,

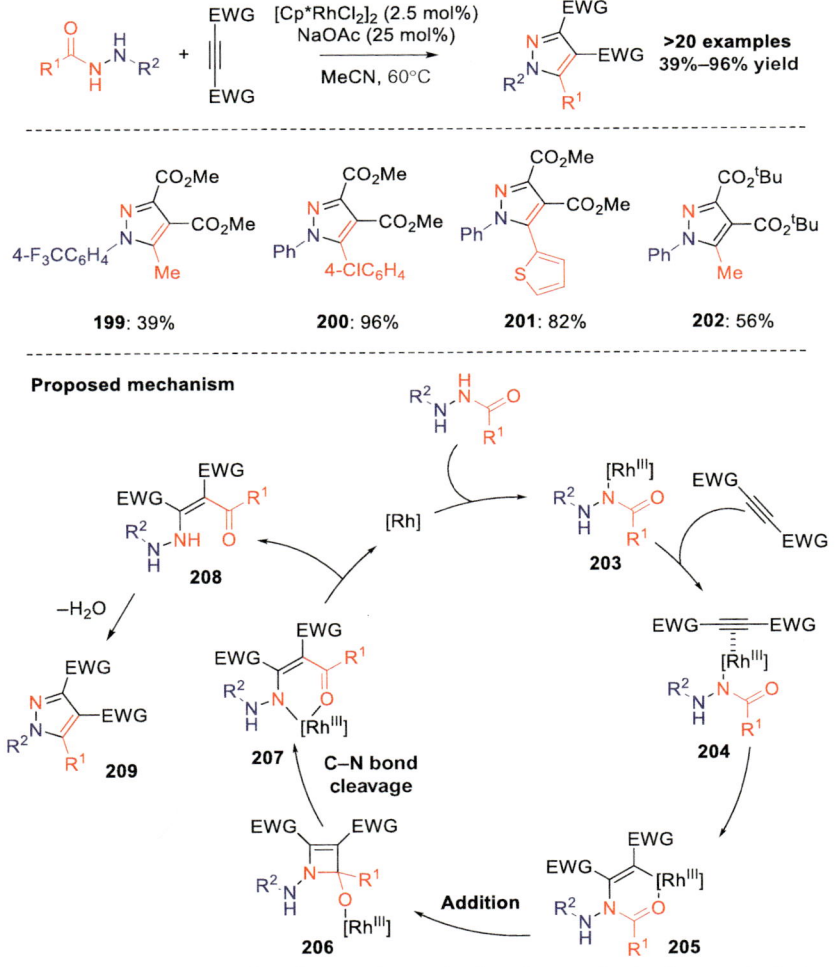

Scheme 47 Rhodium-catalyzed pyrazole synthesis using acylhydrazines and electron-deficient alkynes. *D.Y. Li, X.F. Mao, H.J. Chen, G.R. Chen, and P.N. Liu*, Org. Lett., **16**, 3476 (2014).

leading to intermediate **204**. Intramolecular insertion generates rhodacycle **205**, which then undergoes intramolecular nucleophilic addition to generate the 4-membered intermediate **206**. Ring opening by C–N bond cleavage results in the formation of 6-membered rhodacycle **207** which dissociates to regenerate the catalyst and form **208**, which undergoes subsequent cyclization to form pyrazole **209**.

A particularly important and intriguing metal-promoted pyrazole synthesis was discovered serendipitously by Glorius et al. (2010AGE7790). During investigations into the oxidative cyclization of *N*-aryl enaminones to form indoles (2008AGE7340), these authors found that if acetonitrile is used as solvent the corresponding pyrazole product is formed rather than the target indole. Unlike the vast majority of pyrazole syntheses, the N–N bond is formed during the reaction. The scope of the reaction is quite broad, both electron-poor and electron-rich *N*-aryl groups are tolerated, e.g., **210** and **212** (Scheme 48). Increased steric hindrance did not have

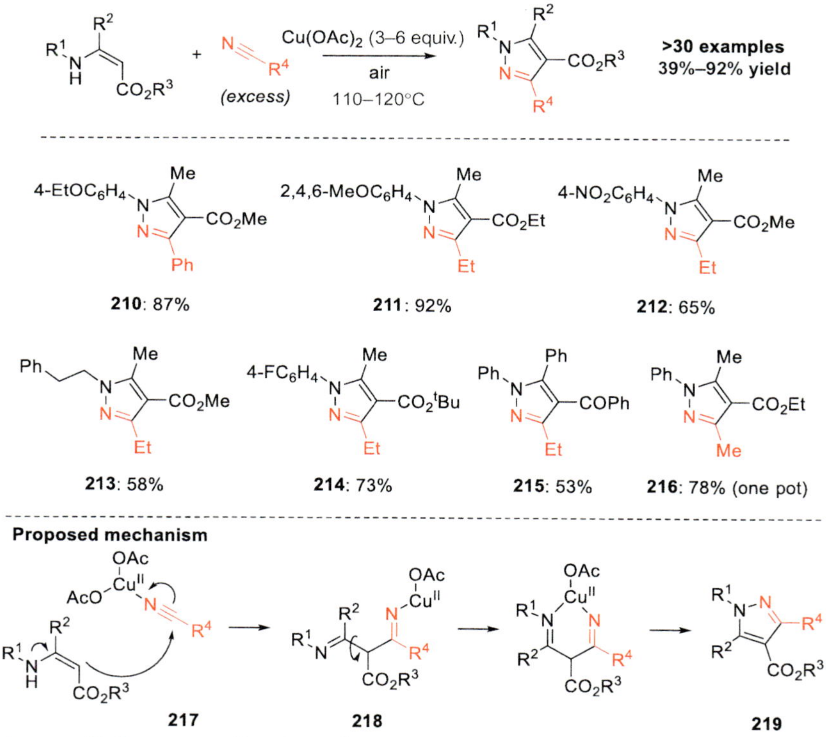

Scheme 48 Copper-mediated oxidative coupling of enaminones and nitriles. *J.J. Neumann, M. Suri, and F. Glorius*, Chem. Int. Ed., **49**, 7790 (2010).

an adverse effect on reactivity, e.g., **211**. Indeed, only in the case of 2,6-diisopropylphenyl was a lower yield observed. A single example of an N-alkyl group was shown to proceed in moderate yield, e.g., **213**. Both alkyl- and aryl-nitriles are effective substrates, and a range of esters and even ketones are tolerated, e.g., **210**, **214**, and **215**. To increase utility, it was demonstrated that the enaminone can be formed in situ and reacts in a one-pot manner without a significant drop-off in yield, e.g., **216**. Limitations of the reaction include the requirement of superstoichiometric copper and the large excess of nitrile required. Nonetheless, the reaction allows the construction of densely substituted pyrazoles from simple feedstocks. The mechanism of the reaction is thought to proceed via Lewis acidic activation of the nitrile by copper (**217**), facilitating attack by the enaminone. The resulting 1,3-diimine **218** then undergoes copper-mediated oxidative coupling to afford the pyrazole **219**.

3. PROPERTIES AND APPLICATIONS
3.1 Biologically Active Pyrazoles
3.1.1 Medicinal Chemistry Applications of Pyrazoles

Pyrazoles are perhaps most renowned for their lucrative commercial success in the field of therapeutics. In 2016, 6 of the top 200 highest-grossing pharmaceuticals contained the pyrazole motif with cumulative sales of over $9 billion (Scheme 49) (2010JCE1348). Of these, apixaban, **220**, marketed by Bristol-Myers Squibb, was the highest grossing with sales of $3.34 billion. Apixaban is an anticoagulant, shown to be at least as efficacious as warfarin with fewer side effects. The drug works by inhibiting factor Xa, preventing prothrombin from being converted to thrombin, which in turn prevents fibrin clot formation (2014COC293). Ibrutinib, **221**, appears twice in the top two hundred list, and is marketed by both Abbvie and Johnson and Johnson. It had cumulative sales of $3.08 billion in 2016. The molecule was originally designed as a tool compound, but was then developed into a therapeutic for the treatment of chronic lymphocytic leukemia (2012BD2590). It targets the B-cell antigen receptor (BCR) signalosome and is an irreversible binder of Bruton's tyrosine kinase (2012BD1182). By blocking BCR signaling, it promotes leukemia cell apoptosis and prevents their proliferation. Sildenafil **222** is one of the most well-known drugs of all time. The phosphodiesterase-5 (PDE-5) inhibitor was originally developed to treat cardiovascular diseases, but has been marketed for the

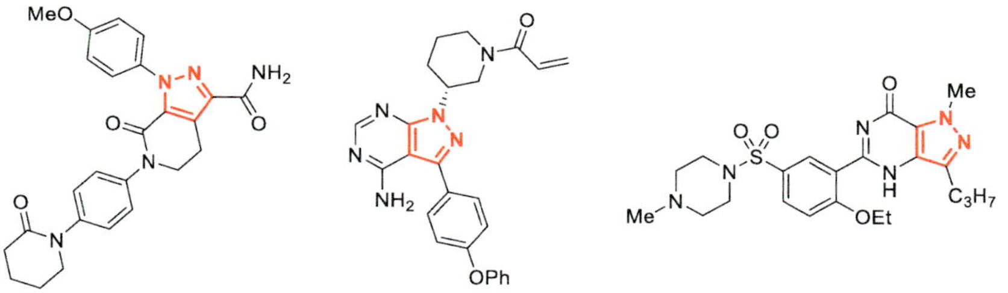

220
#21 Apixaban
$3.34 Billion
Blood disorders

221
#57 + #98 Ibrutinib
$1.83 Billion + $1.25 Billion
Oncology

222
#74 Sildenafil
$1.56 Billion
Sexual health

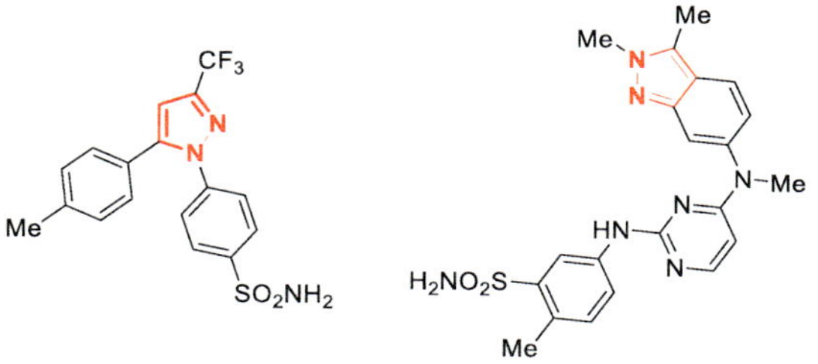

184
#176 Celecoxib
$733 Million
Bone health

223
#180 Pazopanib
$729 Million
Oncology

Scheme 49 Top 200 grossing drugs of 2016 containing a pyrazole motif.

treatment of erectile dysfunction after side effects were noted in clinical trials (2005EODM283). Inhibition of PDE-5 prevents the degradation of cyclic guanosine monophosphate, which after further downstream effects allow increased blood flow to the penis (1999AJCY21).

Celecoxib **184**, marketed by Pfizer, is a nonsteroidal antiinflammatory drug (NSAID) which inhibits the cyclooxygenase-2 (COX-2) enzyme (1997SAR2). Before it lost its patent protection, it grossed over $2 billion per year. The final therapeutic that made it onto the list is the multityrosine kinase inhibitor pazopanib **223**. The compound targets the vascular endothelial growth factor which leads to inhibition of angiogenesis and thus slows tumor growth (2013EOP929). It has been approved for use against soft tissue sarcomas.

The biological areas in which pyrazole-containing compounds have shown promise are almost as diverse as the structures themselves. Recent examples include α7 nicotinic acetylcholine receptor agonists for Alzheimer's disease (2012JMC10277), selective agonists of GPR109a for high cholesterol (2012JMC3644), NS3-4A protease inhibitors for treatment

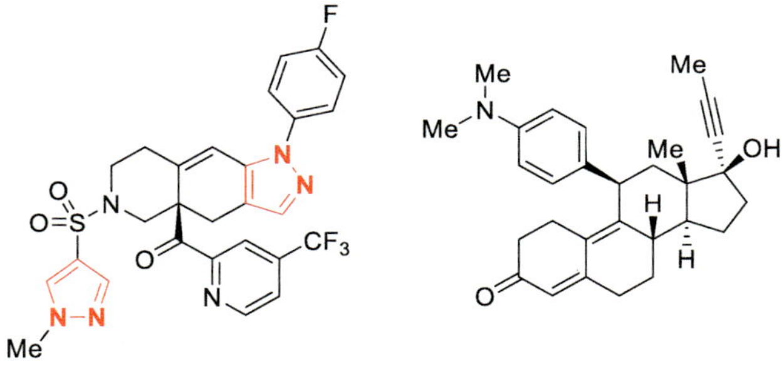

224
Relacorilant

225
Mifepristone

Scheme 50 GR antagonists for the treatment of Cushing's syndrome.

of hepatitis C (2014JMC1770) and tubulin-binding vascular disrupting agents for the treatment of cancer (2016JMC9473). Corcept Therapeutics in collaboration with Sygnature Discovery recently developed a GR antagonist for the treatment of Cushing's syndrome (2017JMC3405). Relacorilant **224** contains two pyrazole structures and is progressing through Phase II clinical trials (Scheme 50). The compound is being developed to replace the current treatment for Cushing's syndrome (mifepristone), which relies on this mechanism. Relacorilant is an improvement on mifepristone **225** as it does not have affinity for the progesterone receptor and will therefore not lead to aborted pregnancies. Recent reviews on pyrazoles in medicinal chemistry give numerous examples and further detail (2016EJM170, 2017NJC16).

3.1.2 Pyrazoles in Crop Protection

Pyrazoles have also had lucrative commercial success in the agrochemical industry. Lamberth's review on pyrazole-based crop protection provides an excellent, in-depth analysis of the area (2007H1467). Pyrazoles have shown impressive diversity in their applications in crop protection with important uses as herbicides, fungicides, and insecticides (Scheme 51). Pinoxaden **226** was developed by Syngenta and is an acetyl-coenzyme A carboxylase (ACCase) inhibitor. Inhibition of ACCase has effects on fatty acid biosynthesis in plants resulting in herbicidal activity (2016SR34066). Pyroxasulfone **227** is a modern, selective herbicide developed by BASF for the control of grasses, sedges, and broadleaf weeds. It is active against susceptible germinating seedlings before (or soon after) they emerge from the soil. The compound works by inhibiting the biosynthesis of very long chain fatty acids which likely play a key role in shoot formation and cell

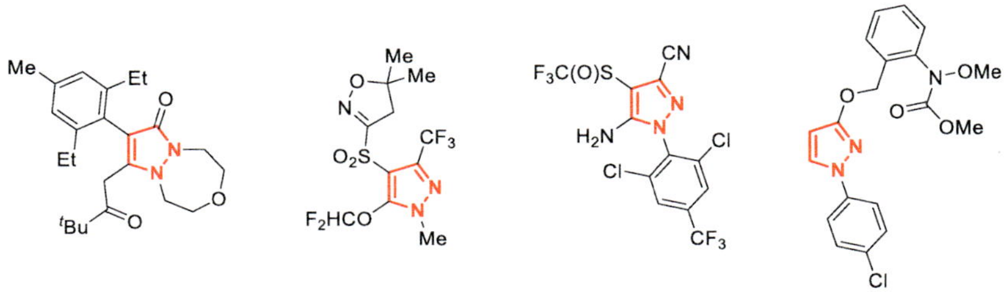

226	**227**	**8**	**228**
Pinoxaden	**Pyroxasulfone**	**Fipronil**	**Pyraclostrobin**
Herbicide	*Herbicide*	*Insecticide*	*Fungicide*

Scheme 51 Pyrazole-based crop protection compounds.

proliferation (2009PBP47). Fipronil **8** was developed by Rhône-Poulenc (now marketed by BASF) and became the first commercially available broad spectrum insecticide from a class called fiproles. This structurally related family are antagonistic inhibitors of γ-aminobutyric acid-gated chloride channel. The channel is located in the membrane of insect nerve fibers and inhibition leads to neuronal cell hyperexcitability (1993PBP47, 1998CRT1529). Also marketed by BASF, pyraclostrobin, **228**, is a fungicide which prevents the production of ATP in the mitochondria of various fungi and some other eukaryotes. The compound binds to cytochrome *b*, disrupting the energy cycle by preventing oxidative phosphorylation (2002PMS649).

3.2 Pyrazoles in Coordination Chemistry

The sp^2 hybridization of the pyrazole N2 atom results in the structures being fundamentally Lewis basic. Chemists have exploited this property to coordinate a wide variety of Lewis acids and undertake a wide variety of chemistries. Furthermore, pyrazolide anions (anions of ^{1}H-pyrazoles) are some of the most versatile ligands known in coordination chemistry. Currently, over 20 different terminal or bridging coordination modes have been identified for these anions (2009JCS(D)2059). These intrinsic properties have led to the widespread application of pyrazoles in coordination chemistry, for example, 1,3-diethyl carboxylate pyrazolide anions have been shown to be amphiphilic receptors for dopamine and amphetamines (2006JA16458). This facilitated the crystallization of stable complexes and in the case of (+)-amphetamine yielded a double-helical supramolecular structure.

In organic chemistry, the chelation properties of pyrazoles have been exploited for metal-catalyzed C–H activation. The functionalization of

C–H bonds as an alternative to traditional crosscoupling methodology has been an exciting growth area of organic chemistry in recent years with numerous reports on the subject and extensive reviews (2009AGE9792, 2011CRV1315, 2014CRV8613, 2017CRV9247). N-Arylpyrazoles have been shown to successfully undergo *ortho*-functionalization using a number of metals, and under an array of conditions, to install a variety of functional groups (Scheme 52). Examples of successful reactions include arylations (2008OL2299, 2010AGE6629), amidations (2013OL3286), and acylations (2013OL5862).

Pyrazoles have also proven to be useful ligands for transition metals due to their Lewis basicity and the ability to tune their properties by modification of appended functional groups. Examples include P,N-ligands based on pyrazole-substituted phosphaferrocene, e.g., **229**, discovered by Ganter et al. (Scheme 53) (2009OM3049). These authors investigated the coordination properties of the ligands toward molybdenum, ruthenium, and palladium. (Allyl)Pd complexes of the ligands were shown to be effective catalysts in asymmetric allylic alkylation reactions. Meyer and coworkers also used pyrazole-based ligands to aid asymmetric allylic alkylations (2010OM1117). The authors prepared a number of methallylpalladium complexes of pyrazolate-based N-donor ligands with chiral oxazoline side arms, dubbed pyrbox's, e.g., **230** (Scheme 53). It was found that the reaction

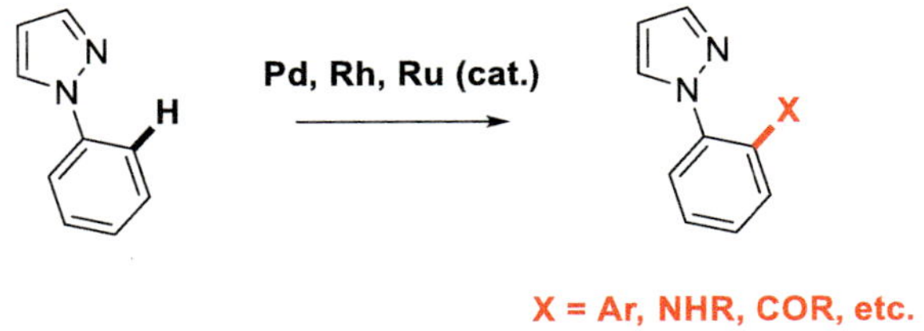

X = Ar, NHR, COR, etc.

Scheme 52 Transition metal-catalyzed functionalization of *N*-arylpyrazoles.

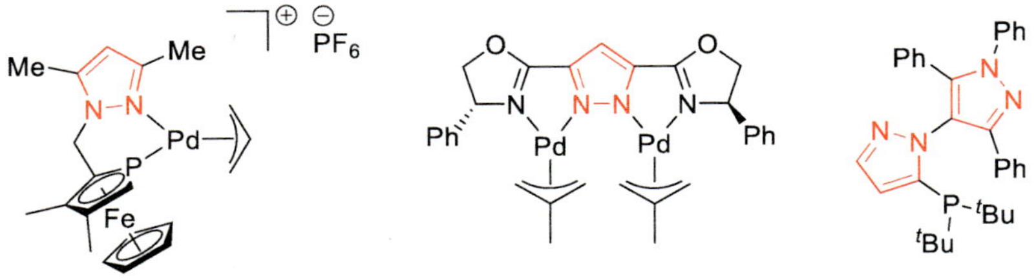

229   **230**   **231**

Scheme 53 Pyrazole-based ligands for catalysis.

enantioselectivity is governed by both the steric and electronic properties of the ligands. Singer et al. enhanced the yields of palladium-catalyzed amination reactions using pyrazole-based ligand **231** (Scheme 53) (2006TL3727).

Pyrazoles have shown great promise in the area of metal–organic frameworks (MOFs), and their contribution has been reviewed in detail (2016CCR1). Indeed, there are over 100 examples of pyrazole-based MOFs described in the literature. A MOF is a coordination network with organic ligands containing potential voids. The coordination network is made up of cross-linked coordination compounds spanning two or three dimensions with repeating coordination entities (2013PAC1715). MOFs have exciting potential in a number of areas including luminescence, drug delivery, electrical conductivity, magnetism, catalysis, and gas storage (2016CCR1). An example of the latter was discovered by Kitagawa et al. who found a silver/bipyrazole-based 3D coordination polymer with interesting properties, e.g., **232** (Scheme 54) (2007AGE889). The host framework was shown to flexibly expand some cavities and contract others to accommodate different guest molecules. This property allowed it to reversibly bind not only water and carbon dioxide but also toluene and benzene. A reasonably diverse set of pyrazole-based ligands have now been shown to form MOFs with a variety of metals including copper (2015CRECF44992), zinc (2012IC5235), cadmium (2013CRECF48280), iron (2013SCI960), and nickel (2013AGE8290).

In a related area, Tidmarsh, Ward, and coworker developed pyrazole-based ligand **233** (Scheme 54) which reacted with silver(I) salts and self-assembled into various infinite coordination networks (2010JCS(D)3805).

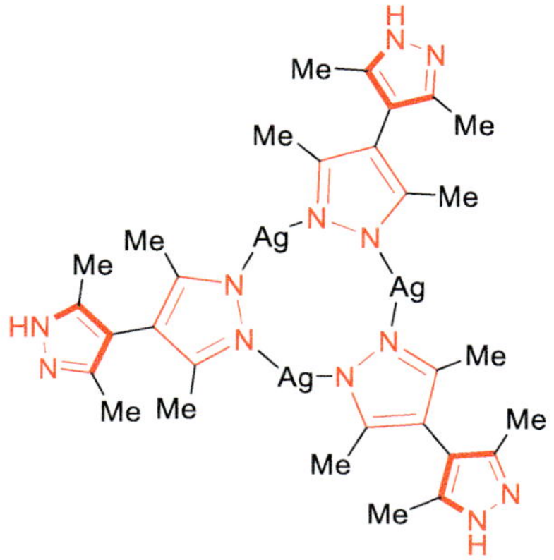

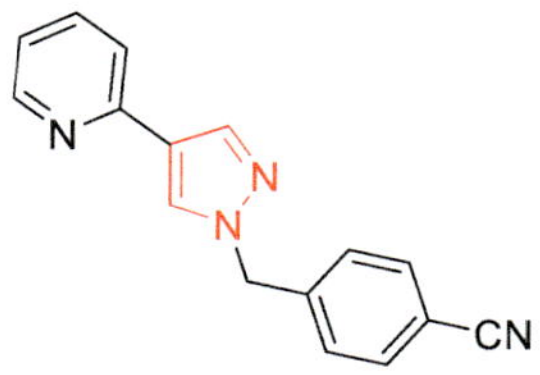

Scheme 54 Pyrazole-based metal–organic framework and pyrazole-based ligand for self-assembly coordination networks.

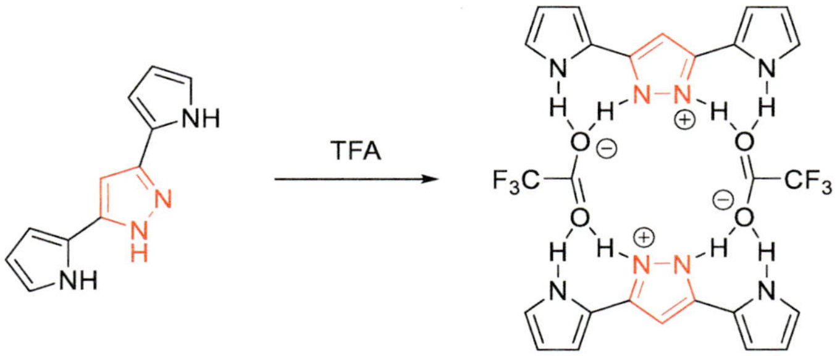

Scheme 55 Anion binding of dipyrrolyl-substituted pyrazoles.

The pendant nitrile group was also found to coordinate Ag(I) as part of the network. Interestingly, when reacted with first-row transition metal dications (cobalt, nickel zinc), the compound formed simple metal trischelates $[ML_3]^{2+}$.

Maeda et al. showed that pyrazoles can also be used as building blocks in supramolecular assemblies not based around metal cations (2007JCS(CC) 1136). Protonated dipyrrolyl-substituted pyrazole **234** was shown to bind trifluoroacetate in the solid state, in a [2+2] binding mode, to afford nanometer- and micrometer-scale architectures **235** (Scheme 55).

3.3 Miscellaneous Applications of Pyrazoles

Pyrazoles have also found use in polymer chemistry with diverse properties. For example, pyrazole-based polymers have been shown to have interesting electroluminescent properties (1999JCS(M)339). In a different application, Hawker, Klinger, and coworkers developed a high-performance thermoset polymer using sydnone cycloaddition chemistry (2016JA6400). The bifunctional sydnone **236** was reacted with a trifunctional alkyne unit (**237**) to form fully aromatic thermosets with pyrazole cross-linking (**238**) (Scheme 56). This material possessed remarkable thermal stability, with $T_{d5\%}$ of 520°C and a weight loss of <0.1% per day at 225°C in air.

Achilefu et al. functionalized the heptamethine cyanine dye, IR820, using Suzuki–Miyaura coupling to form near-infrared dyes which exhibited pH-dependent photophysical properties (2013PAP326). The authors found that an N-unsubstituted pyrazole **239** exhibited pH-dependent fluorescence lifetime changes from 0.93 ns in neutral conditions to 1.27 ns in acidic conditions (Scheme 57). Study of the molecule by time-resolved emission spectra showed fluorophores with two distinct lifetimes at low pH, which

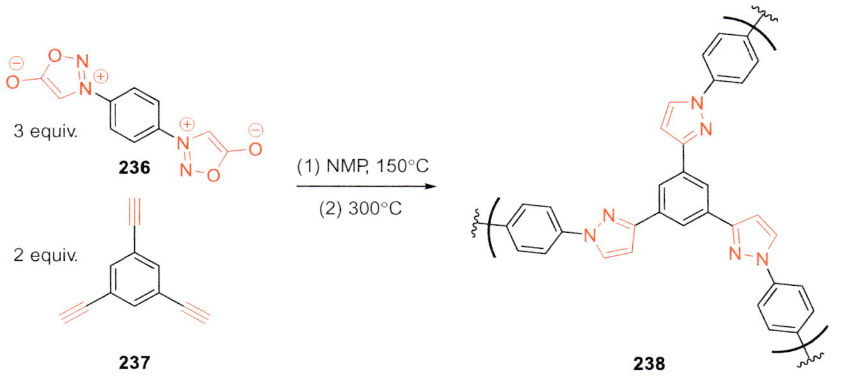

Scheme 56 Preparation of pyrazole-based high performance thermosets. *N.V. Handa, S. Li, J.A. Gerbec, N. Sumitani, C.J. Hawker, and D. Klinger*, J. Am. Chem. Soc., **138**, 6400 (2016).

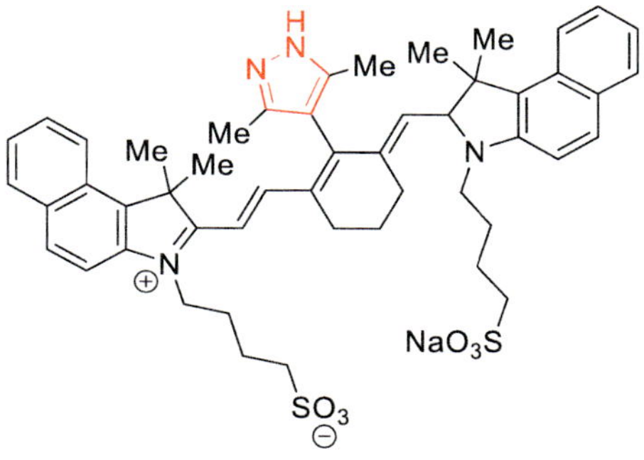

Scheme 57 Pyrazole-containing, pH-dependent near-infrared dye.

corresponded to pH-sensitive and non-pH-sensitive species. N-Substituted pyrazoles did not show pH dependence. However, the N-substituted compounds did not contain *ortho*-substitution relative to the connection point. It was therefore not clear if the lack of pH dependence was due to the N-substitution or orientation of the pyrazole ring.

Pyrazoles have also shown themselves to be amenable to studies of complex physical chemical phenomena. Chi et al. discovered a unique set of pyrazole-based molecules (**240**, **241**) that undergo excited-state intramolecular proton transfer (ESIPT) (Scheme 58) (2003JA10800). The uniqueness of their system is that ESIPT has an appreciably larger energy barrier associated with it, attributed to skeletal reorganization. This energy barrier facilitated studies on the reaction potential energy surface.

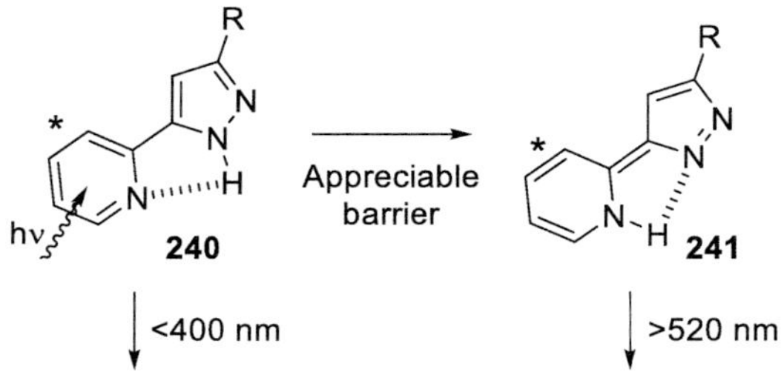

Scheme 58 Excited-state intramolecular proton transfer in pyrazole-based systems.

4. CONCLUDING REMARKS

While pyrazoles have a rich chemical history, the motif continues to enjoy significant attention from research groups. The 5-membered diazole's enduring popularity stems from its continued commercial success in the pharmaceutical and agrochemical industries, as well as its application to numerous other fields. In particular, successful examples in exciting new areas of chemistry such as MOFs, C–H activation chemistry, and high-tech polymers have further enhanced the reputation of pyrazoles. This diversity of applications has inspired numerous innovations in synthetic chemistry for the construction of pyrazoles efficiently, regioselectively, and containing significant complexity. As highlighted in this review, diazo compounds have been shown to be excellent building blocks for the construction of pyrazoles with high regioselectivity and with high structural diversity. However, they are limited to the formation of N–H pyrazoles. Alternatively, *N*-tosylhydrazones have shown potential for the regioselective synthesis of N-substituted pyrazoles. Ynones have also shown themselves to be interesting alternatives to 1,3-dicarbonyl compounds for inherently regioselective condensation reactions to form either N-substituted or N–H pyrazoles. Sydnone cycloadditions have also taken significant steps forward, particularly with the application of copper-catalyzed click chemistry to the reactions. In a similar manner, catalysis featuring other metals has become more prevalent, and unusual mechanistic features have been identified. Overall, interest in pyrazole derivatives continues to be prevalent throughout the chemical literature. In the coming years more applications for the molecules will no doubt be discovered, which will drive further synthetic innovations to meet demand.

REFERENCES

1883BDCG2597	L. Knorr, *Ber. Dtsch. Chem. Ges.*, **16**, 2597 (1883).
1884BDCG2032	L. Knorr, *Ber. Dtsch. Chem. Ges.*, **17**, 2032 (1884).
1884BDCG2756	C. Paal, *Ber. Dtsch. Chem. Ges.*, **17**, 2756 (1884).
1884BDCG2863	L. Knorr, *Ber. Dtsch. Chem. Ges.*, **17**, 2863 (1884).
1896BKW1061	W. Filehne, *Berl. Klin. Wochenschr.*, **33**, 1061 (1896).
1935JA2656	D.C. Hurd and S.C. Lui, *J. Am. Chem. Soc.*, **57**, 2656 (1935).
1935JCS286	D.W. Adamson and J. Kenner, *J. Chem. Soc.*, 286 (1935).
1943RTC485	J. van Alphen, *Recl. Trav. Chim. Pays-Bas*, **62**, 485 (1943).
1950JCS1542	W. Baker, W.D. Ollis, and V.D. Poole, *J. Chem. Soc.*, 1542 (1950).
1960CB1425	R. Hüttel, J. Reidl, H. Martin, and K. Francke, *Chem. Ber.*, **93**, 1425 (1960).
1960CB1433	R. Hüttel, K. Francke, H. Martin, and J. Reidl, *Chem. Ber.*, **93**, 1433 (1960).
1961CB3260	G. Wittig and A. Krebs, *Chem. Ber.*, **94**, 3260 (1961).
1962AGE48	R. Huisgen, R. Grashley, H. Gotthardt, and R. Schmidt, *Angew. Chem. Int. Ed. Engl.*, **1**, 48 (1962).
1963JCS6062	R.C. Cookson and M.J. Locke, *J. Chem. Soc.*, 6062 (1963).
1966T2895	H.-B. Schroeter and D. Neumann, *Tetrahedron*, **22**, 2895 (1966).
1969AGE513	E. Brunn and R. Huisgen, *Angew. Chem. Int. Ed.*, **8**, 513 (1969).
1981S1	O. Mitsunobu, *Synthesis*, **1**, 1 (1981).
1984TL5183	I. Shimizu, Y. Ohashi, and J. Tsuji, *Tetrahedron Lett.*, **25**, 5183 (1984).
1986BCJ483	T. Fuchigami, C.-S. Chen, T. Nonaka, M.-Y. Yeh, and H.-J. Tien, *Bull. Chem. Soc. Jpn.*, **59**, 483 (1986).
1989SC561	S. Ohiria, *Synth. Commun.*, **19**, 561 (1989).
1993PBP47	L.M. Cole, R.A. Nicholson, and J.E. Casida, *Pestic. Biochem. Physiol.*, **46**, 47 (1993).
1993JHC755	R.D. Miller and O. Reiser, *J. Heterocycl. Chem.*, **30**, 755 (1993).
1995TL3087	A.S. Kende and M. Journet, *Tetrahedron Lett.*, **36**, 3087 (1995).
1996CHEC(3)1	J. Elguero, In A.R. Katritzky, C.W. Rees, and E.F.V. Scriven, editors: *Comprehensive Heterocyclic Chemistry II*, Pergamon Press: Oxford, UK: 1996, **Vol. 3**, p 1.
1996SC1441	K. Turnbull and S. Ito, *Synth. Commun.*, **26**, 1441 (1996).
1996SL521	S. Mueller, B. Liepold, J.G. Roth, and H.J. Bestmann, *Synlett*, **6**, 521 (1996).
1997SAR2	J.R. Vane, *Semin. Arthritis Rheum.*, **26**(6 Suppl. 1), 2 (1997).
1998CRT1529	D. Hainzl, L.M. Cole, and J.E. Casida, *Chem. Res. Toxicol.*, **11**, 1529 (1998).
1998JOC5031	Z. Xu and X. Lu, *J. Org. Chem.*, **63**, 5031 (1998).
1999AJCY21	D.J. Webb, S. Freestone, M.J. Allen, and G.J. Muirhead, *Am. J. Cardiol.*, **83**, 21 (1999).
1999JCS(CC)1893	V. Atlan, L. El Kaim, S. Lacroix, and R. Morgentin, *J. Chem. Soc. Chem. Commun.*, 1893 (1999).
1999JCS(M)339	A. Danel, Z. He, G.H.W. Milburn, and P. Tomasik, *J. Chem. Soc. J. Mater. Chem.*, **9**, 339 (1999).
2000AGE3702	K.A. Jørgensen, *Angew. Chem. Int. Ed. Engl.*, **112**, 3702 (2000).
2000SL353	L. El Kaim and S. Lacroix, *Synlett*, **3**, 353 (2000).
2000SL489	V. Atlan, C. Buron, and L. El Kaim, *Synlett*, **4**, 489 (2000).
2000T4139	G. Maas and V. Gettwert, *Tetrahedron*, **56**, 4139 (2000).
2001AGE1430	V.K. Aggarwal, E. Alonso, G. Hynd, K.M. Lyndon, M.J. Palmer, M. Porcelloni, and J.R. Studley, *Angew. Chem. Int. Ed. Engl.*, **40**, 1430 (2001).

2002PMS649	D.W. Bartlett, J.M. Clough, J.R. Godwin, A.A. Hall, M. Hamer, and B. Parr-Dobrzanski, *Pest. Manag. Sci.*, **58**, 649 (2002).
2002TL8319	J.E. Ancel, L. El Kaim, A. Gadras, L. Grimaud, and N.K. Jana, *Tetrahedron Lett.*, **43**, 8319 (2002).
2003JA10800	W.-S. Yu, C.-C. Cheng, Y.-M. Cheng, P.-C. Wu, Y.-H. Song, Y. Chi, and P.-T. Chou, *J. Am. Chem. Soc.*, **125**, 10800 (2003).
2003JOC5381	V.K. Aggarwal, J. de Vincente, and R.V. Bonnert, *J. Org. Chem.*, **68**, 5381 (2003).
2003TL6737	K.Y. Lee, J.M. Kim, and J.M. Kim, *Tetrahedron Lett.*, **44**, 6737 (2003).
2004JCS(CC)394	N. Jiang and C.-J. Li, *J. Chem. Soc. Chem. Commun.*, 394 (2004).
2005EODM283	S.H. Francis and J.D. Corbin, *Expert Opin. Drug Metab. Toxicol.*, **1**, 283 (2005).
2005OL4487	M.S.M. Ahmed, K. Kobayashi, and A. Mori, *Org. Lett.*, **7**, 4487 (2005).
2006AGE7079	R. Martín, M. Rodríguez Rivero, and S.L. Buchwald, *Angew. Chem. Int. Ed. Engl.*, **45**, 7079 (2006).
2006JA16458	F. Reviriego, M.I. Rodríguez-Franco, P. Navarro, E. García-España, M. Liu-González, B. Verdejo, and A. Domènech, *J. Am. Chem. Soc.*, **128**, 16458 (2006).
2006OL2213	V. Nair, A.T. Biju, K. Mohanan, and E. Suresh, *Org. Lett.*, **8**, 2213 (2006).
2006TL3727	R.A. Singer, M. Dore, J.E. Sieser, and M.A. Berliner, *Tetrahedron Lett.*, **47**, 3727 (2006).
2007AGE889	J.-P. Zhang, S. Horike, and S. Kitagawa, *Angew. Chem. Int. Ed.*, **46**, 889 (2007).
2007AGE8656	D.L. Browne, M.D. Helm, A. Plant, and J.P.A. Harrity, *Angew. Chem. Int. Ed.*, **46**, 8656 (2007).
2007H1467	C. Lamberth, *Heterocycles*, **71**, 1467 (2007).
2007JCS(CC)1136	H. Maeda, Y. Ito, Y. Kusunose, and T. Nakanishi, *J. Chem. Soc. Chem. Commun.*, 1136 (2007).
2007OL1125	R. Muruganantham, S.M. Mobin, and I.N.N. Namboothiri, *Org. Lett.*, **9**, 1125 (2007).
2008AGE2359	E.L. Myers and R.T. Raines, *Angew. Chem. Int. Ed. Engl.*, **48**, 2359 (2008).
2008AGE7340	S. Würtz, S. Rakshit, J.J. Neumann, T. Dröge, and F. Glorius, *Angew. Chem. Int. Ed.*, **120**, 7340 (2008).
2008CHEC(4)1	L. Yet, In A.R. Katritzky, C.A. Ramsden, E.F.V. Scriven, and R.J.K. Taylor, editors: *Comprehensive Heterocyclic Chemistry III*, Elsevier: Oxford, UK: 2008, **Vol. 4**, p 1.
2008OL2299	L. Ackermann, R. Vincente, and A. Althammer, *Org. Lett.*, **10**, 2299 (2008).
2009AGE9792	L. Ackermann, R. Vicente, and A.R. Kapdi, *Angew. Chem. Int. Ed. Engl.*, **48**, 9792 (2009).
2009JA7762	D.L. Browne, J.F. Vivat, A. Plant, E. Gomez-Bengoa, and J.P.A. Harrity, *J. Am. Chem. Soc.*, **131**, 7762 (2009).
2009JCS(D)2059	M.A. Halcrow, *J. Chem. Soc. Dalton Trans.*, 2059 (2009).
2009OBC4052	R.S. Foster, J. Huang, J.F. Vivat, D.L. Browne, and J.P.A. Harrity, *Org. Biomol. Chem.*, **7**, 4052 (2009).
2009OL2097	B.S. Gerstenberger, M.R. Rauckhorst, and J.T. Starr, *Org. Lett.*, **11**, 2097 (2009).
2009OM3049	H. Willms, W. Frank, and C. Ganter, *Organometallics*, **28**, 3049 (2009).

2009PBP47	Y. Tanetani, K. Kaku, K. Kawai, T. Fujioka, and T. Shimizu, *Pestic. Biochem. Physiol.*, **95**, 47 (2009).
2009S650	A. Rodriguez and W.J. Moran, *Synthesis*, **4**, 650 (2009).
2009TL3942	A. Rodriguez, R.V. Fennessey, and W.J. Moran, *Tetrahedron Lett.*, **50**, 3942 (2009).
2010AGE3196	K. Mohanan, A.R. Martin, L. Toupet, M. Smietana, and J.-J. Vasseur, *Angew. Chem. Int. Ed.*, **49**, 3196 (2010).
2010AGE6629	P.B. Arockiam, C. Fischmeister, C. Bruneau, and P.H. Dixneuf, *Angew. Chem. Int. Ed.*, **49**, 6629 (2010).
2010AGE7790	J.J. Neumann, M. Suri, and F. Glorius, *Angew. Chem. Int. Ed.*, **49**, 7790 (2010).
2010JCE1348	N.A. McGrath, M. Brichacek, and J.T. Njararson, *J. Chem. Educ.*, **87**, 1348 (2010).
2010JCS(D)3805	H. Fenton, I.S. Tidmarsh, and M.D. Ward, *J. Chem. Soc. Dalton Trans.*, **39**, 3805 (2010).
2010JOC2197	R. Muruganantham and I.N.N. Namboothiri, *J. Org. Chem.*, **71**, 2197 (2010).
2010OM1117	A. Ficks, C. Sibbald, M. John, S. Dechert, and F. Meyer, *Organometallics*, **29**, 1117 (2010).
2010T553	D.L. Browne and J.P.A. Harrity, *Tetrahedron*, **66**, 553 (2010).
2011CRV1315	L. Ackermann, *Chem. Rev.*, **111**, 1315 (2011).
2011EJO3184	A.R. Martin, K. Mohanan, L. Toupet, J.-J. Vasseur, and M. Smietana, *Eur. J. Org. Chem.*, **2011**(17), 3184 (2011).
2011JOC4764	D. Verma, S. Mobin, and I.N.N. Namboothiri, *J. Org. Chem.*, **76**, 4764 (2011).
2011JOC5915	D.J. Babinski, H.R. Aguilar, R. Still, and D.E. Frantz, *J. Org. Chem.*, **76**, 5915 (2011).
2011OL4016	R. Kumar and I.N.N. Namboothiri, *Org. Lett.*, **13**, 4016 (2011).
2012BD1182	M.F.M. de Rooij, A. Kuil, C.R. Geest, E. Eldering, B.Y. Chang, J.J. Buggy, S.T. Pals, and M. Spaargaren, *Blood*, **119**, 2590 (2012).
2012BD2590	S. Ponader, S.-S. Chen, J.J. Buggy, K. Balakrishnan, V. Gandhi, W.G. Wierda, M.J. Keating, S. O'Brien, N. Chiorazzi, and J.A. Burger, *Blood*, **119**, 1182 (2012).
2012IC5235	C. Pettinari, A. Tăbăcaru, I. Boldog, K.V. Domasevitch, S. Galli, and N. Masciocchi, *Inorg. Chem.*, **51**, 5235 (2012).
2012JMC3644	P.D. Boatman, B. Lauring, T.O. Schrader, M. Kasem, B.R. Johnson, P. Skinner, J.K. Jung, J. Xu, M.C. Cherrier, P.J. Webb, G. Semple, C.R. Sage, J. Knudsen, R. Chen, W.L. Luo, L. Caro, J. Cote, E. Lai, J. Wagner, A.K. Taggart, E. Carballo-Jane, M. Hammond, S.L. Colletti, J.R. Tata, D.T. Connolly, M.G. Waters, and J.G. Richman, *J. Med. Chem.*, **55**, 3644 (2012).
2012JMC10277	R. Zanaletti, L. Bettinetti, C. Castaldo, I. Ceccarelli, G. Cocconcelli, T.A. Comery, J. Dunlop, E. Genesio, C. Ghiron, S.N. Haydar, F. Jow, L. Maccari, I. Micco, A. Nencini, C. Pratelli, C. Scali, E. Turlizzi, and M. Valacchi, *J. Med. Chem.*, **55**, 10277 (2012).
2012OL5030	J. Hu, S. Chen, Y. Sun, J. Yang, and Y. Rao, *Org. Lett.*, **14**, 5030 (2012).
2012OL5354	J.D. Kirkham, S.J. Edeson, S. Stokes, and J.P.A. Harrity, *Org. Lett.*, **14**, 5354 (2012).
2013AGE8290	N.M. Padial, E. Quartapelle Procopio, C. Montoro, E. López, J.E. Oltra, V. Colombo, A. Maspero, N. Masciocchi, S. Galli, I. Senkovska, S. Kaskel, E. Barea, and J.A.R. Navarro, *Angew. Chem. Int. Ed. Engl.*, **52**, 8290 (2013).

2013AGE12056	S. Kolodych, E. Rasolofonjatovo, M. Chaumontet, M.-C. Nevers, C. Créminon, and F. Taran, *Angew. Chem. Int. Ed. Engl.*, **52**, 12056 (2013).
2013AGE13324	Z. Chen, Q. Yan, Z. Liu, Y. Xu, and Y. Zhang, *Angew. Chem. Int. Ed. Engl.*, **52**, 13324 (2013).
2013CEJ7555	L. Wang, J. Huang, X. Gong, and J. Wang, *Chemistry*, **19**, 7555 (2013).
2013CRECF48280	V.V. Ponomarova, V.V. Komarchuk, I. Boldog, H. Krautscheid, and K.V. Domasevitch, *CrstEngComm*, **15**, 8280 (2013).
2013EOP929	J. Verweij and S. Sleijfer, *Expert Opin. Pharmacother.*, **14**, 923 (2013).
2013JOC3482	D. Verma, R. Kumar, and I.N.N. Namboothiri, *J. Org. Chem.*, **78**, 3482 (2013).
2013JOC4960	S.V. Kumar, S.K. Yadav, B. Raghava, B. Saraiah, H. Ila, K.S. Rangappa, and A. Hazra, *J. Org. Chem.*, **78**, 4960 (2013).
2013OL3286	S.V. Thirunavukkarasu, K. Raghuvanshi, and L. Ackermann, *Org. Lett.*, **15**, 3286 (2013).
2013OL5862	P.M. Liu and C.G. Frost, *Org. Lett.*, **15**, 5862 (2013).
2013OL5967	G. Zhang, H. Ni, W. Chen, J. Shao, H. Liu, B. Chen, and Y. Yu, *Org. Lett.*, **15**, 5967 (2013).
2013PAC1715	S.R. Batten, N.R. Champness, X.-M. Chen, J. Garcia-Martinez, S. Kitagawa, L. Öhrström, M. O'Keeffe, M.P. Suh, and J. Reedijk, *Pure Appl. Chem.*, **85**, 1715 (2013).
2013PAP326	H. Lee, M.Y. Berezin, R. Tang, N. Zhegalova, and S. Achilefu, *Photochem. Photobiol.*, **89**, 326 (2013).
2013SCI960	Z.R. Herm, B.M. Wiers, J.A. Mason, J.M. van Baten, M.R. Hudson, P. Zajdel, C.M. Brown, N. Masciocchi, R. Krishna, and J.R. Long, *Science*, **340**, 960 (2013).
2014COC293	Y.C. Lau and G.Y. Lip, *Curr. Opin. Cardiol.*, **29**, 293 (2014).
2014CRV8613	F. Chen, T. Wang, and N. Jiao, *Chem. Rev.*, **114**, 8613 (2014).
2014JCS(CC)9376	L. Plougastel, O. Koniev, S. Specklin, E. Decuypere, C. Créminon, D.-A. Buisson, A. Wagner, S. Kolodych, and F. Taran, *J. Chem. Soc. Chem. Commun.*, **50**, 9376 (2014).
2014JCS(CS)1742	S. Wallace and J.W. Chin, *J. Chem. Soc. Chem. Sci.*, **5**, 1742 (2014).
2014JMC1770	B. Moreau, J.A. O'Meara, J. Bordeleau, M. Garneau, C. Godbout, V. Gorys, M. Leblanc, E. Villemure, P.W. White, and M. Llinas-Brunet, *J. Med. Chem.*, **57**, 1770 (2014).
2014JOC2049	R. Harigae, K. Moriyama, and H. Togo, *J. Org. Chem.*, **79**, 2049 (2014).
2014OL576	Y. Kong, M. Tang, and Y. Wang, *Org. Lett.*, **16**, 576 (2014).
2014OL3110	H. Guo, D. Zhang, C. Zhu, J. Li, G. Xu, and J. Sun, *Org. Lett.*, **16**, 3110 (2014).
2014OL3476	D.Y. Li, X.F. Mao, H.J. Chen, G.R. Chen, and P.N. Liu, *Org. Lett.*, **16**, 3476 (2014).
2014OL5108	Z.-J. Cai, X.-M. Lu, Y. Zi, C. Yang, L.-J. Shen, J. Li, S.-Y. Wang, and S.-J. Ji, *Org. Lett.*, **16**, 5108 (2014).
2014OL5940	J.-J. Wen, H.-T. Tang, K. Xiong, Z.-C. Ding, and Z.-P. Zhan, *Org. Lett.*, **16**, 5940 (2014).
2014T5214	M.M.D. Pramanik, R. Kant, and N. Rastogi, *Tetrahedron*, **70**, 5214 (2014).
2015CEJ3257	J. Comas-Barceló, R.S. Foster, B. Fiser, E. Gomez-Bengoa, and J.P.A. Harrity, *Chemistry*, **21**, 3257 (2015).

2015CRECF44992	A. Tăbăcaru, S. Galli, C. Pettinari, N. Masciocchi, T.M. McDonald, and J.R. Long, *CrstEngComm*, **17**, 4992 (2015).
2015CRV9981	A. Ford, H. Miel, A. Ring, C.N. Slattery, A.R. Maguire, and M.A. McKervey, *Chem. Rev.*, **115**, 9981 (2015).
2015JOC2467	A.W. Brown and J.P.A. Harrity, *J. Org. Chem.*, **80**, 2467 (2015).
2015JOC4325	A. Kamal, K.N.V. Sastry, D. Chandrasekhar, G.S. Mani, P.R. Adiyala, J.B. Nanubolu, K.K. Singarapu, and R.A. Maurya, *J. Org. Chem.*, **80**, 4325 (2015).
2015OBC1492	S. Ahmed, A.K. Gupta, R. Kant, and K. Mohanan, *Org. Biomol. Chem.*, **13**, 1492 (2015).
2015OL362	E. Decuypere, S. Specklin, S. Gabillet, D. Audisio, H. Liu, L. Plougastel, S. Kolodych, and F. Taran, *Org. Lett.*, **17**, 362 (2015).
2015OL872	Q. Zhang, L.-G. Meng, K. Wang, and L. Wang, *Org. Lett.*, **17**, 872 (2015).
2015OL1405	D.C. Schmitt, A.P. Taylor, A.C. Flick, and R.E. Kyne Jr., *Org. Lett.*, **17**, 1405 (2015).
2015OL1521	G.C. Senadi, W.-P. Hu, T.-Y. Lu, A.M. Garkhedkar, J.K. Vandavasi, and J.-J. Wang, *Org. Lett.*, **17**, 1521 (2015).
2016AGE12073	H. Liu, D. Audisio, L. Plougastel, E. Decuypere, D.-A. Buisson, O. Koniev, S. Kolodych, A. Wagner, M. Elhabiri, A. Krzyczmonik, S. Forsback, O. Solin, V. Gouverneur, and F. Taran, *Angew. Chem. Int. Ed. Engl.*, **55**, 12073 (2016).
2016CCR1	C. Pettinari, A. Tăbăcaru, and S. Galli, *Coord. Chem. Rev.*, **307**, 1 (2016).
2016EJM170	M.F. Khan, M.M. Alam, G. Verma, W. Akhtar, M. Akhter, and M. Shaquiquzzaman, *Eur. J. Med. Chem.*, **120**, 170 (2016).
2016JA6400	N.V. Handa, S. Li, J.A. Gerbec, N. Sumitani, C.J. Hawker, and D. Klinger, *J. Am. Chem. Soc.*, **138**, 6400 (2016).
2016JMC9473	A. Brown, M. Fisher, G.M. Tozer, C. Kanthou, and J.P.A. Harrity, *J. Med. Chem.*, **59**, 9473 (2016).
2016OBC8486	Y. Shao, J. Tong, Y. Zhao, H. Zheng, L. Ma, M. Ma, and X. Wan, *Org. Biomol. Chem.*, **14**, 8486 (2016).
2016SR34066	X. Xia, W. Tang, S. He, J. Kang, H. Ma, and J. Li, *Sci. Rep.*, **6**, 34066 (2016).
2016TL3146	D. Nair, P. Pavashe, S. Katiyar, and I.N.N. Namboothiri, *Tetrahedron Lett.*, **57**, 3146 (2016).
2017AGE15612	S. Bernard, D. Audisio, M. Riomet, S. Bregant, A. Sallustrau, L. Plougastel, E. Decuypere, S. Gabillet, R.A. Kumar, J. Elyian, M.N. Trinh, O. Koniev, A. Wagner, S. Kolodych, and F. Taran, *Angew. Chem. Int. Ed. Engl.*, **5**, 15612 (2017).
2017CEJ5228	A.W. Brown, J. Comas-Barceló, and J.P.A. Harrity, *Chemistry*, **23**, 5228 (2017).
2017CRV9247	Y. Park, Y. Kim, and S. Chang, *Chem. Rev.*, **117**, 9247 (2017).
2017JOC1688	P. Fricero, L. Bialy, A.W. Brown, W. Czechtizky, M. Méndez, and J.P.A. Harrity, *J. Org. Chem.*, **82**, 1688 (2017).
2017JMC3405	H.J. Hunt, J.K. Belanoff, I. Walters, B. Gourdet, J. Thomas, N. Barton, J. Unitt, T. Phillips, D. Swift, and E. Eaton, *J. Med. Chem.*, **60**, 3405 (2017).
2017NJC16	A. Ansari, A. Ali, M. Asif, and M. Shamsuzzaman, *New J. Chem.*, **41**, 16 (2017).
2017OL1534	S. Panda, P. Maity, and D. Manna, *Org. Lett.*, **19**, 1534 (2017).

2017OL3466 Y. Tu, Z. Zhang, T. Wang, J. Ke, and J. Zhao, *Org. Lett.*, **19**, 3466 (2017).
2017T3160 A.W. Brown and J.P.A. Harrity, *Tetrahedron*, **73**, 3160 (2017).
2018OL198 P. Fricero, L. Bialy, W. Czechtizky, M. Méndez, and J.P.A. Harrity, *Org. Lett.*, **20**, 198 (2018).

FURTHER READING
2012JOC9041 N. Panda and A.K. Jena, *J. Org. Chem.*, **77**, 9401 (2012).

CHAPTER THREE

Recent Developments in the Chemistry of 1,2,3-Thiadiazoles

Yuri Shafran*, Tatiana Glukhareva*, Wim Dehaen[†], Vasiliy Bakulev*,[1]

*Technology for Organic Synthesis, Ural Federal University, Ekaterinburg, Russia
[†]Molecular Design and Synthesis, KU Leuven, Belgium
[1]Corresponding author: e-mail address: v.a.bakulev@urfu.ru

Contents

1. Introduction	110
2. Synthesis of 1,2,3-Thiadiazoles	110
2.1 Monocyclic 1,2,3-Thiadiazoles	110
2.2 Fused 1,2,3-Thiadiazoles	126
2.3 1,2,3-Thiadiazoles Linearly Connected to Another Heterocycle	133
3. Ring Transformations of 1,2,3-Thiadiazoles	144
3.1 Rearrangements of 1,2,3-Thiadiazoles	145
3.2 Cleavage of the Ring	151
4. 1,2,3-Thiadiazoles in Medicine and Agriculture	162
5. Concluding Remarks	163
Acknowledgment	164
References	164

Abstract

The chapter is focused on the synthesis of 1,2,3-thiadiazoles including Hurd–Mori reaction, Wolff approach, and oxidative cyclization of hydrazones bearing thioamide group. It also contains data on the synthesis of fused 1,2,3-thiadiazoles and those linearly connected to other heterocycles. Moreover, a short overview of biological properties is given in a separate section. The data on the chemical properties are limited in the current review to ring-cleavage reactions and rearrangements of 1,2,3-thiadiazoles to other heterocyclic compounds. An analysis of 1,2,3-thiadiazole chemistry for the period 2003–17 is presented. Materials covered in earlier periods are included when they are appropriate for discussions.

Keywords: 1,2,3-Thiadiazoles, Heterocycles, Hurd–Mori reaction, Wolff synthesis, Oxidative cyclization, Transformation, Ring cleavage, Rearrangements, Biological activity

1. INTRODUCTION

More than 10 years have passed since two books were published covering the synthesis, chemical, and biological properties of 1,2,3-thiadiazoles (2003M[1], 2004M[2]). Since then, there has been further progress in the organocatalytic Hurd–Mori reaction (HMR) and oxidative heterocyclizations of 2-hydrazonothioacetamides, and new data have recently been published on 1,2,3-thiadiazole ring transformations, including novel rearrangements and base- and rhodium-catalyzed annulations of 1,2,3-thiadiazoles to other heterocyclic compounds. In this review, data on the synthesis of 1,2,3-thiadiazoles that are monocyclic, fused to other heterocycles, or linearly connected to other rings have been carefully scrutinized and most of the recent publications in this area are summarized, paying attention to both our own work and the work of others (Section 2).

Data on ring transformations are placed in Section 3. Rearrangements are discussed in Section 3.1 and various transformations, including photochemical, thermal, base-, and metal-catalyzed ring degradations, are discussed in Section 3.2. Finally, the data on applications of 1,2,3-thiadiazole derivatives in medicine and agriculture are collected in Section 4. The literature is covered through 2003–17. The materials covered in earlier publications are partially included when relevant to discussions. We hope that this work will be useful to chemists working with heterocyclic compounds and will stimulate further researches on the chemistry of 1,2,3-thiadiazoles.

2. SYNTHESIS OF 1,2,3-THIADIAZOLES

In the following sections the 1,2,3-thiadiazoles are divided into three groups, referring to the topology of 1,2,3-thiadiazole fragment. Traditionally, these groups are monocyclic (Section 2.1) and fused (or condensed) 1,2,3-thiadiazoles (Section 2.2). Although 1,2,3-thiadiazoles directly connected to another heterocycle belong to monocyclic thiadiazoles, because of the high specificity of the synthesis of this type of system, they are discussed separately (Section 2.3).

2.1 Monocyclic 1,2,3-Thiadiazoles

The known methods of formation of the 1,2,3-thiadiazole ring can be conveniently subdivided into four groups:

(i) cyclization of hydrazones with thionyl chloride (Hurd–Mori synthesis);
(ii) cycloaddition of diazoalkanes onto a C=S bond (Pechmann synthesis);
(iii) generation and cyclization of α-diazo thiocarbonyl compounds (Wolff synthesis (WS)); and
(iv) oxidative cyclization of 2-arylhydrazonothiocarbonyl compounds.

2.1.1 Cyclization of Hydrazones With Thionyl Chloride (Hurd–Mori Synthesis)

N^1-Carbonyl- and sulfonyl hydrazones **1** bearing an adjacent methylene or acidic methyl group react with thionyl chloride to form the 1,2,3-thiadiazoles **3**. Because the sulfur atom of the 1,2,3-thiadiazole ring is derived from $SOCl_2$, while the two nitrogen and two carbon atoms come from hydrazones **1**, this synthetic method represents a [1 + 4] approach to 1,2,3-thiadiazoles (Scheme 1).

This reaction was discovered by Hurd and Mori in 1955 and was named afterward as the HMR (1955JA5359). Since then more than 200 publications on the HMR have appeared in the literature. The data for the period 1956–2004 have been the subject of a review by Stanetty et al. (1999THS(3)265) and was more recently also partially reviewed in a book chapter (2008CHECIII(5)468) and in two monographs (2003M[1], 2004M[2]). The further development of the reaction in the period 2003–17 dealt with improvement of the synthetic protocol (2008SC4407, 2010PS1594, 2012JOC9391, 2016JOC271, 2017ACSC4986), with the use of the reaction product in organic synthesis (2005PS1593, 2006RJO1735, 2007RJO630, 2012RCI1999, 2012RJO728, 2013OL40382014RJC1863) and with the synthesis of new derivatives of 1,2,3-thiadiazoles in the framework of research toward new biologically active compounds (2005CHE1186, 2007BML869,

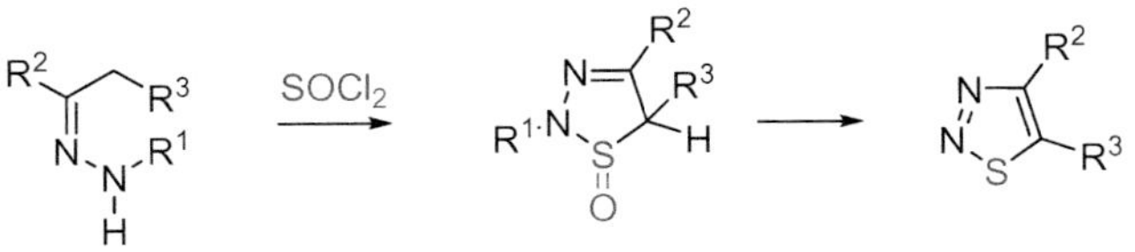

$R^1 = CO_2Alk, CO_2NH_2, Ac, C(=O)H, SO_2Ar, Ts;$
$R^2 = H, Alk, Ar, Het; R^3 = COR, CN, SR, Hal,$
$OR, SO_2R, CS_2SR, cycloalkyl, R^2 + R^3 = cycloalkyl$

Scheme 1

2009BML1089, 2009WOP134110, 2010EJM4920, 2010EJMC6127, 2012JC1276, 2012RCI1999, 2015RJC600, 2015RJC866). Hydrazones bearing carbonyl and sulfonyl groups at N^1 give the best yields of 1,2,3-thiadiazoles (2003M[1], 2004M[2], 2008CHECIII(5)468). Sulfur monochloride or dichloride can also be used as a source of the sulfur atom in 1,2,3-thiadiazoles, though the yields are poor due to side reactions (2004M [2]). The reaction also proceeds using the hydrazones **4** ($R^1 = Ar$). However, in this case 2-aryl-1,2,3-thiadiazolium salts **5** are formed instead of 1,2,3-thiadiazoles **3** (Scheme 2) (1995JCS(P1)2079, 2013IJC427).

The mechanism of the HMR has been carefully studied and has been discussed in several papers (1982JCS(P1)1223, 1998H(48)259, 2013JHC630), in reviews (1999THS(3)265, 2000ANH37, 2003M[1], 2008CHECIII(5)468), and in a book (2004M[2]). Kinetic data and the identification of the intermediates are consistent with a mechanism in which the formation of a thiadiazolinone **2** is the first step. The latter intermediate undergoes aromatization via a Pummerer-type rearrangement followed by cleavage of R^1 from the resulting salt to give the final product (Scheme 1).

The use of ionic liquids (2012JOC9391), solid-phase synthesis (2008SC4407), and microwave and ultrasonic irradiation (2010PS1594) has allowed improvement of the synthetic procedure for the HMR. The advantages of these methods are the ease of work-up, simple reaction conditions, and high purity of the prepared 1,2,3-thiadiazoles. Thus, Kumar et al. have synthesized the ionic liquids **6** and used them with the ketones **7** to obtain a series of 4-aryl- and 4-hetaryl-1,2,3-thiadiazoles **9** in very high yield, via the hydrazones **8** (Scheme 3) (2012JOC9391).

HMR is especially suitable for the synthesis of alkyl-, aryl-, and hetaryl-substituted 1,2,3-thiadiazoles which are stable under the conditions of HMR. A series of various 1,2,3-thiadiazoles with halide (2013OL4038), alkyl (2017CCL372), aryl (2012RJO728, 2014RJC1863), adamantyl (2007RJO630), ester (2013MOL12725), nitro (2007JOC5368), sulfide (2009BMC5920), sulfonyl (2007JHC1165), and other functional groups have been prepared by HMR and also used as reagents for further

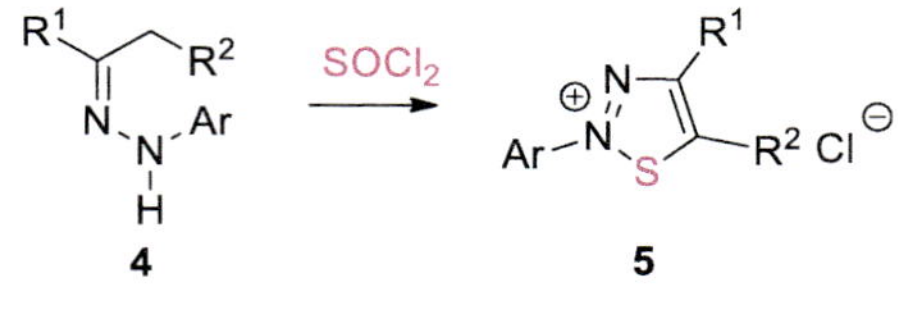

Scheme 2

derivatization giving fused 1,2,3-thiadiazoles (2014CBD223), dendrimers (2007ASJC1783), and polymers (2009S2539). These data published in the last decade, and those reviewed earlier (1999THS(3)265, 2003M[1], 2004M [2], 2008CHECIII(5)468), demonstrate the wide scope of the HMR. The HMR was implemented in an industrial process to prepare the cotton defoliant thidiazuron™. It has become the most powerful method for the synthesis of 1,2,3-thiadiazoles with a wide variety of 4- and 5-substituents.

Although the HMR is indeed the most developed method for the synthesis of 1,2,3-thiadiazoles, and very general in scope, unexpected reactions can occur, due to the use of rather aggressive sulfur-containing reagents such as $SOCl_2$, SCl_2, and S_2Cl_2. These can react with unprotected hydroxy and amino groups; hence, 1,2,3-thiadiazoles with such groups are not available using this methodology. Furthermore, these reagents violently react with water, so this method requires the rigorous extrusion of moisture. Therefore, there has been a demand for the development of a more eco-friendly and safe methodology for the synthesis of 1,2,3-thiadiazoles. Two papers have recently described the synthesis of 1,2,3-thiadiazoles using more benign elemental sulfur in combination with an oxidant. Cheng et al. first demonstrated the use of sulfur in the HMR (Scheme 4) (2016JOC271). Thus, the tetrabutylammonium iodide (TBAI) catalyzed reaction of

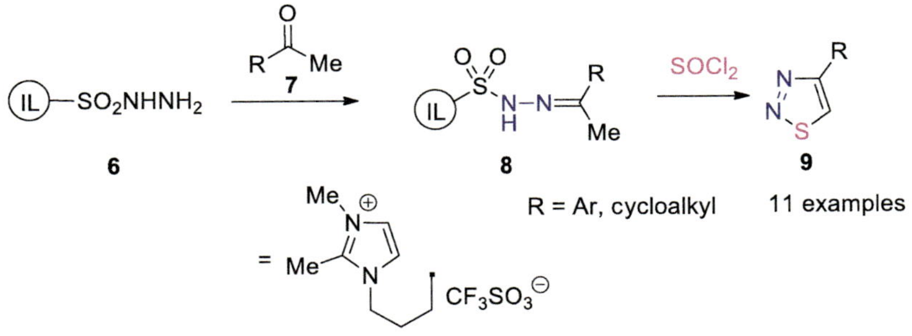

Scheme 3

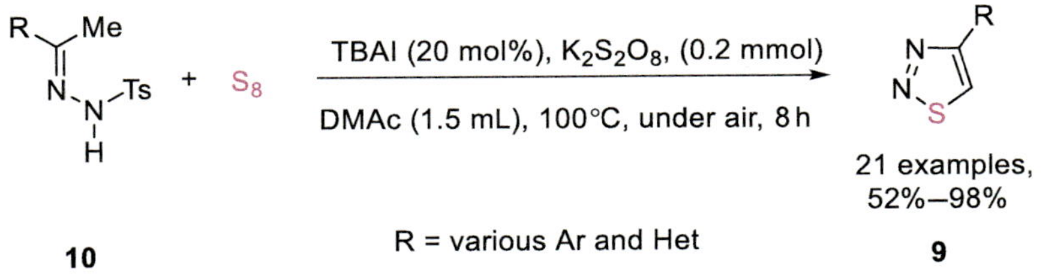

Scheme 4

N-tosylhydrazones **10** in *N,N*-dimethylacetamide (DMAc) with sulfur and an oxidant, preferably $K_2S_2O_8$, at 100°C leads to the formation of the 1,2,3-thiadiazoles **9** in good yields. Other catalysts such as I_2, KI, NH_4I, and CuI also worked to some extent to give 1,2,3-thiadiazoles **9** in moderate yields. The reaction was shown to be very sensitive to the solvent: replacement of DMAc by *N,N*-dimethylformamide (DMF) decreased the yield of the product by 12%, and the reaction did not work at all in other organic solvents such as toluene, dioxane, and acetonitrile. Scheme 4 demonstrates the scope of the reaction. Thiadiazoles bearing various aryl and heteroaryl substituents and alkenyl groups can be prepared in moderate to good yield. However, the *N*-tosylhydrazone of acetone did not enter into this catalyzed reaction with sulfur powder.

To gain insight into the mechanism, Cheng et al. carried out the reaction of the hydrazones **10** with the radical scavenger 2,2,6,6-tetramethyl-1-piperidinyloxy (TEMPO) and additional experiments with proposed intermediates **11** and **14** using both standard conditions and those including radical scavengers. They have shown that the addition of a radical scavenger did not inhibit the reaction (Scheme 5). Furthermore, they found that compounds **11** and **14** under standard reaction conditions afforded thiadiazole **12** in moderate yields. Because disulfide **13** was not detected in the reaction of α-iodo

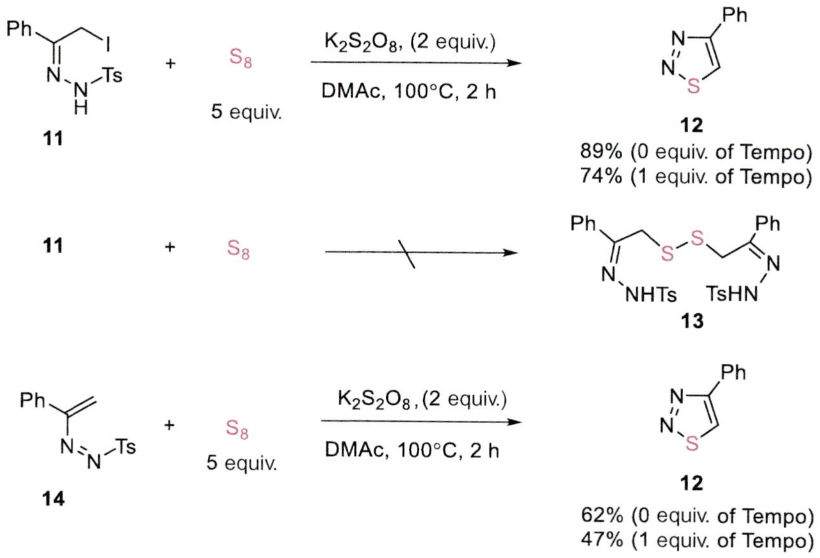

Scheme 5

acetophenone hydrazone **11** with sulfur, the authors ruled out this compound from the list of possible intermediates (2016JOC271).

Based on these experimental results, the mechanism shown in Scheme 6 was proposed. First, oxidation of I^- to I^+ by $K_2S_2O_8$ takes place followed by iodination of the methyl group of hydrazone **10** (R = Ph) to form intermediate **11**. Then, generation of the 1,2-diazadiene **14** occurs via elimination of HI from **11**. HI dissociates to H^+ and I^-. The latter is oxidized to I_2 to close the catalytic cycle. In the second step, the addition of sulfur S_8 to the diene system of **14** generates intermediate **16** (via **15**) followed by cyclization to the thiadiazole **17**. The latter eliminates S_7 and toluenesulfinic acid to afford final product **12**.

The work of Cheng et al. demonstrates a unique catalytic reaction and at the same time a synthesis of 1,2,3-thiadiazoles avoiding the employment of hazardous reagents and without the need for the rigorous exclusion of moisture (Scheme 7) (2016JOC271).

A further development of the catalyzed reaction of tosylhydrazones and elemental sulfur in the presence of ammonium iodide using atmospheric

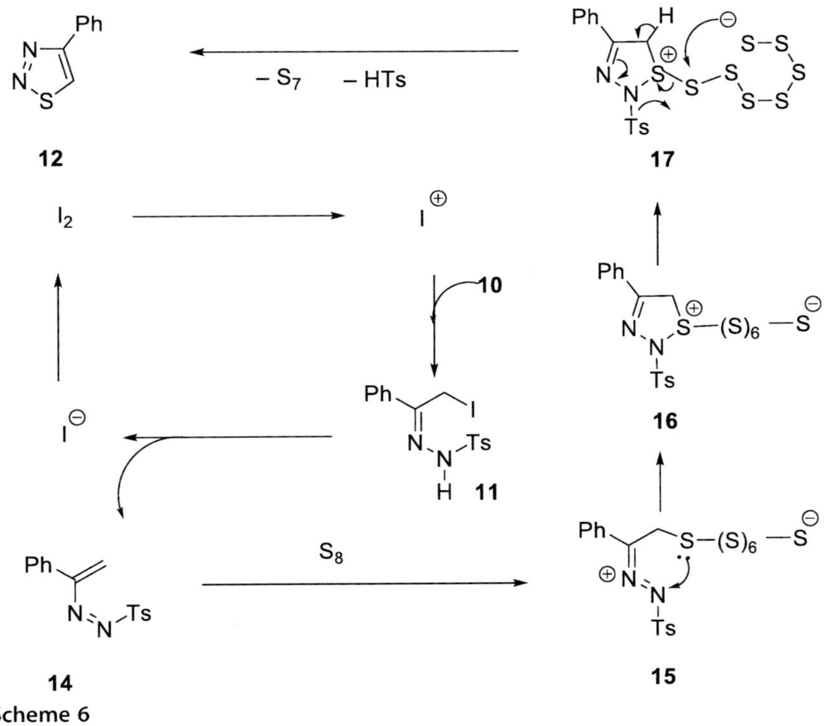

Scheme 6

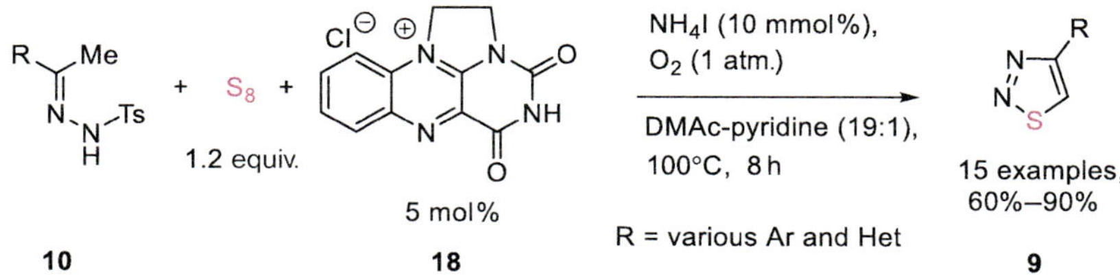

Scheme 7

Scheme 8

oxygen as oxidant instead of $K_2S_2O_8$ has been reported by Iida and coworkers (2017ACSC4986). They have discovered that the application of a flavin-catalyzed aerobic system allows the replacement of $K_2S_2O_8$ by molecular oxygen. The optimal conditions for the reaction involve the use of 1,10-bridged alloxazinium salt **18** and NH_4I as a catalytic system, a mixture of DMAc and pyridine as the solvent, and a temperature of 100°C. These conditions allowed the authors to obtain a variety of 5-substituted 1,2,3-thiadiazoles in 60%–94% yields. Thus, a novel bioinspired two-component redox organocatalytic system has been introduced for the synthesis of 1,2,3-thiadiazoles (2017ACSC4986).

2.1.2 Cycloaddition of Diazoalkanes Onto a C=S Bond (Pechmann Synthesis)

No new data have been published on the synthesis of 1,2,3-thiadiazoles by the Pechmann method during the period 2004–17.

2.1.3 Generation and Cyclization of α-Diazo Thiocarbonyl Compounds (Wolff Synthesis)

The generation and subsequent 1,5-dipolar electrocyclization of α-diazothiocarbonyl compounds **19** represent an efficient method for the synthesis of the 1,2,3-thiadiazoles **20** (Scheme 8). In 1904, Wolff first reported the synthesis of 1,2,3-thiadiazole derivatives by the reaction of NH_4SH with 2-diazo-1,3-dicarbonyl compounds (1904LAC333). Further development of this Wolff method was provided by: (a) the use of other thionation reagents such

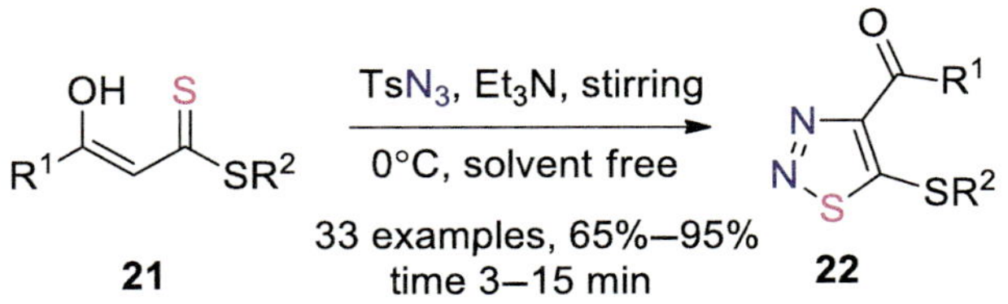

Scheme 9

R¹—C(OH)=CH—C(S)—SR² **21** → (TsN₃, Et₃N, stirring, 0°C, solvent free, 33 examples, 65%–95%, time 3–15 min) → 4-acyl-5-alkylthio-1,2,3-thiadiazole **22**

as phosphorus decasulfide and Lawesson's reagent; (b) the generation of diazothiocarbonyl compounds by diazo transfer reaction of thiocarbonyl compounds with sulfonyl azides or by diazotization reactions of α-aminothiocarbonyl compounds; (c) the use of a rearrangement process of other sulfur-containing heterocyclic compounds, mainly 5-mercapto-1,2,3-triazoles and 1,2,3-triazole-4-carbothioamides. Thus, a variety of 5-alkyl-, aryl-, amino-, and 5-mercapto-1,2,3-thiadiazoles bearing aryl, carbonyl, cyano, phosphoryl, and thiocarbonyl groups at the 4-position, including condensed 1,2,3-thiadiazoles, have been prepared (2003M[1], 2004M[2], 2004RJO870, 2005OBC1835, 2008CHECIII(5)468, 2012RJO1333, 2013GC954, 2014CHE972, 2014TL2430, 2016JHC206, 2017JOC4056).

Diazothiocarbonyl compounds of type **19** are very unstable intermediates; their transformation to the cyclic isomers **20** takes place via a heteroelectrocyclic mechanism with very low activation energy under the conditions for their generation (1995RCR99, 2017JOC4056). Therefore, the literature data on the Wolff method for the synthesis of 1,2,3-thiadiazoles **20** are arranged here according to the method of generation of the intermediate diazothiocarbonyl compounds **19**. These methods are: (i) insertion of a diazo group into thiocarbonyl compounds; (ii) introduction of a thiocarbonyl group into a compound bearing a diazo group; and (iii) simultaneous introduction of diazo- and thiocarbonyl functions.

(i) Insertion of a diazo group onto thiocarbonyl compounds

Reactions of thiocarbonyl active methylene compounds with sulfonyl azides are a convenient and reliable way to generate α-diazothiocarbonyl compounds. The subsequent heterocyclization of the latter leads to the formation of 5-amino- or 5-alkylthio-1,2,3-thiadiazoles (2013GC954, 2017JOC4056). Thus, Singh and coworkers have found that α-enolic dithioesters **21** upon treatment with tosyl azide in the presence of 1 equiv. of Et₃N at 0°C rapidly form the 5-alkylsulfonyl-4-carbonyl-1,2,3-thiadiazoles **22** in excellent yields, accompanied by the formation of tosyl amide (Scheme 9). The latter can be removed by simple extraction with

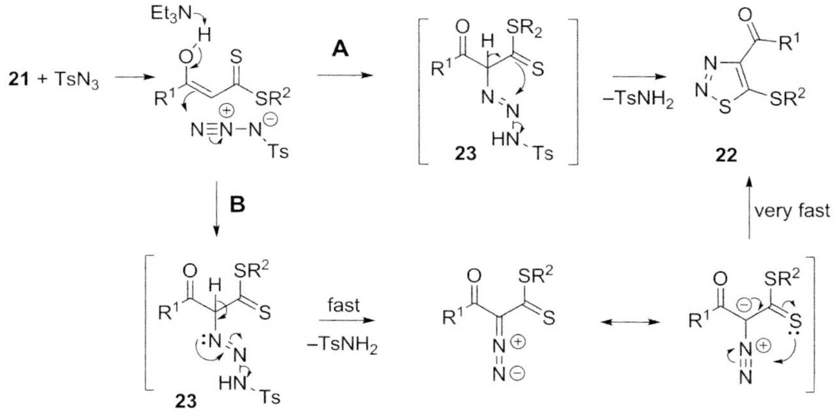

Scheme 10

water. The presence of a base is important for the formation of 1,2,3-thiadiazoles and the reaction did not occur in the absence of a base in dichloromethane. Other readily available organic and inorganic bases afforded 1,2,3-thiadiazoles in reaction of dithioesters **21** with tosyl azide in solvent-free conditions, but in lower yield compared to Et$_3$N and the reactions had to be run for a longer time (2013GC954).

The reaction has a wide scope and dithioesters bearing aryl substituents (R^1) with both electron-donating and electron-withdrawing groups, and heteroaromatic substituents (R^1) have been employed to give good yields of the final products **22**. Various alkyl-substituted dithioesters **21** derived from aliphatic ketones have also been found to react smoothly with TsN$_3$ affording the desired compounds. The reaction has also been shown to tolerate the modification of R^2 with various alkyl groups including benzyl and allyl. The authors proposed a mechanism (Scheme 10, path **A**) where the first step of the formation of triazene **23** is the rate-limiting one of the overall process. Then, the cyclization of thiocarbonyl onto triazene group occurs via cleavage of an N—N bond while extruding tosyl amide to form the final product **22**. Because density functional theory (DFT) calculations have shown a low barrier for the formation of α-diazothioacetamides from triazenes of similar structure and a very low barrier for their cyclization to 1,2,3-thiadiazoles, we propose an alternative two-step route (Scheme 10, path **B**) including transformation of triazene **23** first to diazothiocarbonyl compound followed by heteroelectrocyclization to the final product **22** (Scheme 10) (2017JOC4056).

In contrast to alkyl dithioacetates **21**, the reactions of 2-cyanothioacetamides **24** with sulfonyl azides have been shown to proceed in two possible directions, depending on the structure of the thioamide and the type of solvent and base, to give either 5-amino-1,2,3-thiadiazoles **25** or sodium salts of 5-sulfonylamino-1,2,3-triazole-4-carbothioamides **26** (Scheme 11). In an aprotic solvent, tertiary thioamides **24** in reaction with sulfonyl azides in the presence of bases, such as pyridine, triethylamine, 1,8-diazabicyclo[5.4.0]undec-7-ene (DBU), or sodium ethoxide, function as sulfur-containing building blocks resulting in the formation of the thiadiazoles **25** in high yields. The use of a sodium ethoxide/ethanol system completely switches the reaction direction, resulting in the formation of the sodium salts of 1,2,3-triazole-4-carbothioamide **26** as the only products (2017JOC4056). Interestingly, the primary and secondary 2-cyanothioacetamides when reacting with sulfonyl azides in the presence of any base afford 5-amino-1,2,3-thiadiazole 4-carbonitriles as the only products (1989T7329).

An alternative way to generate α-diazothiocarbonyl compounds is the treatment of α-aminothiocarbonyl compounds with sodium nitrite in the presence of an acid (2004M[2]). Singh and coworkers reported a new method for the synthesis of 5-methylsulfanyl-1,2,3-thiadiazoles bearing a ketone group at the 4-position via a one-pot procedure starting from active methylene α-enolic dithioesters (Scheme 12). This involves nitrosation of the starting dithioesters **21**, reduction of the resulting oximes **27** followed by diazotization of the intermediate α-aminothiocarbonyl compounds **28** to generate diazo compounds **29**. The latter undergo fast cyclization to form final product **22** in good yields (2014TL2430).

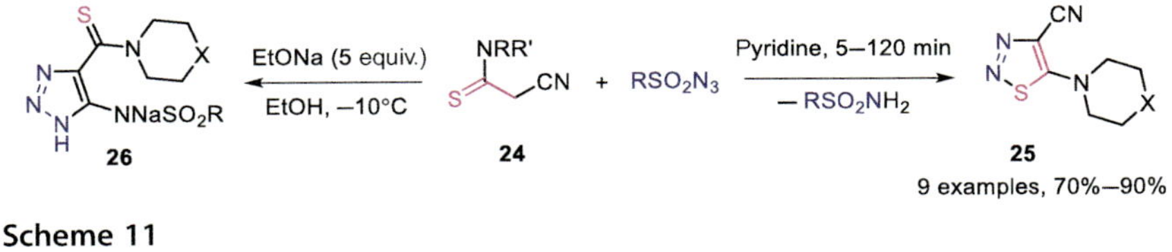

Scheme 11

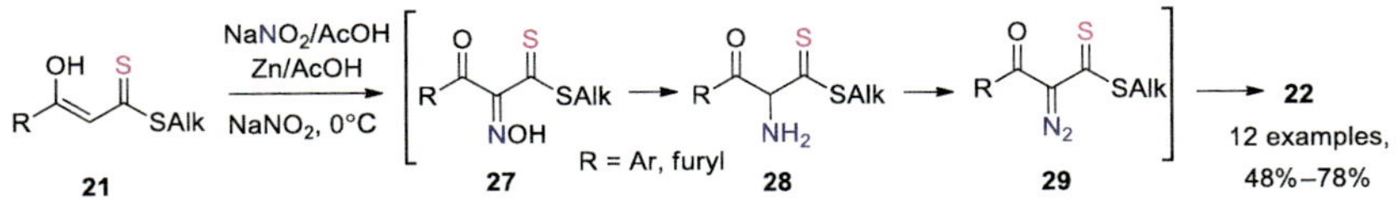

Scheme 12

(ii)
Introducing a thiocarbonyl group into a compound bearing a diazo group
This approach to α-diazothiocarbonyl functionalities includes the reaction of α-diazocarbonyl compounds and diazoacetonitriles. It was developed at the end of the last century and reviewed in an earlier report (2004M [2]). Two patents appeared during the last decade in which improvements of the protocol for the synthesis of 5-amino-1,2,3-thiadiazoles in reactions of diazoacetonitrile with H_2S were presented 2006RUP2276146, 2013CNP103058956).

(iii)
Simultaneous introduction of diazo- and thiocarbonyl functions
This section refers to the rearrangements of 5-mercapto- and 1,2,3-triazole-4-carbothioamides to 5-amino-1,2,3-thiadiazoles. The generation of intermediate α-diazothiocarbonyl compounds **34** and **35** results from the opening of a 1,2,3-triazole ring followed by ring closure onto the thiocarbonyl group to form 5-amino-1,2,3-thiadiazoles. Some years ago, we showed that the equilibrium between the triazoles **30** and **31** and the corresponding thiadiazoles **32** and **33** via diazocompounds (**34** and **35**) is shifted largely in favor of the triazole form and therefore these compounds are not suitable for the synthesis of 1,2,3-thiadiazoles (Scheme 13) (2004RJO1338, 2005CHE542).

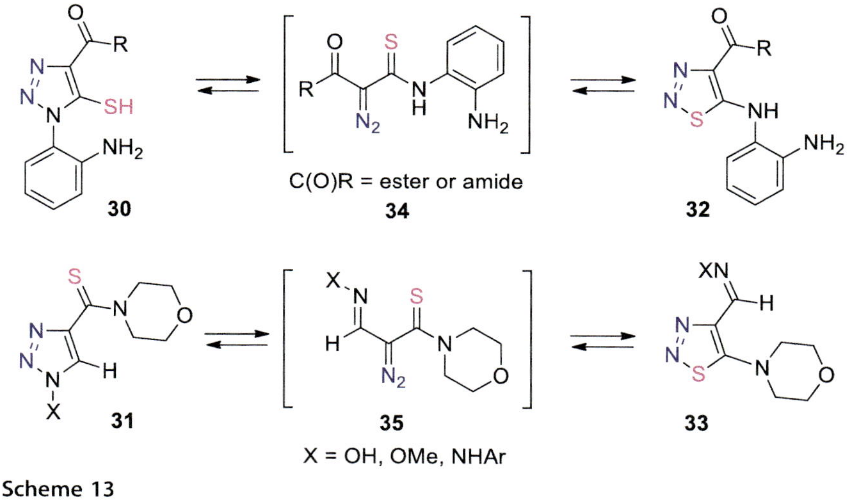

Scheme 13

Twelve years later, we found that 5-sulfonylamino-1,2,3-triazole-4-carbothioamides **26**, prepared by reaction of tertiary cyanothioacetamides with sulfonyl azides in the presence of 2–5 equiv. of sodium ethoxide, are stable only as sodium salts and smoothly rearrange at room temperature in water, or in the presence of acids, to 5-cycloamino-1,2,3-thiadiazole-4-carbamidines **36** as the sole products in excellent yields (Scheme 14) (2017JOC4056, 2016CHE206, 2014CHE972).

2.1.4 Oxidative Cyclization of 2-Arylhydrazonothiocarbonyl Compounds to 2-Aryl-1,2,3-thiadiazolines

Wilkins classified the title reaction in 2008 as "New method for the synthesis of 1,2,3-thiadiazoles from thioanilide derivatives" (2008CHECIII(5)468). Based on a systematic study of the cyclization of 2-hydrazonothioacetamides **37** to thiadiazole-5-imines **38** and **39**, which was carried out by Belskaya and coworkers, and the literature data cited in this chapter, we classify this approach as "Oxidative cyclization of 2-arylhydrazonothiocarbonyl compounds to 2-aryl-1,2,3-thiadizolines" (2010ARK275, 2004RJO818, 2007T3042, 2011RCB896, 2013T7423, 2016EJM245).

Gewald and Hain in 1975 first reported the synthesis of 2-aryl-1,2,3-thiadiazoline-5-imines **38a** by reaction of 2-arylhydrazonothioacetamides **37** with bromine in acetic acid (Scheme 15) (1975JPR317). Some years later, systematic studies of the oxidative heterocyclization reaction (OHR) of 2-hydrazonothioacetamides **37** using a large series of starting thioamides and with variations of oxidant, including bromine, iodine, and chlorosulfonyl isocyanate (CSI), were carried out by Belskaya et al. Thus a variety of 5-imino **38b** and 5-alkyliminothiadiazoles **38c**, bearing aryl substituents with both electron-withdrawing and electron-releasing groups at position 2 and cyano, ester, and amide functions at position 4, were successfully prepared as their salts with hydrogen halides (Scheme 15) (2004RJO818, 2011RCB896). Gil et al. prepared thiadiazolium salts of type

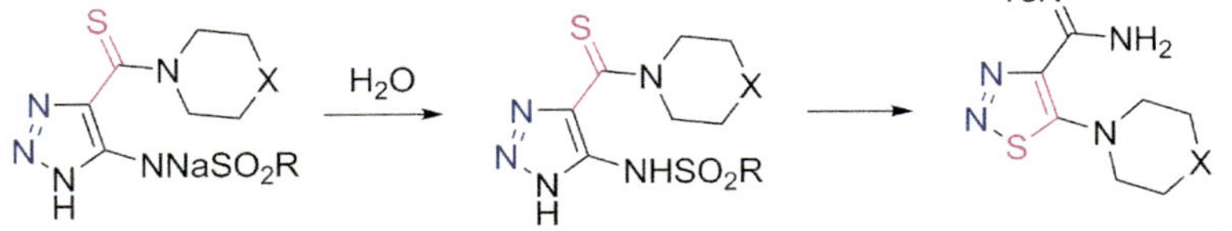

Scheme 14

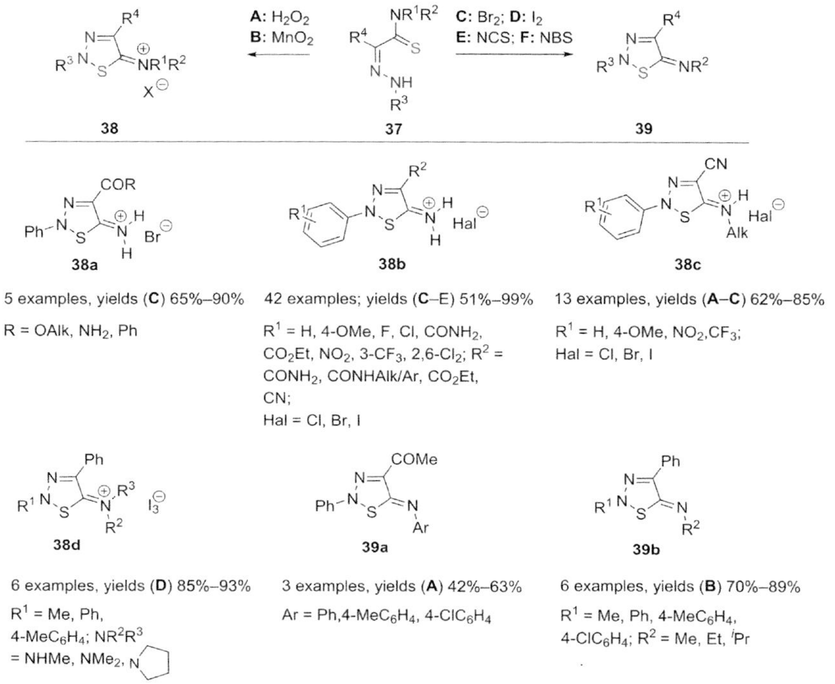

Scheme 15

38 — **A:** H₂O₂; **B:** MnO₂ ← **37** → **C:** Br₂; **D:** I₂; **E:** NCS; **F:** NBS → **39**

38a: 5 examples, yields (**C**) 65%–90%
R = OAlk, NH₂, Ph

38b: 42 examples; yields (**C–E**) 51%–99%
R¹ = H, 4-OMe, F, Cl, CONH₂, CO₂Et, NO₂, 3-CF₃, 2,6-Cl₂; R² = CONH₂, CONHAlk/Ar, CO₂Et, CN;
Hal = Cl, Br, I

38c: 13 examples, yields (**A–C**) 62%–85%
R¹ = H, 4-OMe, NO₂, CF₃;
Hal = Cl, Br, I

38d: 6 examples, yields (**D**) 85%–93%
R¹ = Me, Ph, 4-MeC₆H₄; NR²R³ = NHMe, NMe₂, N(pyrrolidine)

39a: 3 examples, yields (**A**) 42%–63%
Ar = Ph, 4-MeC₆H₄, 4-ClC₆H₄

39b: 6 examples, yields (**B**) 70%–89%
R¹ = Me, Ph, 4-MeC₆H₄, 4-ClC₆H₄; R² = Me, Et, ⁱPr

38d when they carried out the reaction in the presence of an excess of iodine (1996PS89, 2010ARK275). Furthermore, the use of mild oxidants such as MnO₂ or H₂O₂ allowed Gil et al. and Zaleska et al. to prepare stable thiadiazoles **39a** and **39b**, respectively, as free bases (1996PS89, 2003S2559). Basing on an X-ray analysis for hydrazone **37** (NR¹R² = NHPh, R³ = Ph, R⁴ = COMe), Zaleska et al. have shown that the Z-configuration of this compound with a planar S=C—C=N—N—H system favors interaction between the N and S atoms to form 1,2,3-thiadiazole ring (2003S2559).

The first example of an OHR where atmospheric oxygen was used for the synthesis of 2-aryl-1,2,3-thiadiazolines **40** was discovered by Belskaya and coworkers when they treated thioamides **41** with anhydrides **42** or acyl chloride **43** in pyridine. The authors proposed a mechanism for the formation of the 5-acylimino-2,5-dihydro-1,2,3-thiadiazoles **40** involving an oxidative heterocyclization of intermediates **44** following the initial acylation of the starting thioamides **41** (Scheme 16) (2013T7423).

The structure of compounds **40** has been confirmed by an X-ray diffraction study of compound **40** (Ar=4-MeOC$_6$H$_4$, R^1=pyrrolidinylcarbonyl, R^2=Me). This was the first report of an X-ray analysis for 5-acylimino-2,5-dihydro-1,2,3-thiadiazoles, providing a strong proof of the structure of this type of thiadiazoles (Fig. 1). The X-ray data and DFT calculations confirm the existence of thiadiazoles **40** in the Z-configuration and have rationalized this by an S⋯O bond between the thiadiazole ring and acyl group due to n → σ* orbital overlap in the thiadiazoles (Fig. 2) (2013T7423).

The OHR approach has also been applied to the synthesis of fused 1,2,3-thiadiazoles (see also Section 2.3.1). Thus, oxidation of the 3-hydrazono-1,3-dihydroindoles **45** with halogens or *N*-bromosiccinimide (NBS) has been used to prepare thiadiazolo-indoles **46** in good to excellent yields (Scheme 17) (2007T3042).

Scheme 16

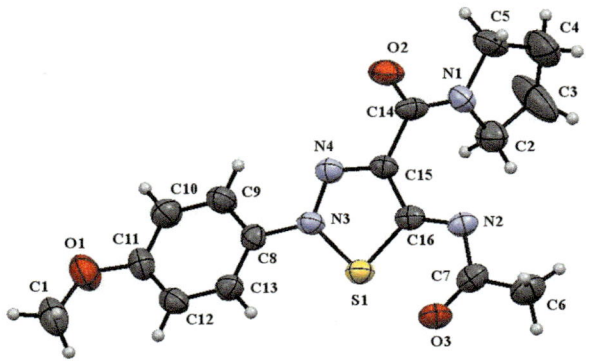

Fig. 1 X-ray diffraction data for **40** (Ar-4-MeOC$_6$H$_4$, R^1 = pyrrolidinylcarbonyl, R^2 =Me).

The analysis of X-ray data of the starting hydrazone **45** (R = ^{t}Bu) confirms the Z-(cis) configuration of this compound (Fig. 3). The hydrogen bond between the S-atom of the thioamide group and the N-atom of the hydrazone fragment was proposed to favor the formation of the N—S bond of 1,2,3-thiadiazole ring (2007T3042).

Fig. 2 S⋯O interaction due to n → σ* orbital overlap in thiadiazoles **40**.

HHal	HCl	HCl	HCl	HCl	HBr	HBr	HBr	HI	HI
R	Me	4-EtOC$_6$H$_4$	4-BnOC$_6$H$_4$	4-ClC$_6$H$_4$	Me	Bn	Ph	Me	Ph
Yield (%)	85	60	69	51	93	60	99	55	57

Scheme 17

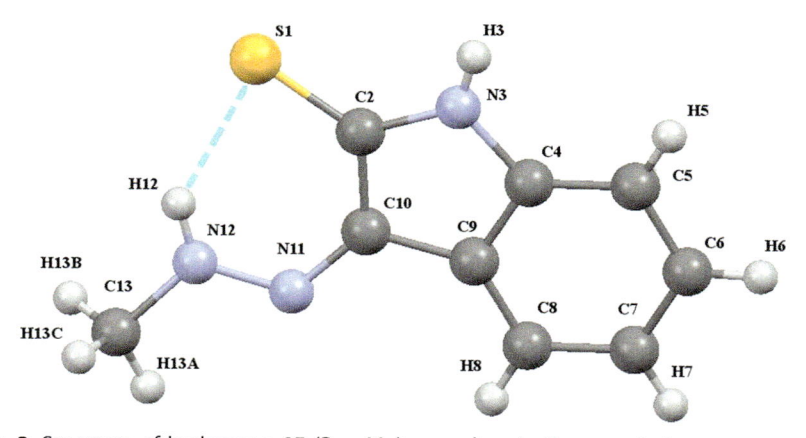

Fig. 3 Structure of hydrazone **45** (R = Me) according to X-ray analysis.

The radical nature for the oxidative heterocyclization of arylhydrazonothiocarbonyl compounds to 2-aryl-1,2,3-thiadiazolines was recently shown by Li and coworkers in the reaction of 2-(indol-2-one-3-yl)-2-aroylthioacetamides **47** with arylhydrazines in the presence of I_2/O_2 as oxidation source (Scheme 18). Based on this reaction, they have elaborated a new and efficient method for the synthesis of 2-aryl-1,2,3-thiadiazolin-5-ylidenes **48** and demonstrated the scope of the reaction by the synthesis of 30 compounds of type **48**. Indeed, the reaction tolerated the substitution of both aryl groups by electron-withdrawing and electron-releasing substituents providing thiadiazoles **48** in 66%–92% yield. The derivatization of indole with F, Cl, Br, and Me was also tolerated.

To gain deeper insight into the mechanism, the authors carried out the reaction of thiolate **47** (R=H, Ar^1 and Ar^2=Ph) in the presence of the radical scavengers 2,2,6,6-tetramethyl-1-piperidinyloxy (TEMPO) or butylhydroxytoluene (BHT) (2.0 equiv.) under optimal conditions. No formation of thiadiazole **48** was observed but rather complex mixtures of unidentified products were formed, confirming the radical process for the reaction. Furthermore, the authors found that thiadiazoles **48** are not formed under an argon atmosphere, indicating that O_2 plays a key role in the reaction. On the basis of this experiment, the authors proposed the mechanism outlined in Scheme 19. Thiocarbonyl compound **47** reacts with aryl hydrazines to afford aryl hydrazones **A**. The latter under treatment with iodine undergo an S—I bond formation generating then intermediates **B**, which via homolysis of the S—I bond yield radicals **C**. An attack of intermediate radical **C** on the nitrogen atom of the hydrazono group provides thiadiazole radicals **D**. In turn, radicals **D** eliminate H•, which couples with I• to form HI and the final 1,2,3-thiadiazoles **48**. Iodine was recovered from the reaction of HI with O_2 (2016OL1258).

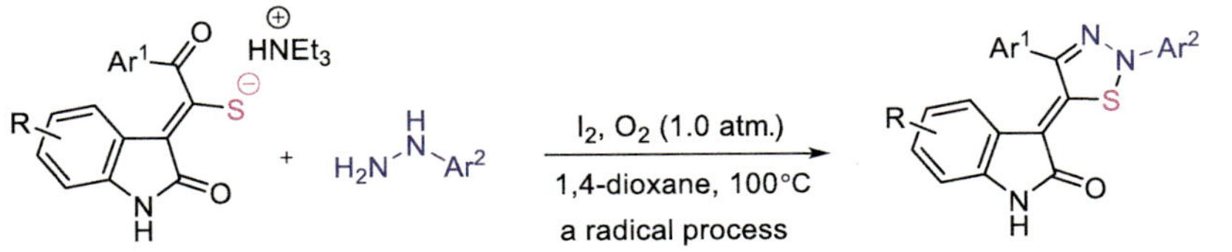

R = H, Me, Cl, Br
Ar^1 = Ph, 4-MeC$_6$H$_4$, 4-MeOC$_6$H$_4$, 4-ClC$_6$H$_4$, 3-ClC$_6$H$_4$, 2-ClC$_6$H$_4$, 4-BrC$_6$H$_4$, 4-NO$_2$C$_6$H$_4$,
Ar^2 = Ph, 4-MeC$_6$H$_4$, 4-MeOC$_6$H$_4$, 3-MeOC$_6$H$_4$, 4-ClC$_6$H$_4$, 3-ClC$_6$H$_4$, 2-ClC$_6$H$_4$, 4-BrC$_6$H$_4$

Scheme 18

2.2 Fused 1,2,3-Thiadiazoles

Fused polycyclic systems, where 1,2,3-thiadiazoles are annelated to other rings, have been attractive synthetic targets in the last decade due to their importance for organic electronics, medicinal chemistry, and agricultural applications. In this section, the synthetic data are divided into two parts, based on whether the 1,2,3-thiadiazole is annelated to an existing ring or alternatively where a new ring is fused to a 1,2,3-thiadiazole. Within these two parts, the reactions are arranged according to the type of a ring condensed to the 1,2,3-thiadiazoles.

2.2.1 Annelation of 1,2,3-Thiadiazole to a Heterocycle

The overwhelming majority of fused 1,2,3-thiadiazoles are prepared by annelating the 1,2,3-thiadiazole ring to an existing ring. Here again, the most popular method is the HMR (see Section 2.1), in which the α,β-C—C bond of the hydrazone **49** is part of a ring (Scheme 20; LG=leaving group) (2003FA63, 2003IJB189, 2003JCC1934, 2003JHC149, 2003JHC427, 2004RJO1092, 2004RJO90, 2005EJMC728, 2005IJB783, 2005MOL367,

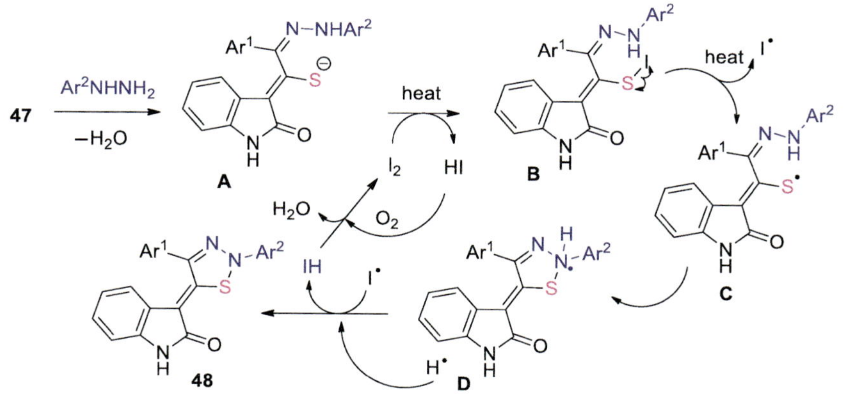

Scheme 19

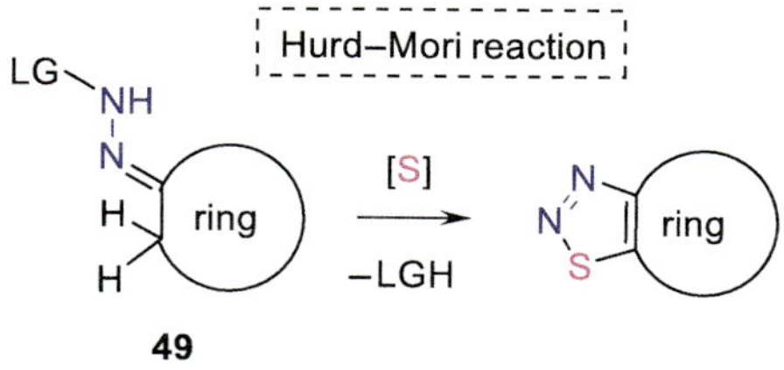

Scheme 20

2005MOL818, 2005USP250783, 2006CHE681, 2006RJO1735, 2006S2573, 2007MCR392, 2007SL2513, 2008JME3005, 2010EJM4920, 2010PS1594, 2010T5472, 2012BML7237, 2012JFA346, 2012JME3122, 2014CBD223, 2016CNP105294598, 2017BMC2336, 2017EJM48, 2017EP3133075).

Several condensed 1,2,3-thiadiazoles have also been prepared by WS (see Section 2.3) (2003BML3513, 2005BMC1615, 2005OBC1835, 2005T6634, 2008M1067, 2011S2154, 2011WOP29459, 2013JME2110, 2016CM6390, 2017USP110666) via generation and subsequent cyclization of heteroaromatic o-diazo-thiones **50** (Scheme 21).

In the reviewed period (2003–17) all benzo[d][1,2,3]thiadiazoles have been prepared by annelating a 1,2,3-thiadiazole ring to a benzene core (2003BML3513, 2003JHC427, 2005BMC1615, 2005EJMC728, 2005T6634, 2008M1067, 2011S2154, 2012JFA346, 2012WOP47617, 2013JME2110, 2016CM6390, 2017USP110666. In this way, an interesting example of a bi-1,2,3-thiadiazole **51** was prepared via WS by diazotization of compound **52**, bearing a double set of amino and mercapto groups both situated in o-position to each other (Scheme 22) (2017USP110666).

2-Amino-4,5-difluorobenzenethiol **53** was prepared by hydrazinolysis of 5,6-difluoro-2-methylbenzo[d]thiazole **54** and used without isolation

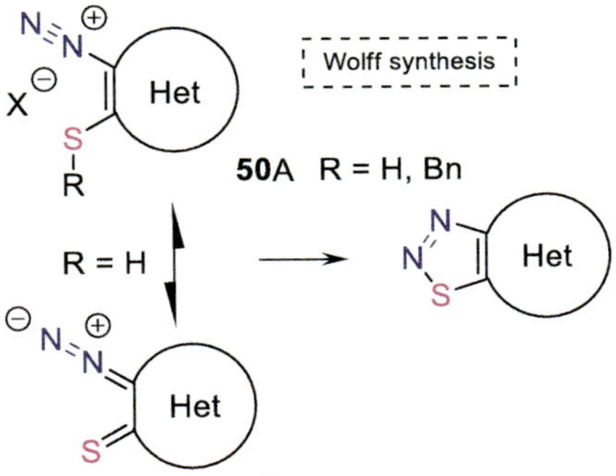

Scheme 21

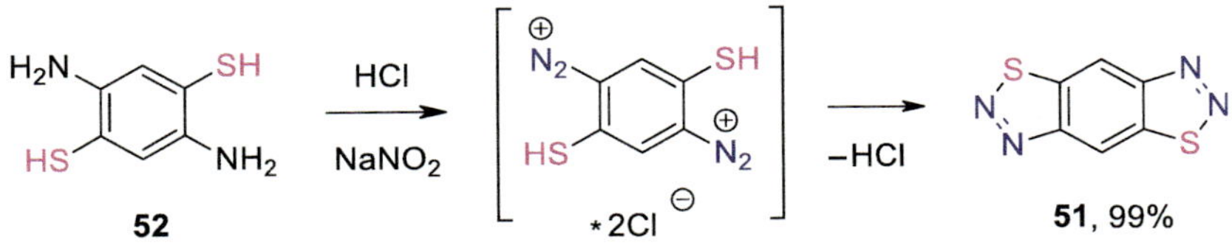

Scheme 22

or identification in a WS leading to 5,6-difluorobenzo[*d*][1,2,3]thiadiazole **55** (Scheme 23) (2016CM6390). Compound **55** was further used for the synthesis of semiconducting polymers.

Acidic cleavage of the resin-immobilized triazenes **56**, bearing an arenethiol moiety, provided high yields of benzo[*d*][1,2,3]thiadiazoles **57** and easy separation of the solid-phase residue from the reaction products (Scheme 24) (2005OBC1835). Quite reasonably, the authors assume that thiadiazoles **57** are formed via the intermediate diazothiones **56A** ↔ **56B**.

One very advantageous implementation of the HMR employing resin-bound acylhydrazine **58** leads to cyclohexene fused 1,2,3-thiadiazole **59** (Scheme 25) (2008SC4407). The advantage of this method is that the product is not contaminated with the leaving group, and resin-bound chloroanhydride **60** can be recycled.

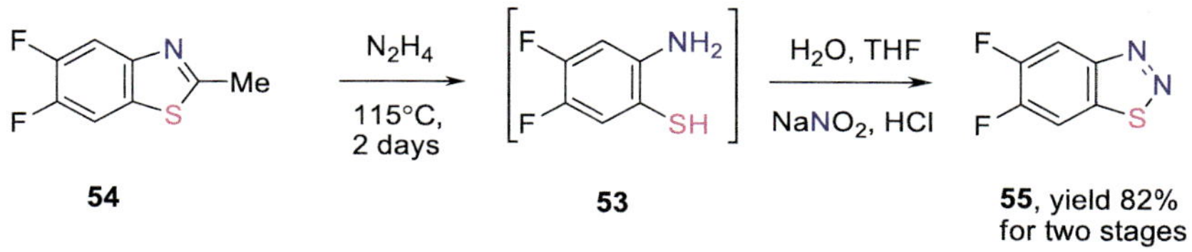

Scheme 23

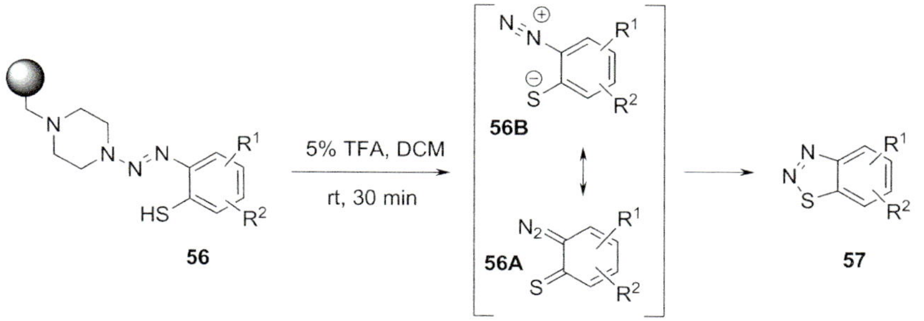

Scheme 24

R^1, R^2 (yield of **57**, %): H, H (62); 4-Cl, 6-Cl (10); 6-NO$_2$ (37); 6-CO$_2$Me (34); 6-CO$_2$Et (41); H, 6-Me (42); 4-Me, 6-Me (56); 4-SH (traces); 4-CO$_2$Me (15)

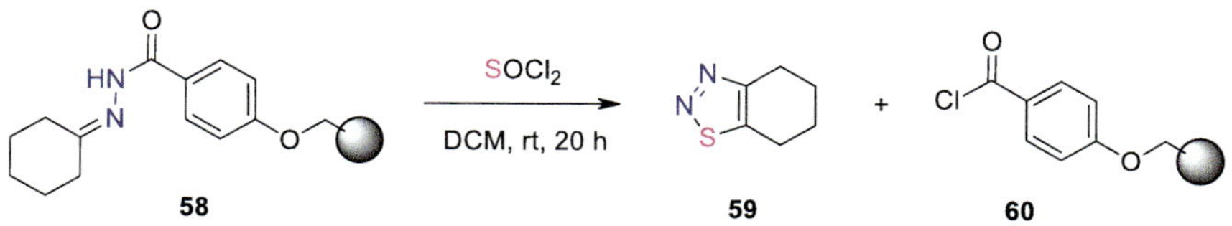

Scheme 25

Synthesis of the cyclohexanone fused thiadiazole **61** has been described (2017BMC2336, 2017EP3133075) (Scheme 26); the reaction is accompanied by cleavage of the protecting 1,3-dioxolane ring to give the keto group.

The classical HMR starting from semicarbazones (2003JCC1934, 2004RJO90, 2014CBD223) and thiosemicarbazones (2003JCC1934) has been successfully used for modification of naturally occurring compounds. By this means, 1,2,3-thiadiazoles annelated to terpenoids (**62** and **63**) (2004RJO90), cholesterol (**64**) (2003JCC1934), and 18β-glycyrrhetinic acid (**65**) (Scheme 27) (2014CBD223) have been prepared. In the synthesis

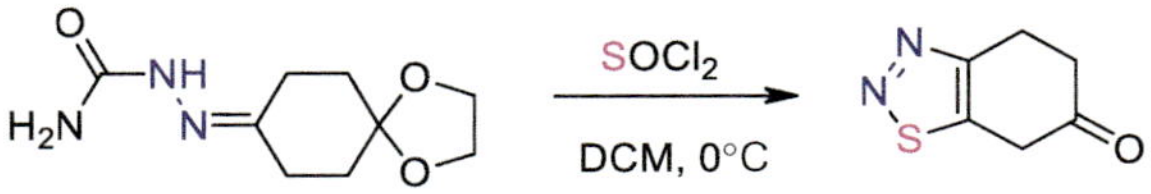

Scheme 26

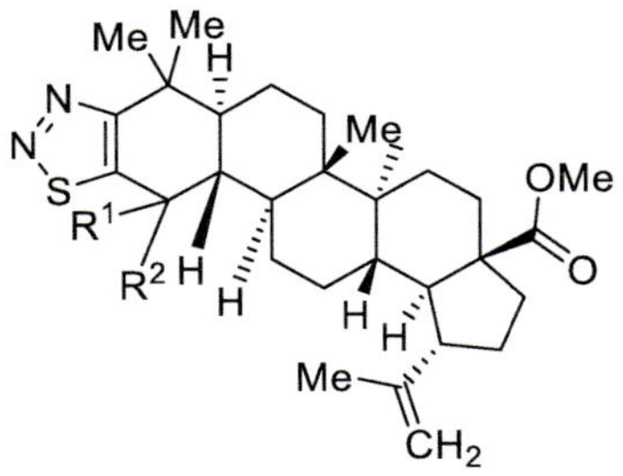

R^1, R^2 = H, H (**62a**), 72%
R^1+R^2 = =S=O (**62b**), 12%

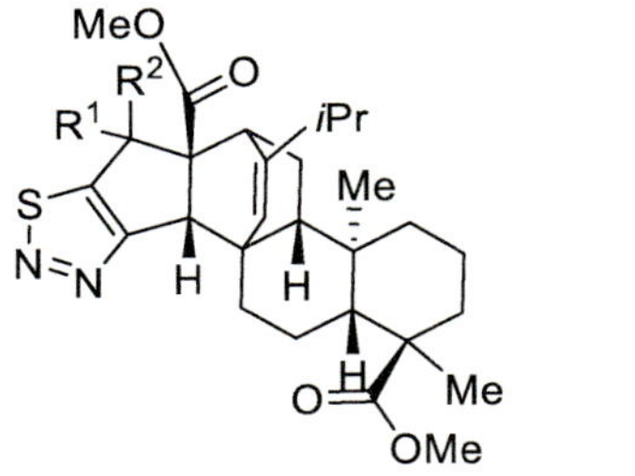

R^1, R^2 = H, H (**63a**), 75%
R^1 + R^2 = =S=O (**63b**), 10%

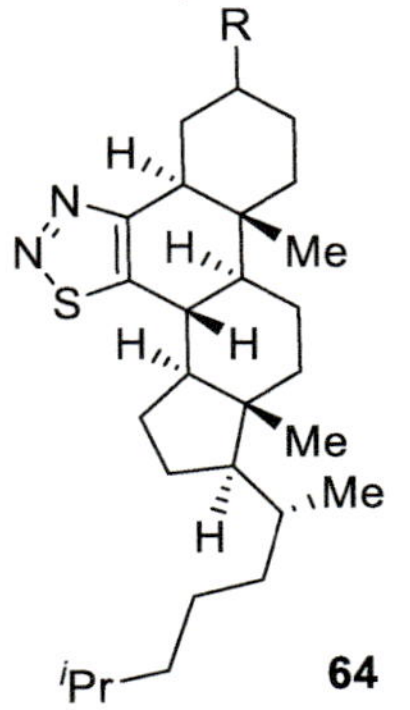

R	Yield, %
H	66/73
Cl	70/64
Ac	75/69

64

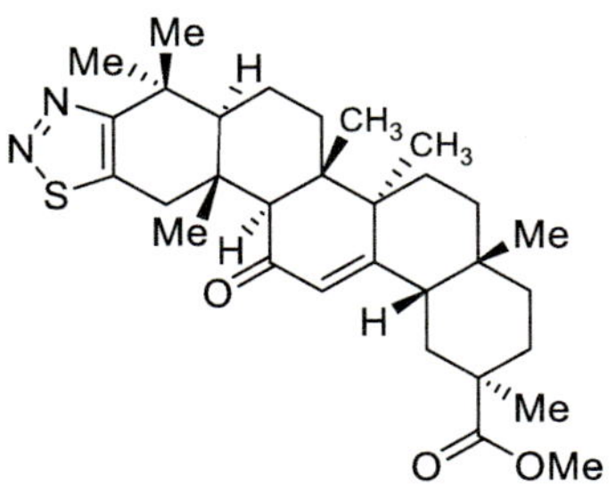

65, 84% from ketone

Scheme 27

of the terpenoid derivatives **62** and **63**, formation of the sulfanone products **62b** and **63b**, as minor admixtures to main products (**62a** and **63a**, respectively), was detected (2004RJO90). It is worthy of note that in all cases (Scheme 27) treatment with thionyl chloride retains the stereochemistry of the precursors, which opens a route to physiologically active species.

Mihovilovic and coworkers (2005MOL367) have demonstrated that the disappointingly low yields of pyrrolo[2,3-*d*][1,2,3]thiadiazoles **66** prepared by HMR can be drastically improved by introducing an electron-withdrawing methoxycarbonyl function in position N^1 of the starting 5-(2-ethoxycarbonyl-2-hydrazinyl-1-yliden)-pyrrolidines **67**. This substituent is easily and selectively removed under mild conditions with formation of the N-unsubstituted pyrrole **68** (Scheme 28).

In further studies, Mihovilovic and coworkers (2010T5472) proved the favoring effect of electron-withdrawing substituents in starting hydrazonopyrrolidine **69**. Thus, without a substituent in compound **69**, the yield of the pyrrolo[2,3-*d*][1,2,3]thiadiazole **70a** was only 29% (Scheme 10). Conversely, substitution of hydrogen by methoxycarbonyl at position 3 in the hydrazonopyrrolidine **69** leads to the *N*-Boc-substituted pyrrolo[2,3-*d*][1,2,3]thiadiazole **70b** in high yield. Both Boc and methoxycarbonyl are easily removed to form the unprotected compounds **71a,b** (Scheme 29).

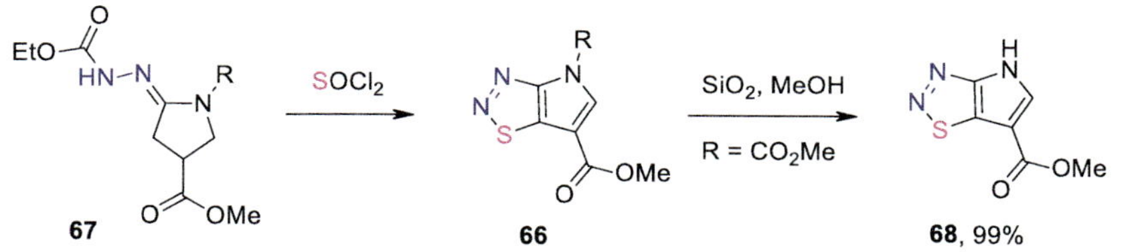

R (yield of **66**, %): Bn (25); Me (14); CO₂Me (94)

Scheme 28

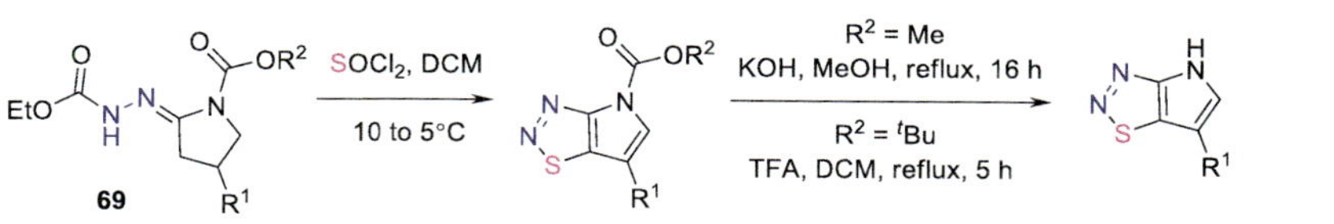

70a: R¹ = H, R² = Me, 29%;
70b: R¹ = CO₂Me, R² = ᵗBu, 82%

71a: R¹ = H, 79%;
71b: R¹ = CO₂Me, 98%

Scheme 29

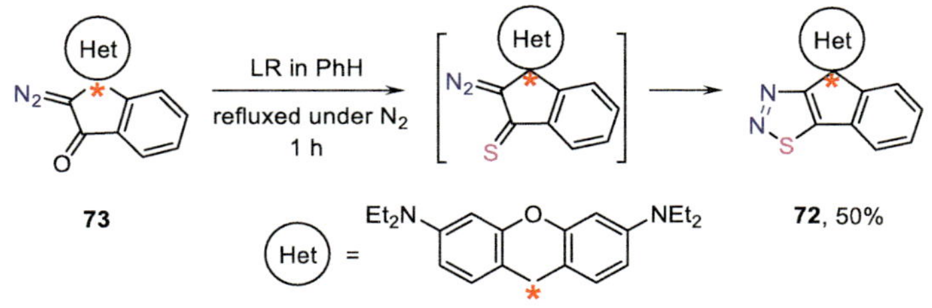

Scheme 30

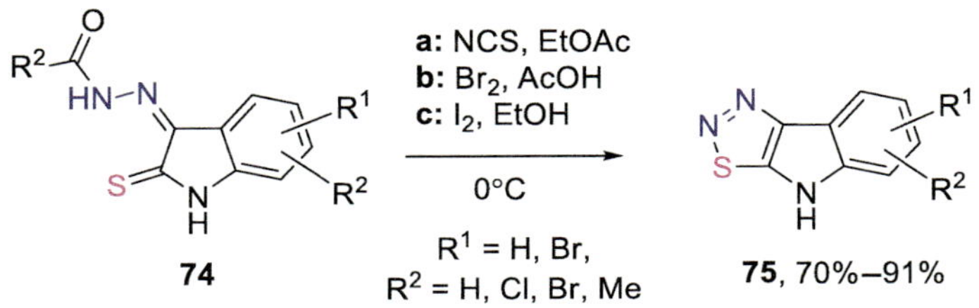

Scheme 31

For the preparation of 4H-indeno[2,1-d][1,2,3]thiadiazole **72**, the authors used a WS starting from 2-diazo-2,3-dihydro-1H-inden-1-one **73** and Lawesson's reagent (LR) where diazo and (thio)carbonyl groups are situated in o-positions to each other (Scheme 30) (2011WOP29459).

The products under invention, including compound **72**, are claimed to be novel photoactivatable fluorescent dyes incorporating small photoactive groups with improved properties, such as the capability for easy and effective photoactivation (uncaging) which does not generate toxic substances that may interfere with biochemical processes in cells or tissues (2011WOP29459).

Belskaya and coworkers have employed an oxidative heterocyclization of 3-hydrazones of indoline-2-thione **74**, after treatment with N-chlorosuccinimide (NCS), bromine or iodine, for the preparation of a series of [1,2,3]thiadiazolo[5,4-b]indoles **75** (Scheme 31) (2007T3042, 2016EJM245). These authors consider this method to be much more advantageous than the Wolff method (2016EJM245).

2.2.2 Annelation of a Heterocycle to 1,2,3-Thiadiazole

This opposite way of constructing fused systems, starting from mononuclear 1,2,3-thiadiazoles, is far less represented in the literature, but a number of examples have been described.

The amino and carboxy groups in the thiadiazole **76** condense giving the 4,5-dihydro-6*H*-pyrrolo[3,4-*d*][1,2,3]thiadiazole **77** in moderate yield on reaction with water-binding reagents (Scheme 32) (2007WOP36733, 2009USP131431).

Another example of the formation of a cyclic amide ([1,2,3]thiadiazolo [4,5-*c*]quinolin-4-one **78**) is observed after Suzuki coupling of methyl 5-bromo-1,2,3-thiadiazole-4-carboxylate **79** with the boronic acid **80** (Scheme 33) (2011WOP25859). Among other compounds, quinolone **78** is claimed to possess diverse biological activities.

The oxidation of the thiadiazoles **81** by iodine, following treatment of the reaction mixture with sodium hydride, unexpectedly led to the formation of the benzo[4,5]thieno-[3,2-*d*][1,2,3]thiadiazoles **82** (2013OL4038). Formation of compounds **82** by oxidative aromatization of the proposed adduct **83** represents a rare example of the oxidative nucleophilic substitution of a hydrogen atom with the participation of a sulfur nucleophile. A plausible mechanism for the reaction, as proposed by the authors, is presented in Scheme 34, but electrophilic or radical mechanisms, as a result of thiolate oxidation, are also not excluded.

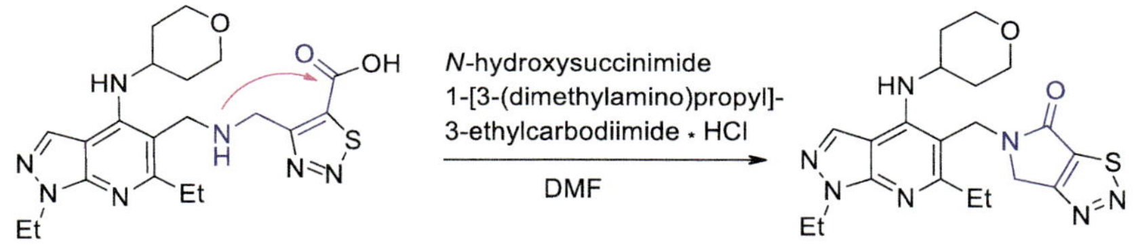

76 **77**, 49%

Scheme 32

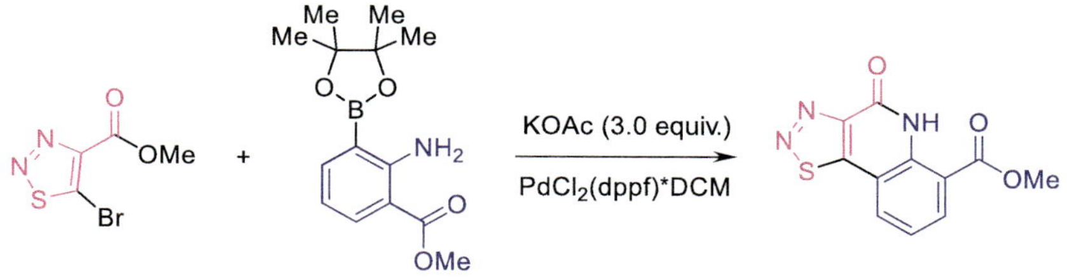

79 (1.0 equiv.) **80** (1.4 equiv.) **78**

Scheme 33

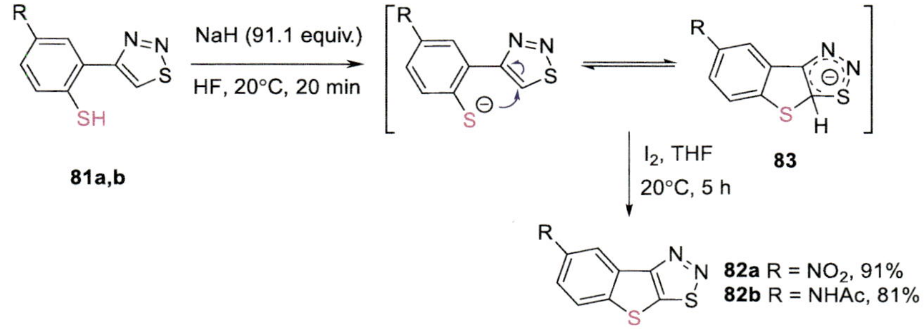

Scheme 34

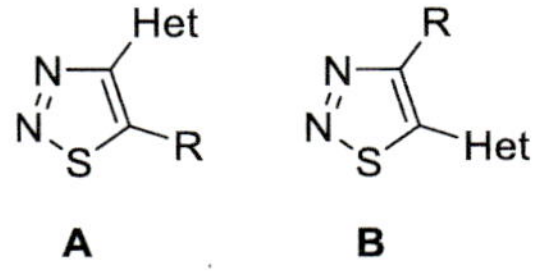

Scheme 35

2.3 1,2,3-Thiadiazoles Linearly Connected to Another Heterocycle

In this section the methods for synthesis of 1,2,3-thiadiazoles directly conjugated to other heteroaromatic systems by means of a single bond are reviewed. For better perception, we divide these compounds into two types (**A** and **B**) that differ in the position of connection of the 1,2,3-thiadiazole (positions 4 and 5, respectively) with a heterocycle (Het) (Scheme 35).

Within each type we distinguish reactions when thiadiazole is introduced to an existing heterocycle ("1,2,3-thiadiazole to Het") and when these compounds are constructed in the opposite manner ("Het to 1,2,3-thiadiazole").

2.3.1 Introduction of 1,2,3-Thiadiazo-4-yl to a Heterocycle

For the "1,2,3-thiadiazole to Het" constructions of type **A**, only variations of the HMR (see Section 2.1), where a starting hydrazone bears a Het function, have been used (Scheme 36).

The HMR has proved to be suitable for the preparation of 1,2,3-thiadiazoles bearing 5-membered heterocycle (furan-2- **84** and -3-yl **85** as well as thiene-2- **86** and -3-yl **87** and **88**). As for six-membered heterocycles, only pyridin-2-, -3-, and -4-yl (compounds **89**, **90**, **91**, and **92**) have

84 R¹ = H, CO₂Alk, CONAlk₂, R² = H, CO₂Alk

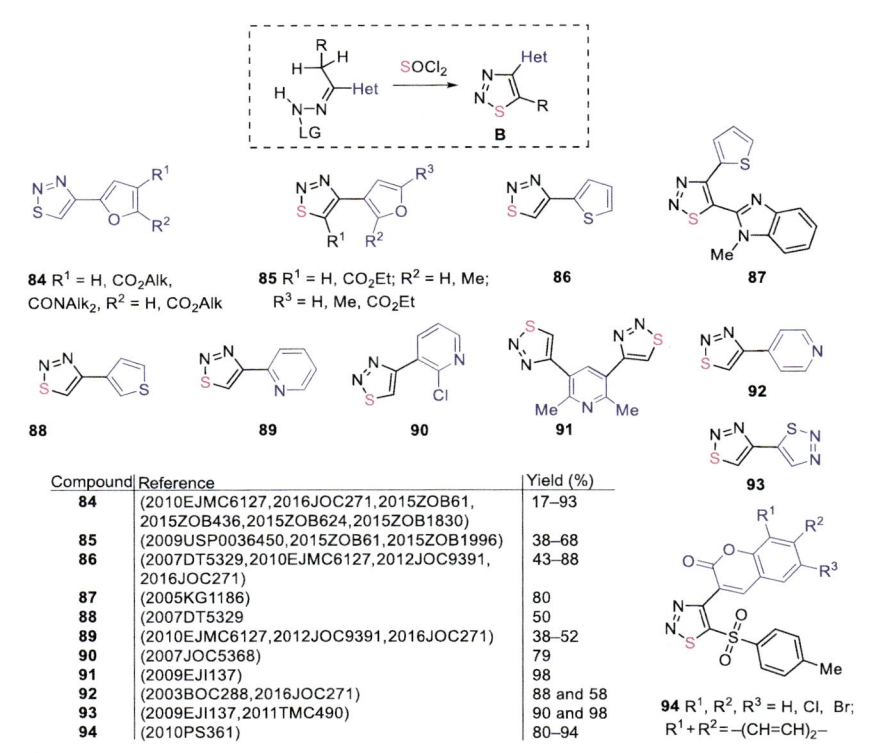

85 R¹ = H, CO₂Et; R² = H, Me; R³ = H, Me, CO₂Et

86

87

88

89

90

91

92

93

94 R¹, R², R³ = H, Cl, Br; R¹+R² = –(CH=CH)₂–

Compound	Reference	Yield (%)
84	(2010EJMC6127,2016JOC271,2015ZOB61, 2015ZOB436,2015ZOB624,2015ZOB1830)	17–93
85	(2009USP0036450,2015ZOB61,2015ZOB1996)	38–68
86	(2007DT5329,2010EJMC6127,2012JOC9391, 2016JOC271)	43–88
87	(2005KG1186)	80
88	(2007DT5329)	50
89	(2010EJMC6127,2012JOC9391,2016JOC271)	38–52
90	(2007JOC5368)	79
91	(2009EJI137)	98
92	(2003BOC288,2016JOC271)	88 and 58
93	(2009EJI137,2011TMC490)	90 and 98
94	(2010PS361)	80–94

Scheme 36

been introduced. Symmetric compound **91** is an interesting example of a bi-1,2,3-thiadiazol-4-yl prepared by an HMR in which the two 1,2,3-thiadiazole rings are formed simultaneously (2009EJI137). Another bi-1,2,3-thiadiazol-4-yl **93** has been prepared in a specific implementation of HMR, when both constructed and existing heteroaromatic rings were 1,2,3-thiadiazol-4-yl, and which can both be formally attributed to reactions of "1,2,3-thiadiazole to Het" type and "Het to 1,2,3-thiadiazole" type, or to neither. In the reviewed period, the synthesis of compound **93** was mentioned in two reports. In one, compound **93** was prepared in the absence of light, which raised the yield to 90% (2011TMC490). The authors used compound **93** as a starting material for the preparation of mixed ruthenium complexes (2011TMC490). An even higher yield (98%) was achieved when the synthesis of bi-1,2,3-thiadiazole **93** was carried out under an inert atmosphere while cooling with ice (2009EJI137). In this work, compound **93** was used as a source of alkynethiolate for the synthesis of dinuclear bis(alkynethiolate)gold(I) derivatives. In the earlier publication, the reported yield of compound **93** was 85% (at ambient temperature) (1997JHC605).

Heterocycles with more than one nucleus in position 4 of the 1,2,3-thiadiazole ring are represented only by 3-[benzo]chromen-2-ones **94** (2010PS361). An unusual leaving group in the starting hydrazone [2-(quinolin-8-yloxy)acetyl] was used in the synthesis, leading to the monosubstituted 1,2,3-thiadiazoles **84**, **86**, and **89** in moderate yields (35%, 43%, and 38%, respectively) (2010EJMC6127). The same furan-2-yl **84** and thien-2-yl derivatives **86** were also prepared by an improved HMR protocol consisting of replacing the thionyl chloride by elemental sulfur in the presence of tetrabutylammonium iodide as a catalyst (2016JOC271) (see Section 2.1 for details). This resulted in an increase in the yields of up to 60% for thien-2-yl **86** and 50% for 4-pyridyl **89**. On the other hand, the yield of the 4-pyridyl derivative **92** by this method was 58% compared to the 88% achieved in the standard HMR (2003BOC288).

2.3.2 Introduction of a Heterocycle Into Position 4 of 1,2,3-Thiadiazole

A few examples of "Het to 1,2,3-thiadiazole" methodology for type **A** molecules (Scheme 35) are known. Thus, the hydroxamic acid **95** in the presence of carbonyldiimidazole **96** cyclizes to give the 1,2,4-oxadiazolone **97**, which has been used as an intermediate in the synthesis of N-hydroxyamidinoheterocycles (modulators of indoleamine 2,3-dioxygenase) (Scheme 37) (2007USP185165).

The 1,3,4-thiadiazol-5-yl-substituted 1,2,3-thiadiazoles **98** have been prepared by treating the carbohydrazides **99** with carbon disulfide under basic conditions (Scheme 38) (2008EP1970377; cf. (2013BML5821, 2014PS1895).

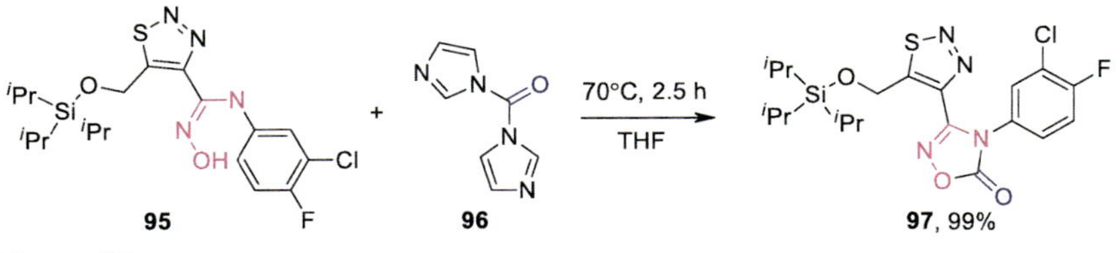

Scheme 37

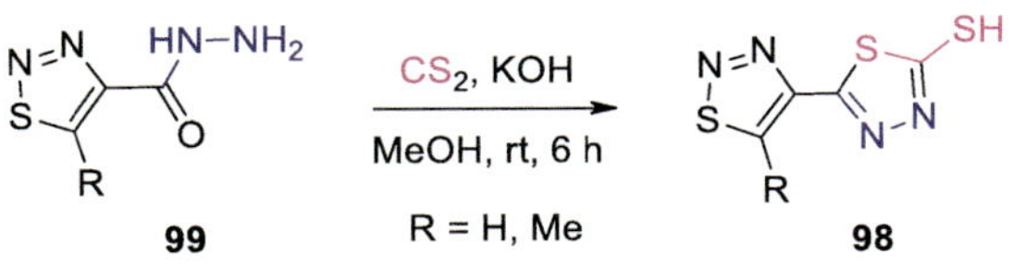

Scheme 38

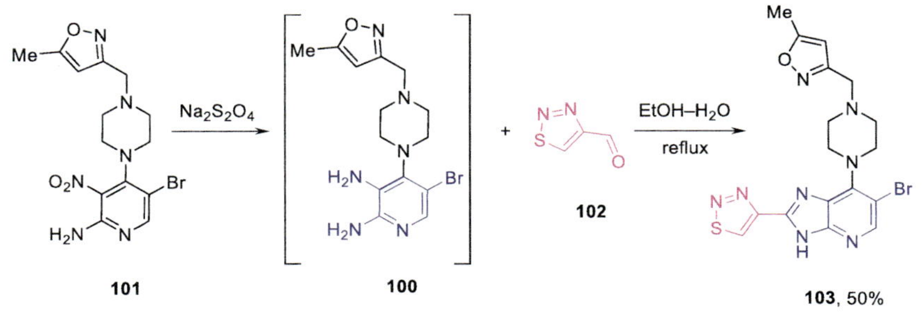

Scheme 39

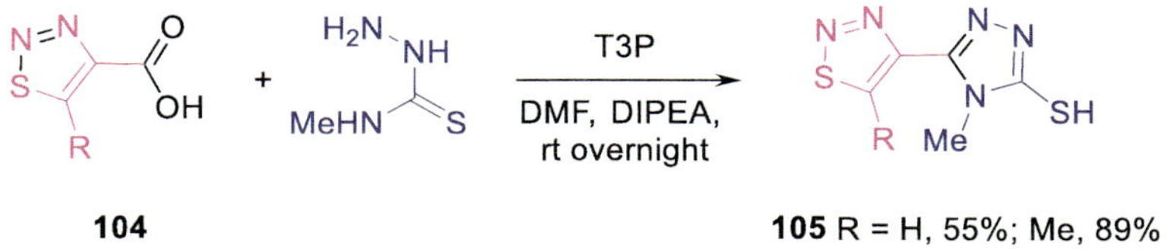

Scheme 40

The o-diamine **100**, generated by reduction of the nitro derivative **101**, reacts with 1,2,3-thiadiazole-4-carbaldehyde **102** to form the (imidazo[4,5-b]pyridin-2-yl)-1,2,3-thiadiazole **103** (Scheme 39) in moderate yield (2012JME8271).

The thiadiazole-4-carboxylic acids **104** condense with N-methylthiosemicarbazide upon treatment with propylphosphonic anhydride (T3P) in the presence of N,N-diisopropylethylamine (DIPEA), with formation of (1,2,3-thiadiazol-4-yl)-1,2,4-triazoles **105** (Scheme 40) (2016JME8549, 2016WOP67043).

2.3.3 Introduction of 1,2,3-Thiadiazole-5-yl to a Heterocycle

As is usual for "Het to 1,2,3-thiadiazole" assembling, the most popular method of formation of the 1,2,3-thiadiazole ring in compounds **106–112** is the HMR (Scheme 41).

With the aim of creating polo-like kinase-2 (PLK2) inhibitors, which are potential drugs against Parkinson's disease, the racemic compound **112** was resolved into enantiomers (R)-**112** and (S)-**112** using chiral HPLC (2011WOP079118, 2013CMC1295). Though thiadiazole **112** was the most promising of the compounds examined, it was found to be not equally potent in cell or enzyme activity compared to previously synthesized inhibitors (2011WOP079118, 2013CMC1295).

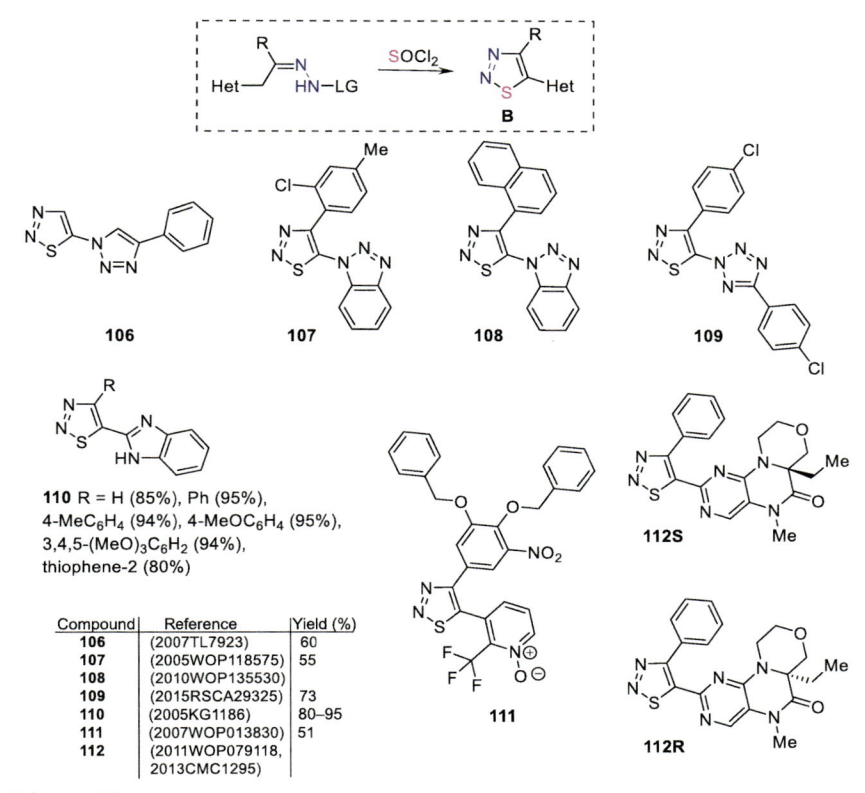

Recent Developments in the Chemistry of 1,2,3-Thiadiazoles

Scheme 41

Compound	Reference	Yield (%)
106	(2007TL7923)	60
107	(2005WOP118575)	55
108	(2010WOP135530)	
109	(2015RSCA29325)	73
110	(2005KG1186)	80–95
111	(2007WOP013830)	51
112	(2011WOP079118, 2013CMC1295)	

110 R = H (85%), Ph (95%), 4-MeC$_6$H$_4$ (94%), 4-MeOC$_6$H$_4$ (95%), 3,4,5-(MeO)$_3$C$_6$H$_2$ (94%), thiophene-2 (80%)

Scheme 42

1,2,3-Thiadiazoles attached to furan-3-yl (**113**, X=O) and thien-3-yl (**113**, X=S) rings have been synthesized by thionation of the corresponding 2-diazo-1,3-dicarbonyl compounds **114** (Scheme 42) (2016OL5408). It is reasonable to surmise that the reaction proceeds with the intermediacy of a diazothione **114A**, and thus the process in general is an example of a WS (see Section 2.3 for details).

2.3.4 Introduction of a Heterocycle Into Position 5 of 1,2,3-Thiadiazoles

The isoxazole ring in the spirocyclic compound **115** was formed as a result of the cyclocondensation of the γ-fluoro ketone **116** and hydroxylamine hydrochloride (Scheme 43) (2010WOP043893).

Other 1,2,3-thiadiazol-5-yl-substituted isoxazoles **117** have been prepared from the corresponding enamines **118** by interaction with hydroxamoyl chlorides **119** (Scheme 44) (2013CHE1880, 2016CHE743). The enamines **118** are also useful materials for building other "1,2,3-thiadiazole–Het"

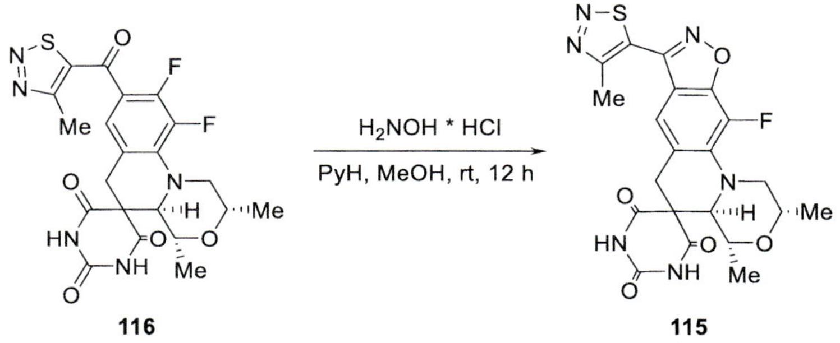

Scheme 43

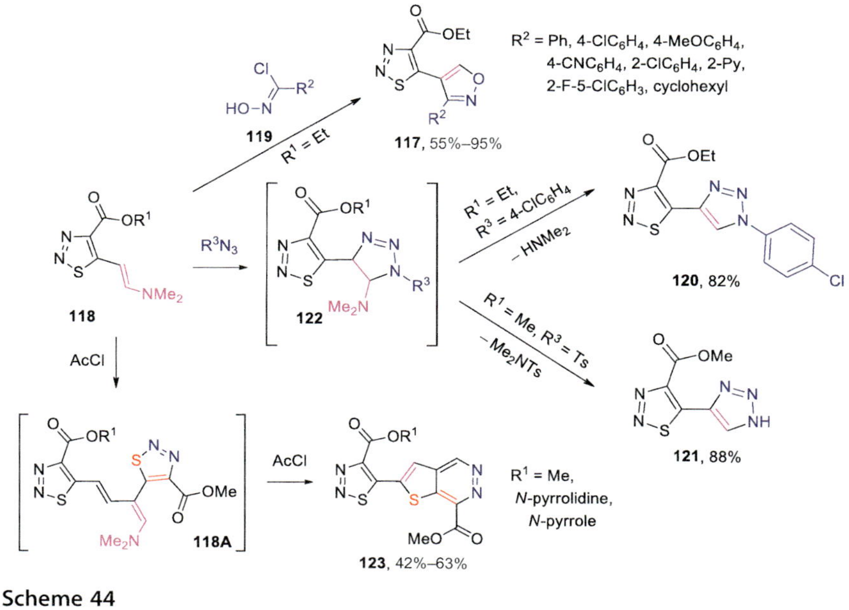

Scheme 44

system. Thus, cycloaddition of enamines **118** to aryl- and sulfonyl azides leads to N^1-aryl (**120**) and N-unsubstituted (**121**) 1,2,3-triazoles as final products, respectively (Scheme 44) (2014EJO3684). As is commonly adopted, the reaction proceeds with intermediacy of a 4,5-dihydro-1H-1,2,3-triazol-4-ine **122**, which undergoes aromatization via elimination of either dimethylamine or N,N-dimethyltosylamide.

A novel and quite unusual rearrangement of the enamines **118** into thieno [2,3-d]pyridazines **123** was discovered by Rozin et al. (Scheme 44) (2015TL1545) (see also Section 5). This involves the intermediacy of a dienamine **118A** formed after treatment of the enamine with acetyl chloride.

Pyrazole **124** has been prepared by standard synthesis from the β-diketone **125** and hydrazine (Scheme 45) (2007BML4694).

N^1-(1,2,3-Thiadiazol-5-yl)-pyrazoles **126**, with an alternative type of conjunction between 1,2,3-thiadiazole and pyrazole, have been prepared from acetophenone (1,2,3-thiadiazol-5-yl)hydrazones **127** by treatment with Vilsmeier–Haack complex (Scheme 46) (2017CHE236). These recently prepared compounds **126** are unique building blocks, containing formyl and ester growing points in the molecule, and have good potential to be involved in numerous syntheses.

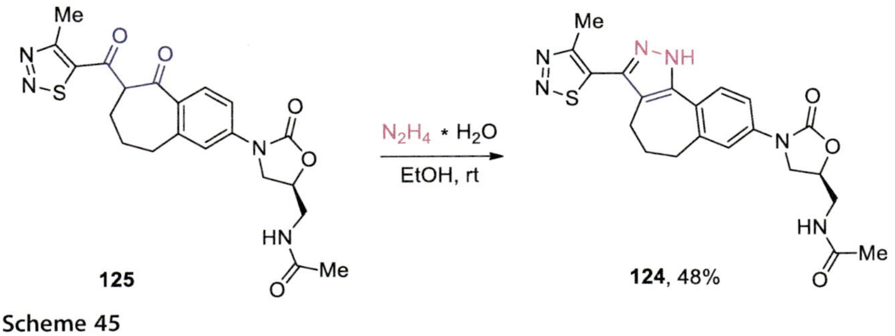

Scheme 45

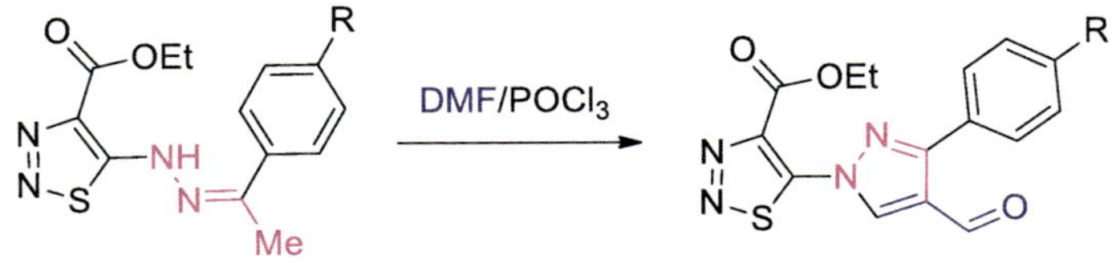

Scheme 46

1,2,3-Thiadiazole **128**, containing a tetrahydropyrrolopyridine fragment, was formed in a Paal–Knorr reaction of the diketone **129** with ammonium acetate (Scheme 47) (2015WOP022073). Annelated pyrroles were found to be useful for the inhibition of the calcium release-activated calcium channel (2015WOP022073).

Interestingly, heating the aldimine **130** at reflux in trifluoroacetic acid (TFA) results in the formation of a pyridine ring annelated to both pyrazole and benzene rings in the tricyclic product **131**, apparently after a Bischler–Napieralski-type cyclization followed by oxidative aromatization under the reaction conditions (Scheme 48) (2005OBC932).

The 5-(thiazol-2-yl)-1,2,3-thiadiazole **132** has been assembled by a Hantzsch synthesis employing the thioamide **133** and ethyl 4-chloroacetoacetate (Scheme 49) (2003WOP082862).

The 1,2,3-thiadiazole-4-carbohydrazide **134** is a fruitful source of several types of heteroaryl-thiadiazoles. Thus, depending upon the reaction conditions,

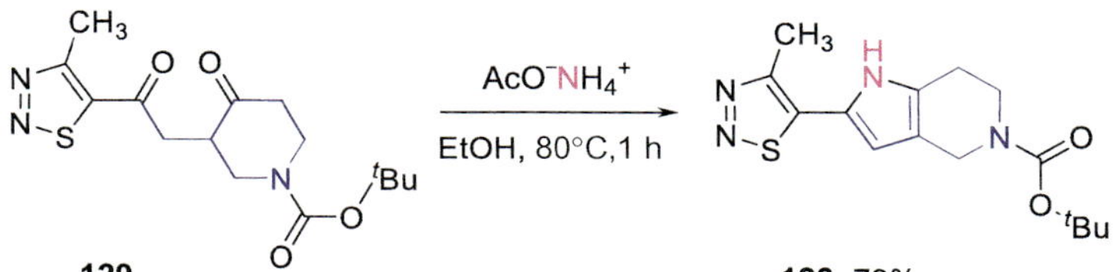

129 **128**, 79%

Scheme 47

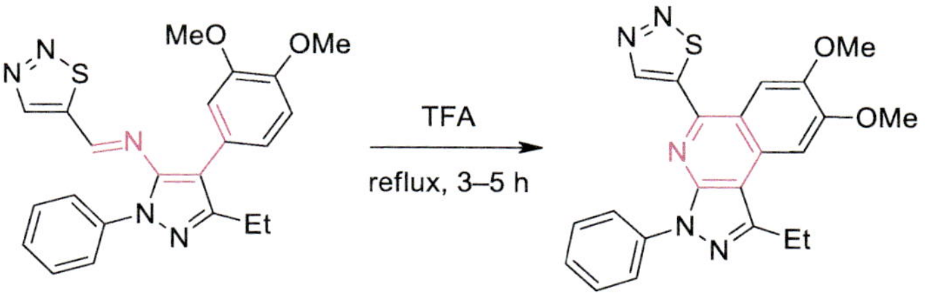

130 **131**, 69%

Scheme 48

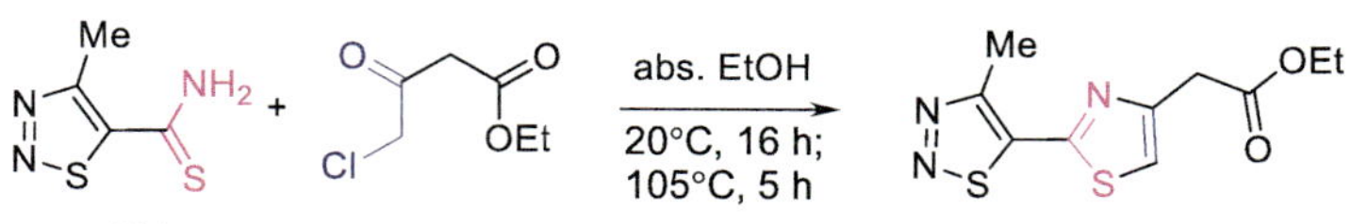

133 **132**

Scheme 49

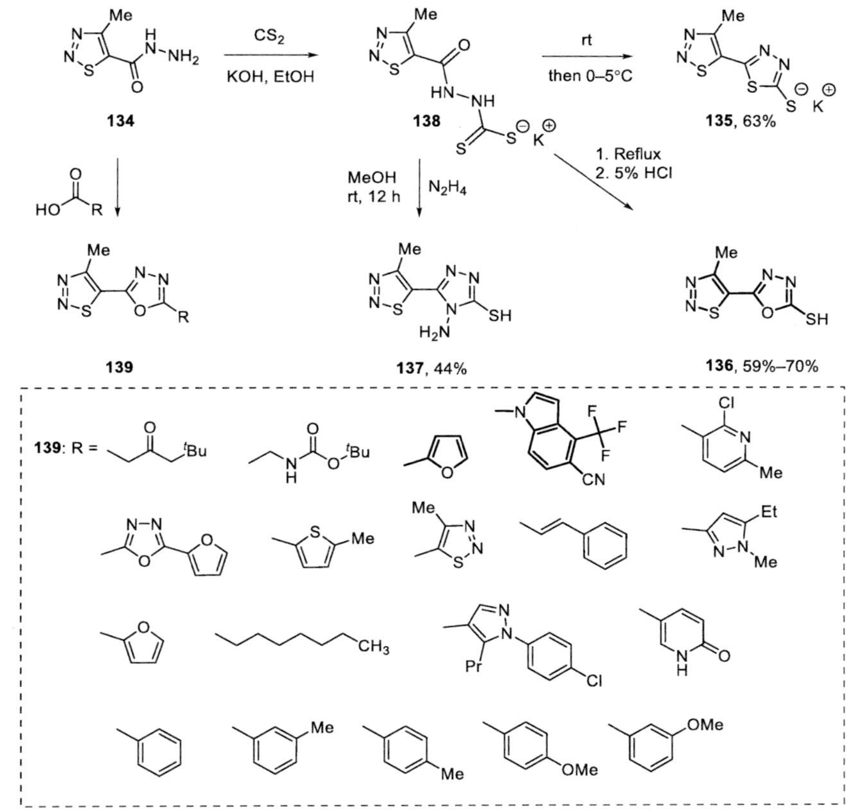

Scheme 50

N-unsubstituted hydrazide **134** by combination with carbon disulfide cyclizes either to the 5-mercapto-1,3,4-thiadiazole **135** (Scheme 50) (2013BML5821) or to the 5-mercapto-1,3,4-oxadiazole **136** (Scheme 50) (2013BML5821, 2014PS1895). N^1-Amino-5-mercapto-1,3,4-triazole **137** is obtained by reaction of the hydrazide **134** with carbon disulfide in the presence of hydrazine (Scheme 50) (2008WOP011045, 2010JFA2630). Most probably, all these reactions are occurring via the intermediate **138**, although it was detected only in some of the reports (2008WOP011045, 2010JFA2630). The intermediate **138** when kept in ethanol converts into the mercapto-1,3,4-thiadiazole **135** (2013BML5821), treatment with hydrochloric acid affords mercapto-1,3,4-oxadiazole **136** (2013BML5821, 2014PS1895), and hydrazine causes its cyclization into mercapto-1,3,4-triazole **137** (Scheme 50) (2008WOP011045, 2010JFA2630).

The same hydrazide **134** can be converted into various 5-substituted 2-(1,2,3-thiadiazol-5-yl)-1,3,4-oxadiazoles **139** by combination with carboxylic acids, with yields varying from 19% to 51% (Scheme 50) (2008WOP042571, 2009JFA4279, 2009T9989, 2009USP0036450).

The derivative **140** of the 1,2,4-oxadiazole ring has been synthesized by carbonyldiimidazole-mediated condensation of the 1,2,3-thiadiazole-5-carboxylic acid **141** with the hydroxamic acid **142** (Scheme 51) (2007WOP017261).

The synthesis of 5-mercapto-1,3,4-triazoles **143** by intramolecular cyclocondensation of the semicarbazide derivatives **144** in aqueous NaOH is described in several reports (Scheme 52) (2010HAC131, 2010HAC521, 2013CCL1134, 2013LDDD2, 2014JHC690, 2014PSS379, 2016JEIMC481).

Conversely, in sulfuric acid the thiosemicarbazides **144** cyclize to the isomeric 5-amino-1,3,4-thiadiazoles **145** (Scheme 53) (2010HAC521).

(1,2,3-Thiadiazol-5-yl)-benzoxazinone **146** has been obtained by intramolecular cyclization of the amide and the carboxy groups in compound **147**; this reaction is activated by 2-chloro-1-methylpyridinium iodide **148** in THF. Among the other compounds investigated, benzoxazinone **146** was found to show an excellent controlling effect against wheat powdery mildew (*Blumeria graminis*) (Scheme 54) (2007EP1852428).

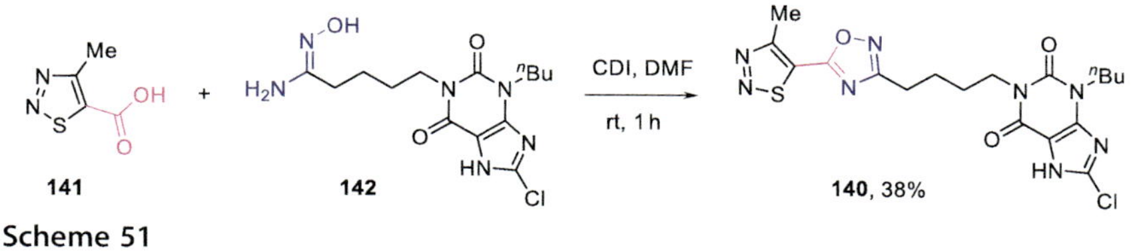

Scheme 51

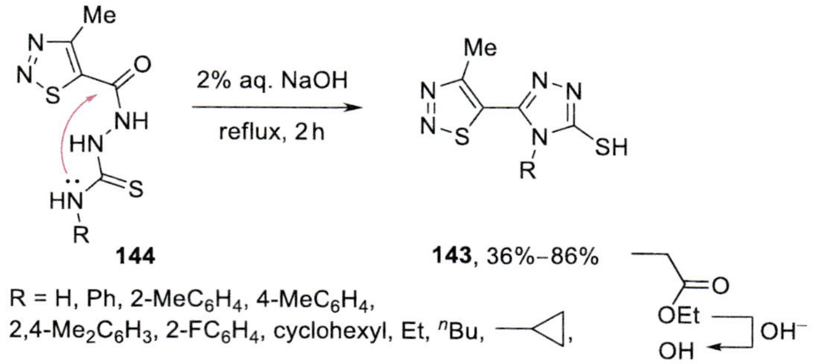

Scheme 52

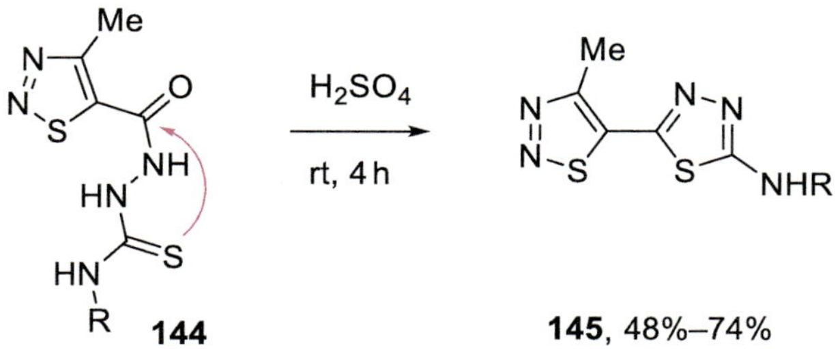
Me
O
NH
HN
S
HN
R
144
H_2SO_4
rt, 4 h
Me
N-N
S
NHR
145, 48%–74%
R = Ph, 2-MeC6H4, 2-FC6H4, cyclohexyl, Et, nBu

Scheme 53

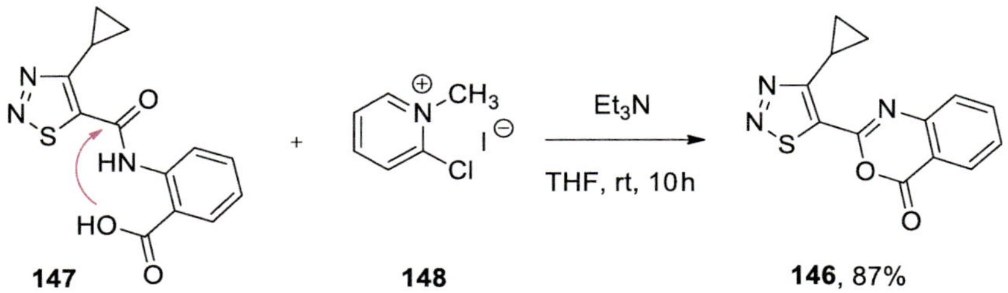
O
HN
HO
147
+
CH3
Cl
148
Et3N
THF, rt, 10h
N
O
O
146, 87%

Scheme 54

The *N*-(1,2,3-thiadiazol-5-yl)substituted benzimidazoles **149** (X=CH, yield 68%–81%) and benzotriazoles **149** (X=N, yield 80%–87%) have been synthesized by reacting *o*-phenylenediamines **150** with triethyl orthoformate and nitrous acid, respectively. Similar products (benzimidazoles **149**; X=CH, yield 65%–74%, and benzotriazoles **149**; X=N, yield 74%–94%) were obtained when mercaptotriazoles **151** were used in these reactions instead of *o*-phenylenediamines **150**. Compounds **150** and **151** are easily interconverted, the actual form being determined by the pH of the medium. Whether thiadiazole **150** or triazole **151** is taken as a starting material, the result of the reaction is almost the same and depends solely on the reaction conditions (Scheme 55) (2004RJO870).

Compound **152**, where the 1,2,3-thiadiazol-5-yl group is attached to the 2-position of a benzimidazole, has been prepared by intramolecular cyclization of compound **153**, in which the *o*-phenylenediamine fragment and 1,2,3-thiadiazole are separated by a carbonyl, by heating in acetic acid (Scheme 56) (2004USP002524).

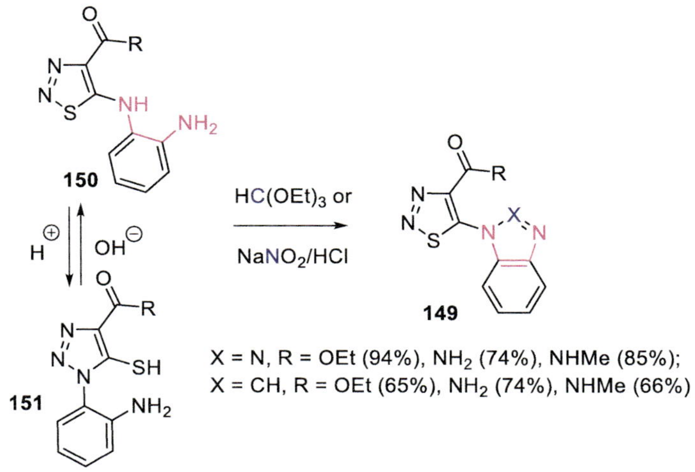
150
151
149
HC(OEt)3 or
NaNO2/HCl
X = N, R = OEt (94%), NH2 (74%), NHMe (85%);
X = CH, R = OEt (65%), NH2 (74%), NHMe (66%)

Scheme 55

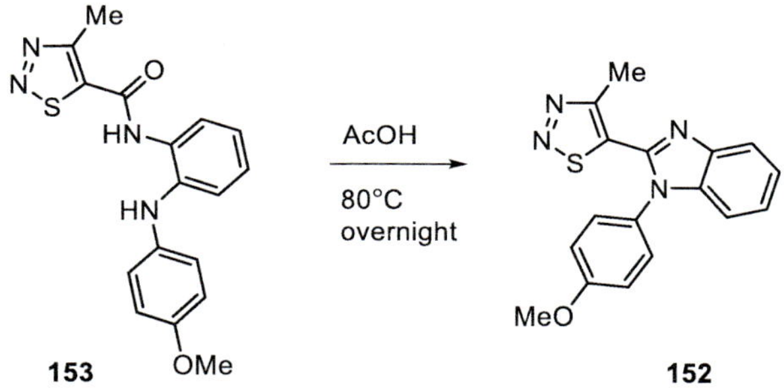
AcOH
80°C
overnight
153
152

Scheme 56

3. RING TRANSFORMATIONS OF 1,2,3-THIADIAZOLES

Ring transformations of 1,2,3-thiadiazoles are mediated by highly reactive species such as α-diazoketones, α-thiovinylcarbenes, alkinthiolates, and thioketenes. By means of these ring transformations, many five- and six-membered O-, N-, and S-containing heterocycles, as well as acyclic products, have been obtained. A review covering the subject was published in 2003 (2003CHE679), and some examples of the transformations were mentioned in several reviews and monographs (2013RJO479, 2000ANH37, 2004M[2]).

3.1 Rearrangements of 1,2,3-Thiadiazoles

The weak N—S bond of 1,2,3-thiadiazoles makes them prone to rearrange via α-diazothiocarbonyl (DT) intermediates that are in equilibrium with the thiadiazoles. Both diazo and thiocarbonyl groups are capable of interaction with other nucleophilic and electrophilic centers within the molecule, leading to further cyclizations (Scheme 57). The known rearrangements of 1,2,3-thiadiazoles can be divided into three main groups depending on the number of chain atoms involved in the cyclization: one for Dimroth-, two for Cornforth-, and three for L'abbé-type rearrangements.

3.1.1 Dimroth Rearrangements

Dimroth-type rearrangements of 1,2,3-thiadiazoles to 1,2,3-triazoles are initiated by deprotonation of NH in position 5 of the original heterocycle under the influence of a base (Scheme 58). Sometimes, acidification of 1,2,3-triazol-5-thiolates affords 5-mercapto-1,2,3-triazoles (2004RJO870); however, in other cases the process cannot be stopped at this stage and further protonation leads to reverse rearrangement back to 5-amino-1,2,3-thiadiazoles.

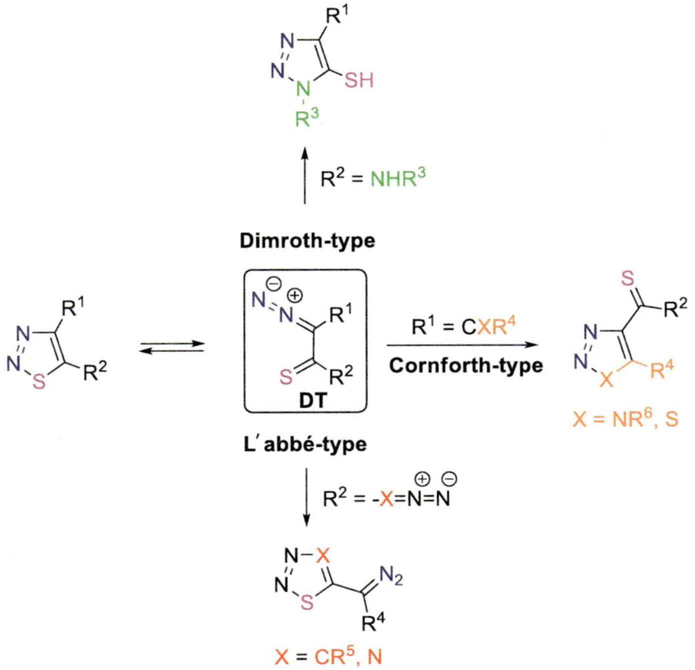

Scheme 57

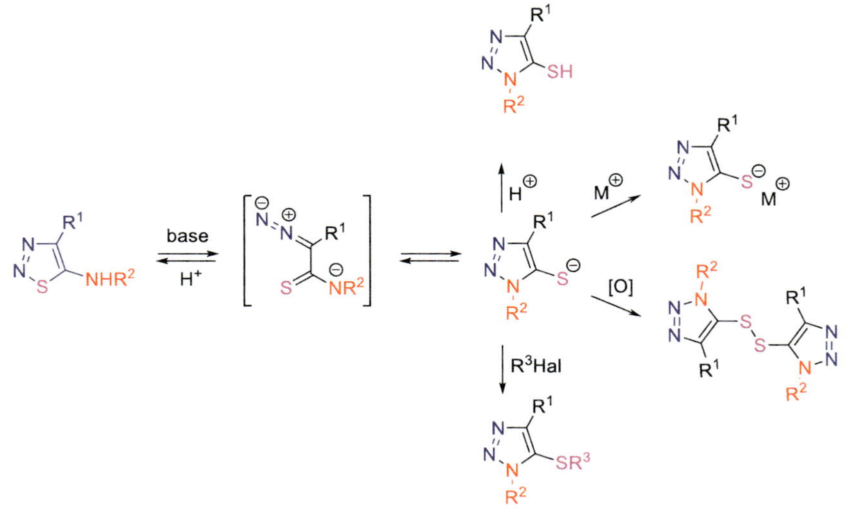

Scheme 58

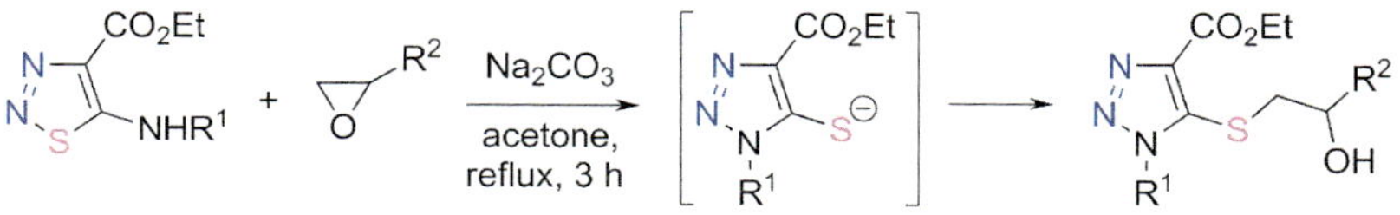

Scheme 59

R^1, R^2 (yield of **156**, %): Ph, CH_2Cl (65); 4-MeC_6H_4, CH_2Cl (85); 4-$MeOC_6H_4$, CH_2Cl (53); Bn, CH_2Cl (62); 4-$EtOC_6H_4$, Me (80); 3-MeC_6H_4, Me (82)

5-Mercapto-1,2,3-triazole-5-thiols can be stabilized against this rearrangement by oxidative dimerization of the thiol group or by its covalent protection, e.g., by alkylation, (hetero)arylation, oxidation to a disulfide, etc. (Scheme 58) (2004M[2]).

Morzherin and coworkers have described an example in which intermediate 1,2,3-triazol-5-ylthiolates **154**, resulting from Dimroth rearrangement of 1,2,3-thiadiazoles **155** in the presence of sodium carbonate, are alkylated by propylene oxide or epichlorohydrin with formation of 5-hydroxyethylthio-1,2,3-triazoles **156** as final products (Scheme 59) (2011PAC715).

Stabilization of 1,2,3-triazoles by heteroarylation is illustrated by transformation of 5-chloro-1,2,3-thiadiazoles **157** into (1,2,3-thiadiazol-5-yl)-1,2,3-triazol-5-ylsulfide **158**, via the isomeric intermediates **159** ↔ **160** (Scheme 60) (2008CHE233).

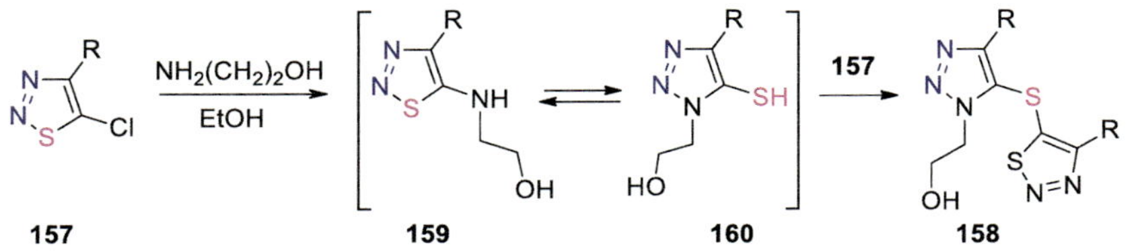

Scheme 60

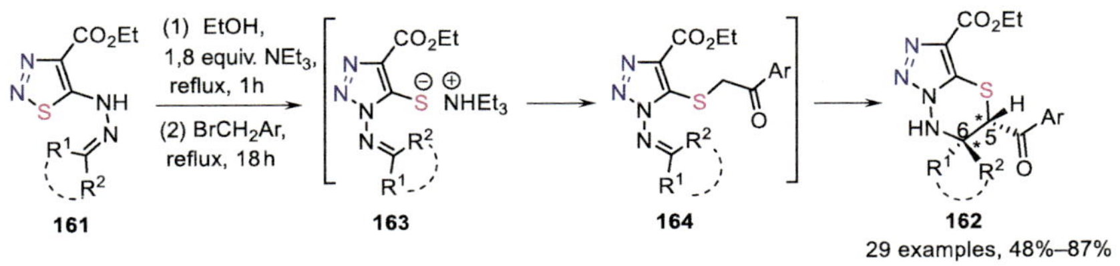

Scheme 61

R¹ = H, Me; R² = Ar, Me; for spiro-compounds CR¹R² = cyclopentane, cyclohexane, cycloheptane

Not only 5-amino- but also 5-hydrazino(hydrazono)-1,2,3-thiadiazoles easily undergo Dimroth rearrangement. Again, the process is initiated by deprotonation by a base. To prevent backward rearrangement, besides the previously mentioned ones, other reactions such as annelation of the heterocycle involving both the thiol and hydrazino(hydrazono) group in condensation with bifunctional reagents are used.

Thus 1,2,3-thiadiazolylhydrazones **161** when treated with α-bromoacetophenones in the presence of triethylamine give 5H-[1,2,3]triazolo[5,1-b][1,3,4]thiadiazines **162** (Scheme 61). The first stage of this transformation is a Dimroth rearrangement. Further alkylation of the 1,2,3-thiadiazol-5-thiolates **163** with α-bromoacetophenone and intramolecular cyclocondensation of the alkylation products **164** into bicyclic triazoles **162** take place (2015CHE589, 2017JHC137). For hydrazones of acetophenones and benzaldehydes ($R^1 \neq R^2$), the process is diastereoselective, providing only one pair of enantiomers (5R6S and 5S6R) (2017JHC137).

Dimroth rearrangement of the 5-hydrazino-1,2,3-thiadiazole **165** in ethanol in the presence of triethylamine leads to the 1-amino-1,2,3-triazole-5-ylthiolate **166**. Subsequent cyclocondensation with the quinoline derivative **167** or the 5-chloro-4-formylpyrazole **168** affords [1,2,3]triazolo[5′,1′:2,3][1,3,4]thiadiazepino[7,6-b]quinoline **169** and 5H-pyrazolo[4,3-f][1,2,3]triazolo[5,1-b][1,3,4]thiadiazepine **170**, respectively (Scheme 62) (2013CHE350).

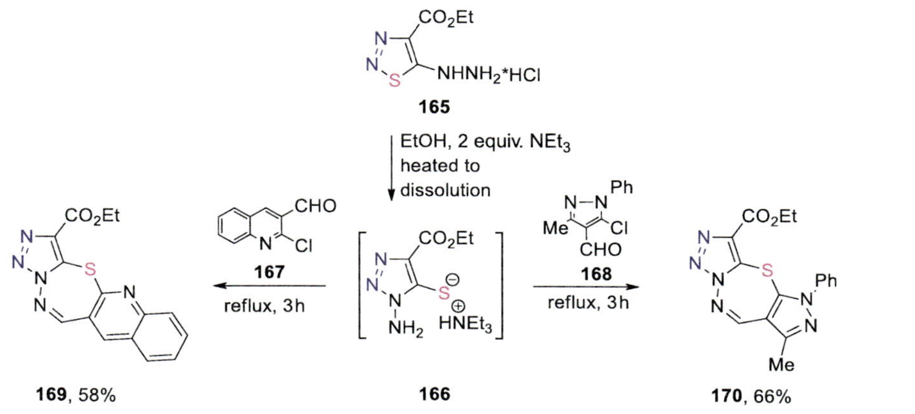

Scheme 62

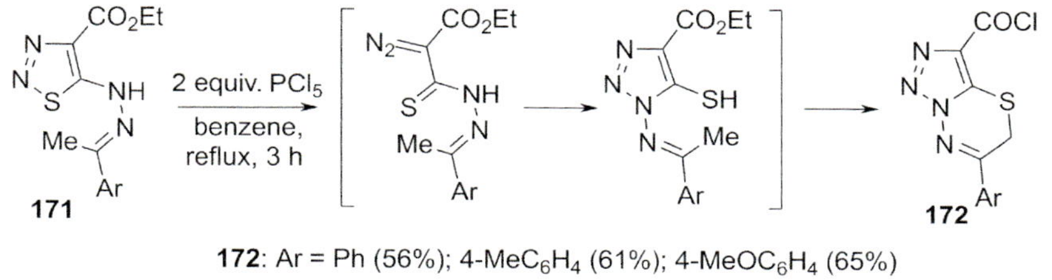

169, 58%   **166**   **170**, 66%

Scheme 63

172: Ar = Ph (56%); 4-MeC$_6$H$_4$ (61%); 4-MeOC$_6$H$_4$ (65%)

It has been shown that rearrangement of 5-amino- and 5-hydrazino-1,2,3-thiadiazoles to 5-mercapto-1,2,3-triazoles can also take place under nonbasic conditions. One of the first examples of such a process was the transformation of 5-amino-1,2,3-thiadiazoles into bi-1,2,3-triazol-5-yldisulfides by reaction with halogen-containing oxidizing agents such as sulfuryl chloride, chlorine, and bromine (Scheme 58) (1993CHE724).

In recent years, one more example of a Dimroth rearrangement in nonbasic media has been reported. It was found that interaction of the 1,2,3-thiadiazolylhydrazones of acetophenones **171** with phosphorus pentachloride in benzene or toluene leads to 5H-[1,2,3]triazolo[5,1-b][1,3,4]thiadiazines **172**. Possibly, the first stage of the process is a Dimroth rearrangement of the thiadiazole ring into a triazole. This is followed by the intramolecular cyclocondensation with formation of thiadiazine ring accompanied with transformation of the ester group to acid chloride (Scheme 63). The mechanism of the reaction has not been studied (2011RCB981).

3.1.2 Cornforth Rearrangement

Cornforth rearrangement is possible for 1,2,3-thiadiazoles bearing C=N- or C=S-containing substituents in position 4 of the ring. Thus, 4-carbimino-1,2,3-thiadiazoles **173**, formed by condensation of 1,2,3-thiadiazol-4-ylcarbaldehydes **174** with primary amines, are labile and highly reactive species that rearrange to 5H-1,2,3-triazol-4-ylcarbothioamides **175** under the conditions needed for their generation (2004RCB1311, 2010RCB867). The 1,2,3-triazole ring of **175** originates from the 1,5-dipolar cyclization of diazo and imino groups (Scheme 64). An NMR study has established that in solvents like chloroform, DMSO and pyridine 1,2,3-triazol-4-ylcarbothioamides **175** bearing hydroxy, methoxy, or 2,4-dinitrophenylamino groups in position 1 of the ring are in equilibrium with the isomeric 4-imino-1,2,3-thiadiazole form **173**, with predominance of the 1,2,3-triazole form **175**. Rearrangement to triazolylcarbothioamides **175** with alkyl or aryl substituents in position 1 is irreversible (2005CHE542).

Unlike 4-carbimino-1,2,3-thiadiazoles, 4-carbamidino-1,2,3-thiadiazoles **176** rearrange only in basic media, e.g., in the presence of sodium methoxide or DBU. The process results in salts of 5-sulfonamido-1,2,3-triazole-4-carbothioamides **177**. Treatment of the salts **177** with water, dilute hydrochloric acid, or even silica gel causes their rearrangement to the starting 4-amidino-1,2,3-thiadiazoles **176** (Scheme 65) (2017JOC4056, 2017JPB (145)315).

Amides of 5-methyl-1,2,3-thiadiazol-4-carboxylic acid **178** are thionated upon treatment with diphosphorus decasulfide to form 5-methyl-1,2,3-thiadiazol-4-ylcarbothioamides **179** that are in equilibrium with rearranged 4-thioacetyl-5-amino-1,2,3-thiadiazoles **180**. The equilibrium position depends on the solvent employed; an increase of solvent polarity shifts the equilibrium in favor of the thioketones **180** (Scheme 66) (2012RJO1338).

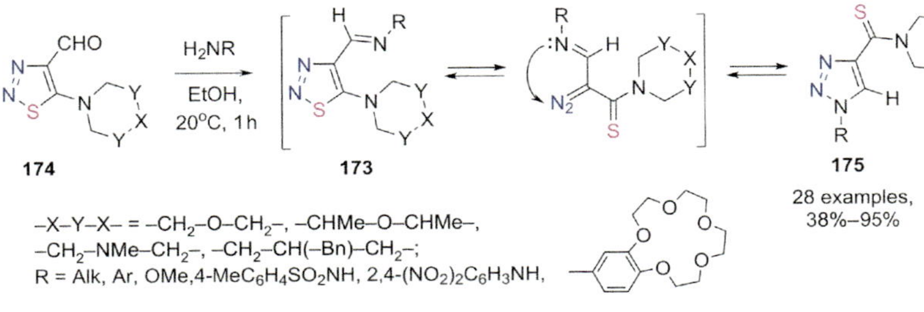

Scheme 64

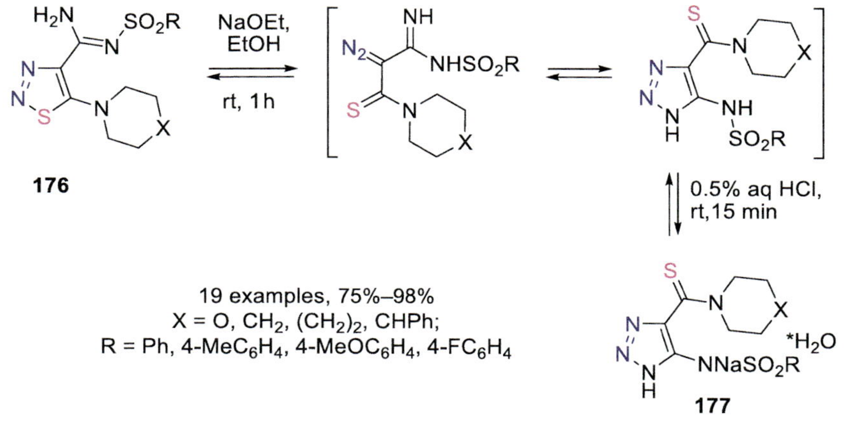

Scheme 65

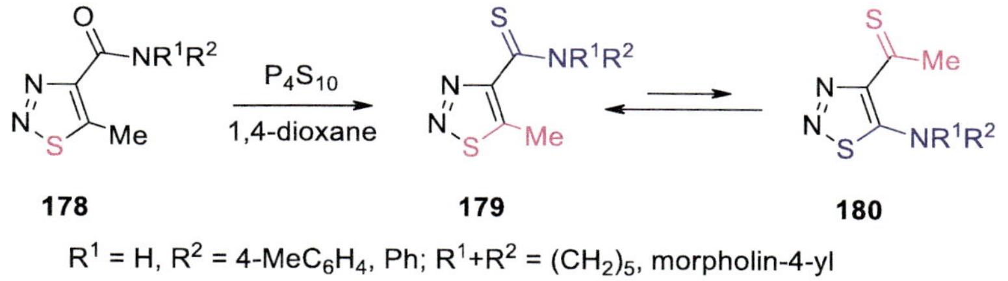

19 examples, 75%–98%
X = O, CH$_2$, (CH$_2$)$_2$, CHPh;
R = Ph, 4-MeC$_6$H$_4$, 4-MeOC$_6$H$_4$, 4-FC$_6$H$_4$

Scheme 66

R^1 = H, R^2 = 4-MeC$_6$H$_4$, Ph; R^1+R^2 = (CH$_2$)$_5$, morpholin-4-yl

3.1.3 L'abbé Rearrangement

L'abbé-type rearrangements, involving three atoms from the side chain at position 5 of the thiadiazole ring, were reviewed in 2004 (2004M[2]). No new examples of L'abbé rearrangements have been published in the period of 2003–17.

3.1.4 Rearrangement of 5-Vinyl-1,2,3-thiadiazoles to 4,5-Dihydropyridazine-4-thiols

A novel rearrangement of 5-vinyl-1,2,3-thiadiazoles to pyridazine-4-thiols has been demonstrated to take place via a tandem transformation of β-(1,2,3-thiadiazole-5-yl)enamines **181** to 2-(1,2,3-thiadiazol-5-yl)thieno[2,3-*d*] pyridazine **182** after treatment with acetyl chloride. A plausible mechanism for this conversion is illustrated in Scheme 67. The opening of the 1,2,3-thiadiazole ring at position 2 of dieneamines **183**, obtained after condensation of two enamine molecules, is proposed to occur with formation of a diazo thione **184** bearing also an enamine moiety. The intermediate **184**

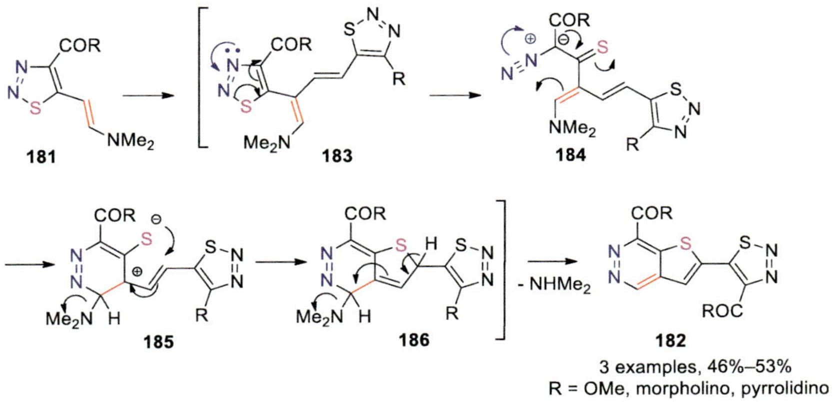

Scheme 67

then undergoes cyclization to generate a pyridazine **185**. In its turn, thiolate anion attacks the double bond to afford a 2,4-dihydrothienopyridazine **186**. The latter can eliminate dimethylamine to form the final thieno[2,3-*d*] pyridazine **182**. Remarkably, only the 1,2,3-thiadiazole ring which is in direct conjugation with enamine double bond is involved in the transformation (2015TL1545).

3.2 Cleavage of the Ring

Factors such as high temperatures, UV irradiation, or action of catalysts cause elimination of nitrogen from α-diazothiones, the open-chain isomers of 1,2,3-thiadiazoles. Further transformations lead to different species (e.g., thiirane, thioketene, thiocarbene, or alkynethiolate) depending on the structure of the starting 1,2,3-thiadiazole and reaction conditions (2003CHE679, 2004M[2], 2013RJO479). These highly reactive intermediates are capable of versatile inter- and intramolecular reactions, leading to a wide spectrum of products.

3.2.1 Photochemical Cleavage of the Ring

Previously, it was found that the main products resulting from the photodissociation of 1,2,3-thiadiazoles are thiirenes **187**, thioketenes **188**, and ethynethiols **189** (Scheme 68). Two possible mechanisms have been proposed to date. The first is a stepwise mechanism via a thiocarbene intermediate. The second mechanism surmises the formation of a 1,3-diradical intermediate. Intermediacy of 1,3-diradicals and thiocarbenes as precursors for thiirenes or thioketenes is only hypothetical (2013PPS895).

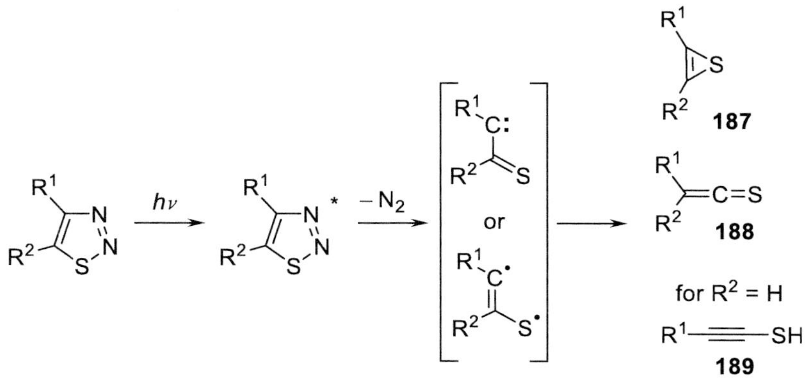

Scheme 68

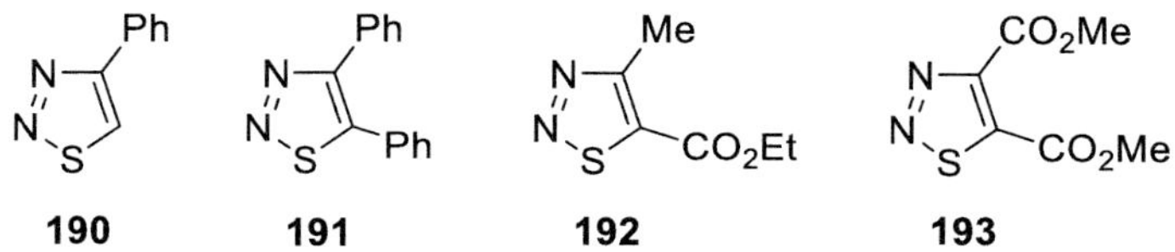

190, 191, 192, 193

Scheme 69

In the reviewed period, a series of papers on the study of photochemical cleavage of 4-phenyl-1,2,3-thiadiazole **190** (2013PPS895), 4,5-diphenyl-1,2,3-thiadiazole **191** (2013PPS895), 4-methyl-5-carboethoxy-1,2,3-thiadiazole **192** (2011JPC(A)14300, 2013PPS895), and dimethyl 1,2,3-thiadiazole-4,5-dicarboxylate **193** (2013JPC(A)4551) have been published (Scheme 69).

Experiments with the 1,2,3-thiadiazoles **190–193** were performed in acetonitrile solution by means of ultraviolet–visible and IR transient absorption spectroscopies. It was discovered that thiirene is an intermediate in all cases. Its quantum yield is in the range of 0.3–0.5. Formation of thioketene **188**, with quantum yield around 0.9, was detected only for phenyl-, diphenyl-, and carboethoxy-1,2,3-thiadiazoles **190–192** (2013PPS895). The quantum yield of diphenylthioketene is lower than that of monophenylthioketene due to the lower migrational ability of phenyl in comparison to that of hydrogen in an ultrafast Wolff rearrangement. In the photocleavage of dimethyl 1,2,3-thiadiazole-4,5-dicarboxylate **193**, the sole intermediate was thiirene. The authors explain this phenomenon as being due to the ester groups present in the molecule, which block Wolff rearrangement (2013JPC(A)4551).

The formation of the final photolysis products can be rationalized by cycloaddition of thiirene to thioketene, and dimerization of thiirene and thioketene. Thus photolysis of 4-phenyl-1,2,3-thiadiazole **190** leads to the 1,3-dithiole **194** (*cis-* and *trans-*isomers). Photolysis of 4,5-diphenyl-1,2,3-thiadiazole **191** gives a complex mixture of products: 4,5-diphenyl-2-(diphenylmethylene)-1,3-dithiole **195**, tetraphenylthiophene **196**, diphenylacetylene **197**, and, presumably, phenyl(2-phenylethynyl)sulfane **198** (Scheme 70) (2013PPS895).

For photolysis of 4-methyl-5-carboethoxy-1,2,3-thiadiazole **192**, the stable photoproducts 1,3-dithiethane **199**, 1,3-dithiol **200**, and the thiophene isomers **201** were detected by NMR spectroscopy (Scheme 71) (2011JPC(A)14300).

Tetra(carbomethoxy)thiophene **202** has been shown to form in the photochemical transformation of dimethyl 1,2,3-thiadiazole-4,5-dicarboxylate **193**. This is attributed to the dimerization of an intermediate thiirene (Scheme 72) (2013JPC(A)4551).

In their works (2013PPS895, 2011JPC(A)14300, 2013JPC(A)4551), Burdzinski et al. have proposed an "excited-state concerted rearrangement mechanism" for these cleavages of the 1,2,3-thiadiazoles **190–193**. In these transformations, occurring in a concerted manner that does not involve

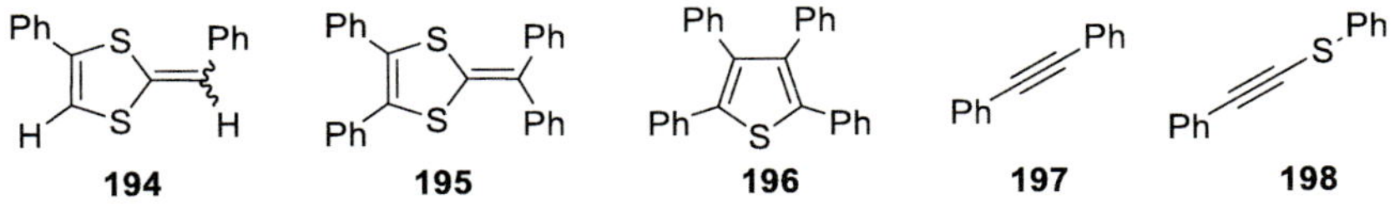

Scheme 70

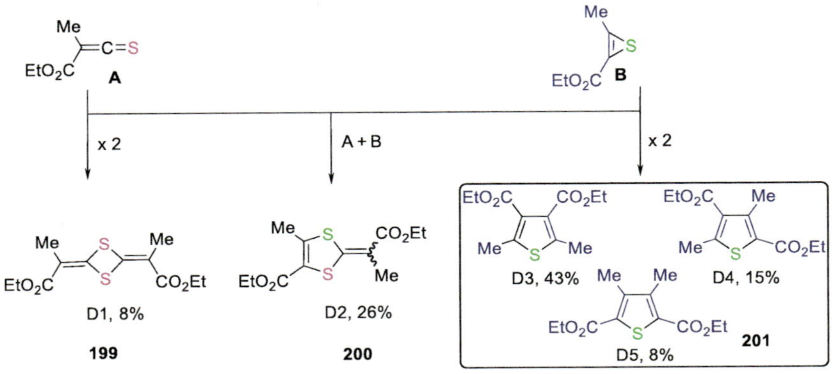

Scheme 71

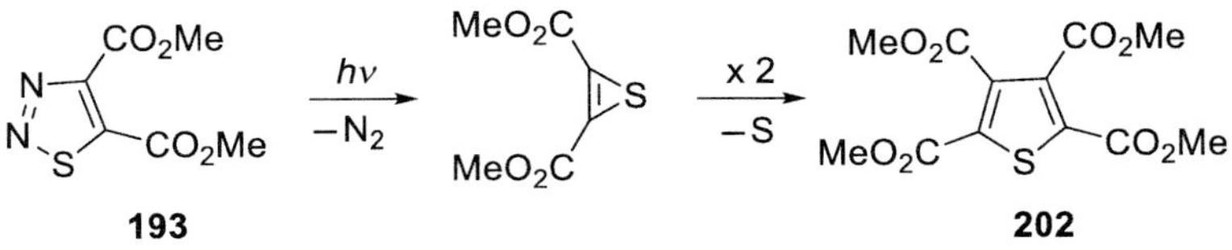

$$\underset{\textbf{193}}{\overset{N}{\underset{S}{\parallel}}\!\!\!\!\!\overset{CO_2Me}{\underset{CO_2Me}{\diagdown}}} \xrightarrow[-N_2]{h\nu} \underset{MeO_2C}{\overset{MeO_2C}{\diagdown}}\!\!\!\!\!\overset{S}{\triangle} \xrightarrow[-S]{\times 2} \underset{\textbf{202}}{\overset{MeO_2C}{\underset{MeO_2C}{\diagdown}}\!\!\!\!\!\overset{CO_2Me}{\underset{S}{\diagdown}\!\!\!\!\!\overset{CO_2Me}{\diagdown}}}$$

Scheme 72

diradical thiocarbene intermediates, thiirenes and thioketenes are formed in excited singlet states.

Liu et al., on the basis of quantum (complete active space self-consistent field second-order perturbation theory) and molecular mechanics studies on unsubstituted 1,2,3-thiadiazole, have calculated that initially the S2 state is populated, followed by ultrafast decay of the excited state of the S—N bond ($S_2 \to S_1 \to S_0$), which finally leads to a very "hot" intermediate with broken S—N bond. Consequent thermal rearrangements of thioketene, thiirene, and ethynethiol are implemented in the concerted asynchronous mode (2017JCP224302).

3.2.2 Thermal Decomposition Reactions of 1,2,3-Thiadiazoles

As a result of their thermal decomposition, 1,2,3-thiadiazoles convert to the same intermediates as in their photochemical cleavage, i.e., thiocarbenes, thiirenes, and thioketenes. Intermolecular combination thereof also affords ethynedithiols, thiophenes, etc. Introduction of compounds capable of interaction with the intermediates leads to various cycloaddition products. One of the characteristic examples of such interactions is cycloaddition to carbon disulfide with formation of 1,3-dithiol-2-thiones, or with acetylenes furnishing thiophenes (2004M[2]).

When the starting 1,2,3-thiadiazole bears a substituent that can interact with the intermediate carbene or thioketene, the transformation can be accompanied with an intramolecular cyclization. One example is the transformation of 1,2,3-thiadiazoles **203** in *n*-butanol at reflux temperature with the loss of N_2 to afford 5-ethoxyfurans **204**. The proposed mechanism includes the generation of intermediate thiocarbenes **205** which cyclize to give the final ethyl 2-ethoxy-5-thioamidofuran-3-carboxylates **204** (Scheme 73). The reaction is, to date, the only example of the synthesis of monocyclic furans from 1,2,3-thiadiazoles (2006MC76).

It is known that thermal decomposition of 2-substituted 1,2,3-thiadiazole derivatives proceeds in a different way. Thus, 2-aryl-1,2,3-thiadiazol-5-imines eliminate sulfur to form hydrazones, while 2-aryl-2,

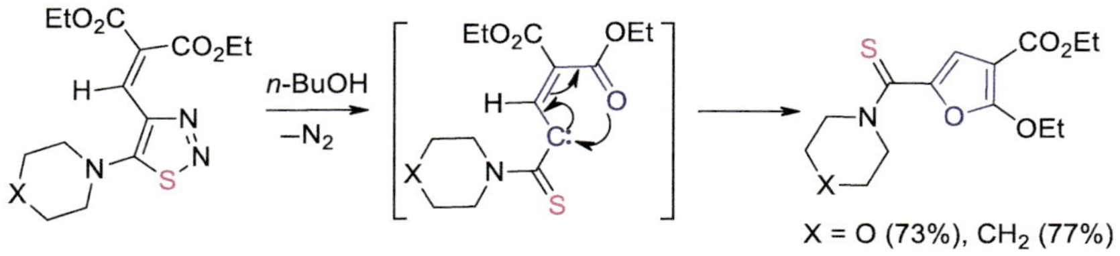

Scheme 73

203 → [205] → 204 X = O (73%), CH$_2$ (77%)

Scheme 74

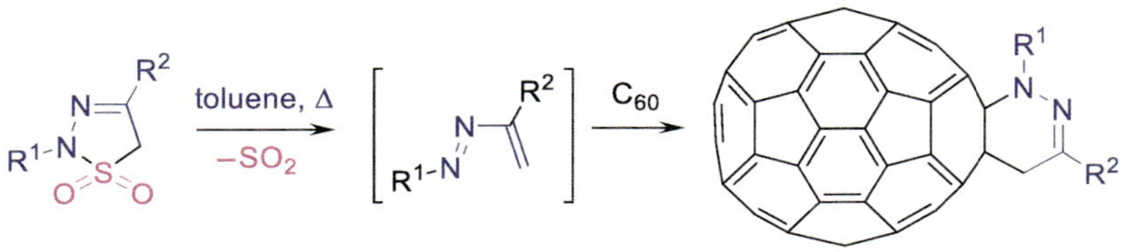

206 → [207] → 208

208: R^1, R^2 = Ph, Ph (47%); Ph, 4-ClC$_6$H$_4$ (48%); Ph, 4-MeOC$_6$H$_4$ (48%); Ph, Me (36%); 4-ClC$_6$H$_4$CH$_2$, Ph (27%)

Scheme 75

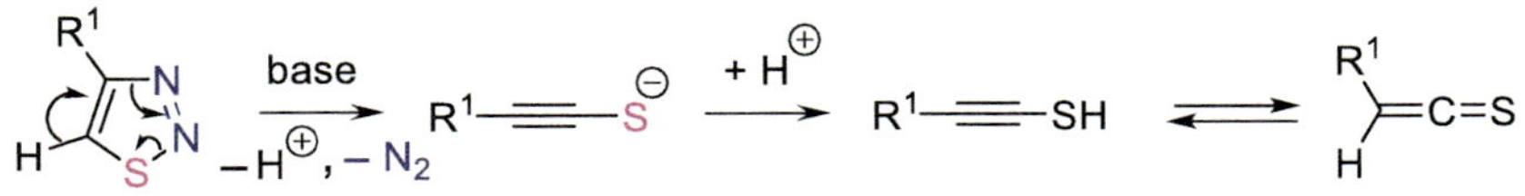

209 → 210 → 211 ⇌ 212

5-dihydro-1,2,3-thiadiazol-1,1-dioxides **206** extrude SO$_2$ affording α-1,2-diazabutadienes **207** (2004M[2]). The latter undergo cycloaddition to a double bond of [60]fullerene to give the tetrahydropyridazine derivatives **208** (Scheme 74) (2006TL4129).

3.2.3 Transformations of 5H-1,2,3-Thiadiazoles Under Basic Conditions

In the presence of potassium *tert*-butoxide, potassium methoxide, or sodium carbonate, 5*H*-1,2,3-thiadiazoles **209** undergo ring cleavage. The process is initiated by deprotonation at position 5 of the ring and is accompanied by elimination of nitrogen. This results in the generation of the highly reactive alkynethiolates **210**, which after protonation and subsequent tautomerization of the alkynethiols **211** gives thioketenes **212** (Scheme 75).

Alkynethiolates **210** can be isolated from the reaction mixtures such as salts, metal complexes, and S-alkylated products (2004M[2]). In the presence of potassium *tert*-butoxide, the bi-1,2,3-thiadiazoles **213** afford bi-alkynthiolates **214**, which after interaction with mononuclear complexes of general formula [AuClL] give the bis(alkynethiolate)gold(I) derivatives **215** and **216** (Scheme 76) (2009JIC259).

Using bis(diphenylphosphanyl)methane (DPPM) and 1,2-bis(diphenylphosphino)ethane (DPPE), tetranuclear macrocyclic complexes **217** and **218** have been synthesized (Scheme 77) (2009JIC259).

Decomposition of the 5*H*-1,2,3-thiadiazoles **219** initiated by potassium *tert*-butoxide leads to the alkynethiolates **220**, which upon alkylation

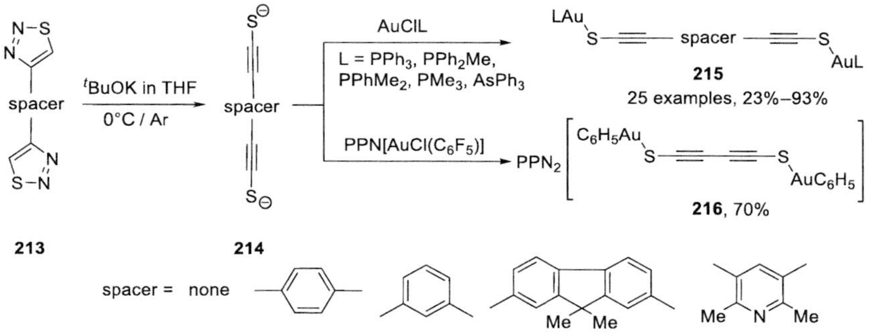

Scheme 76

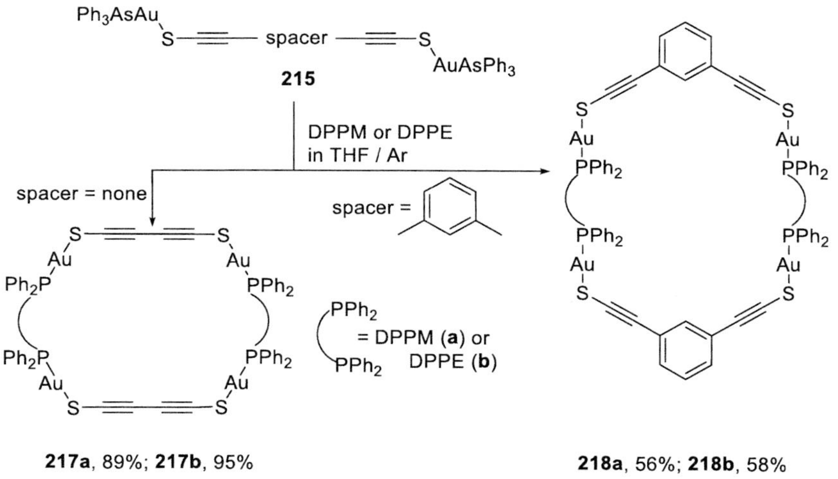

Scheme 77

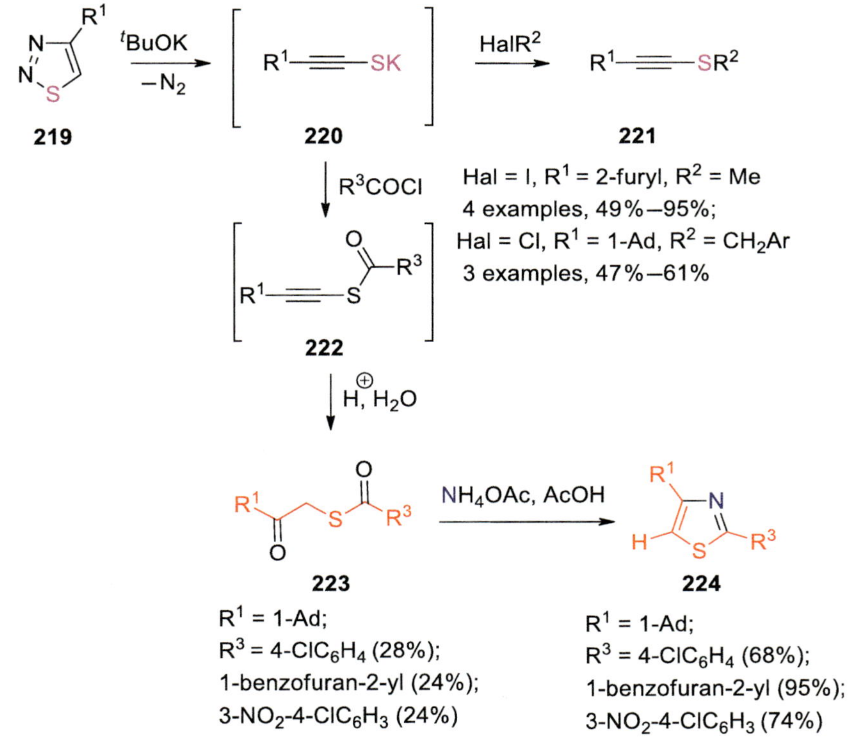

Scheme 78

afford S-adamantyl- and S-furylsulfanylacetylenes **221** (Scheme 78) (2017RJG259, 2010RJO1214, 2011RJO1777). It has been shown that acylation of adamantylethynethiolate **220** ($R^1 = 1$-Ad) with acid chlorides leads to the thioesters **222**, which are easily hydrolyzed to give acylthioles **223**. Compounds **223** react with anhydrous ammonium acetate in acetic acid to furnish 4-(1-adamantyl)-2-aryl-1,3-thiazoles **224** (Scheme 78) (2011RJO1878).

Alkynethiolates **220** that are formed by the decomposition of the 5H-1,2,3-thiadiazoles **219** can transform after acid treatment to the thioketenes **225** (Scheme 79). In turn, the latter are capable of reacting with various nucleophiles to form acyclic and heterocyclic compounds. For example, the reaction of thioketenes **225** with amines leads to the thioacetamides **226** (2008RJO774, 2011RJO1191). Androsov et al. have described the transformation of 5H-1,2,3-thiadiazoles **219** to the thioesters **227** by treatment with sodium methoxide, acting both as a base and as a nucleophile

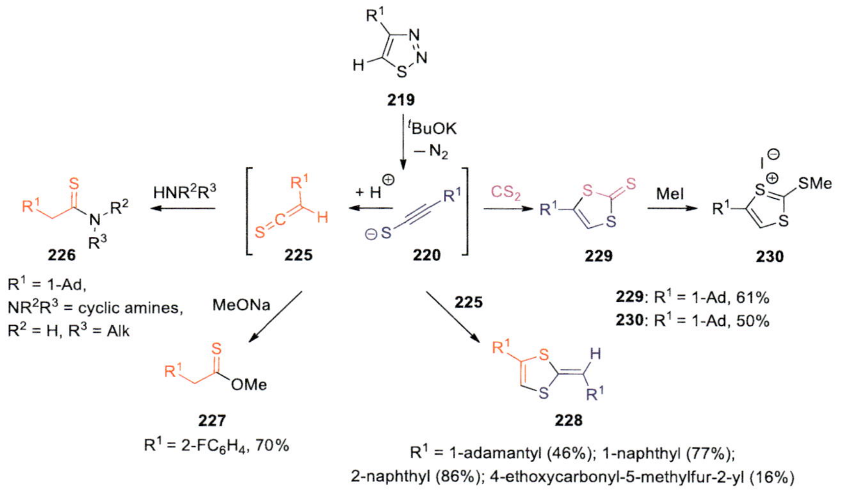

Scheme 79

(2007RJO1870). In the absence of an external nucleophile, the thioketenes **225** react with the alkynethiolates **220**, affording dithiafulvenes **228** (2017RJG259, 2016RGC1762, 2011RJO1191, 2010RJO1214). When the reaction mixture, containing intermediate alkynethiolates **220**, is treated with an excess of carbon disulfide, 1,3-dithiol-2-thiones **229** are formed. Subsequent alkylation with methyl iodide gives 2-methylsulfanyl-1,3-dithiolium iodides **230** (Scheme 79) (2008RJO1089).

When the starting 5*H*-1,2,3-thiadiazole contains a nucleophilic group (OH or NH_2), the thioketenes, generated after base-catalyzed cleavage, are capable of intramolecular cyclization. Thus, transformation of 4-aryl-1,2,3-thiadiazoles **231**, bearing a hydroxyl group in the *ortho*-position of the aryl ring, under the action of the relatively weak base potassium carbonate leads to 2-benzofuranthiolates **232** (X=O). Acidification of the resulting thiolates **232** affords benzo[*b*]furan-2-thiols **233** (2009RJO1727), whereas alkylation furnishes 2-alkylthiobenzofurans **234** (2002RJO1510). The synthesis of derivatives of benzo[*b*]furan-2-thiols by this approach has also been reported (Scheme 80) (2012RJO728, 2004RJO1691).

In contrast to 4-(2-hydroxyphenyl)-1,2,3-thiadiazole **231**, a stronger base (potassium *tert*-butoxide) is required to open and transform the ring of 4-(2-aminophenyl)-1,2,3-thiadiazole **235** to 2-methylsulfanylindole **236** (Scheme 80) (2003RJO284).

The decomposition of 4-(2-halophenyl)-1,2,3-thiadiazoles **237**, initiated by potassium carbonate in DMF, presents a convenient method for

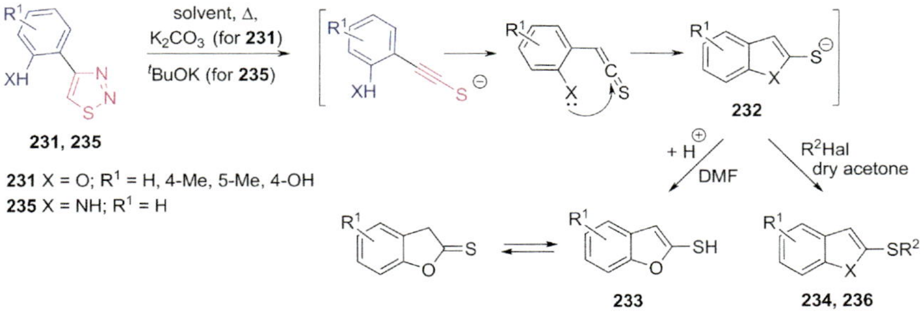

231 X = O; R^1 = H, 4-Me, 5-Me, 4-OH
235 X = NH; R^1 = H

233 X, R^1: O, H, (68%); O, 5-Me (32%); O, 4-Me (55%);
234 X, R^1, R^2: O, H, Me (56%); O, H, $C_{16}H_{31}$ (92%); O, H, CH_2Ph (91%);
O, 4-OH, $C_{16}H_{31}$ (46%); O, 5-OH, $C_{16}H_{31}$ (97%);
236 X, R^1, R^2: NH, H, Me (85%)

Scheme 80

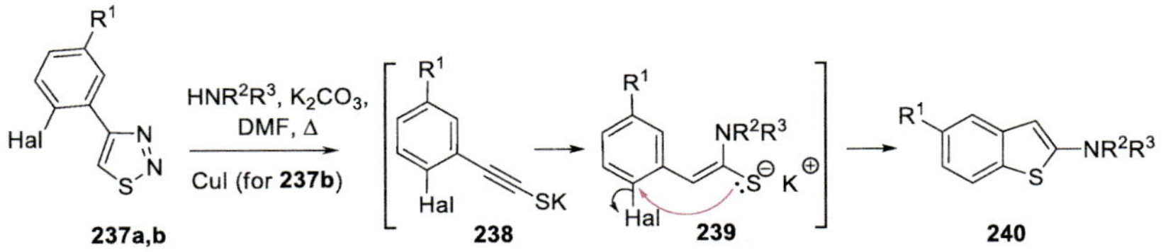

237a R^1 = NO_2; Hal = Cl
237b R^1 = H; Hal = Br

240a R^1, NR^2R^2: NO_2, $N(CH_2CH_2)_2O$ (49%); NO_2, $N(CH_2)_5$ (17%)
240b R^1, NR^2R^2: H, $N(CH_2CH_2)_2O$ (82%); H, $N(CH_2)_5$ (49%)

Scheme 81

the generation of 2-halophenylethenylthiolates **238**, which in the presence of amines give 2-(2-halophenyl)-1-aminoethenylthiolates **239**. The latter are capable of cyclization via intramolecular substitution of the halogen in the benzene ring to give 2-dialkylaminobenzo[*b*]thiophenes **240**. Nucleophilic substitution of the halogen requires activation by a nitro group in the *para*-position (2014RJG2405, 2010T1040, 2007JOC5368) or catalysis by copper iodide (Scheme 81) (2015RJO1040).

It should be noted that 4-substituted-5*H*-1,2,3-thiadiazoles, the starting material for all transformations described in this section, are readily available from the corresponding methyl ketones by HMR (see Section 2.1).

3.2.4 Rhodium-Catalyzed Transannulations of 1,2,3-Thiadiazoles

Gevorgyan et al. have suggested that α-diazothione, a minor open-chain tautomer of 1,2,3-thiadiazole, can be trapped by rhodium to afford the thiavinyl carbenoid. The latter has received increased attention as highly useful precursor for a variety of organic transformations. In this respect, a rhodium-catalyzed transannulation reaction of 1,2,3-thiadiazoles **241** with monosubstituted alkynes, leading to a mixture of isomeric 2,3,4- and 2,3,5-trisubstituted thiophenes **242** and **243**, has been studied. Optimization of the reaction conditions has revealed that the highest yields and highest regioselectivity are observed when the reaction is carried out in an inert atmosphere at 60°C in chlorobenzene and in the presence of [Rh(COD)Cl]$_2$ as catalyst and 1,1′-bis(diphenylphosphino)ferrocene (DPPF) as ligand. Rhodium α-thiovinylcarbenes **244** react with monosubstituted acetylenes mostly with formation of 2,3,5-trisubstituted thiophenes **243** (Scheme 82) (2016OL1804).

Reactions of ethyl 1,2,3-thiadiazole-4-carboxylates **245** with monosubstituted alkenes take place regioselectively to form 2,3,4-trisubstituted 4,5-dihydrothiophenes **246** in 70%–99% yield. The structure of the latter product has been confirmed by X-ray analysis. The dihydrothiophenes **246** can be oxidized to aromatic thiophenes by treatment with 2,3-dichloro-5,6-dicyanobenzoquinone (DDQ). Furthermore, the authors elaborated a one-pot two-step method for the synthesis of thiophenes **247** from easily available thiadiazoles **245** by reaction with alkenes and DDQ (Scheme 83) (2016OL5408).

Lee et al. managed to connect alkene or alkyne moieties to the 1,2,3-thiadiazole ring through different linkers, thus obtaining derivatives of the type **248** and **249**, respectively. This allowed them to carry out intramolecular additions of the intermediate rhodium carbenoids onto double or triple bonds to form bicyclic thiophenes **250** (Scheme 84) (2017JOC1437, 2016OL5408).

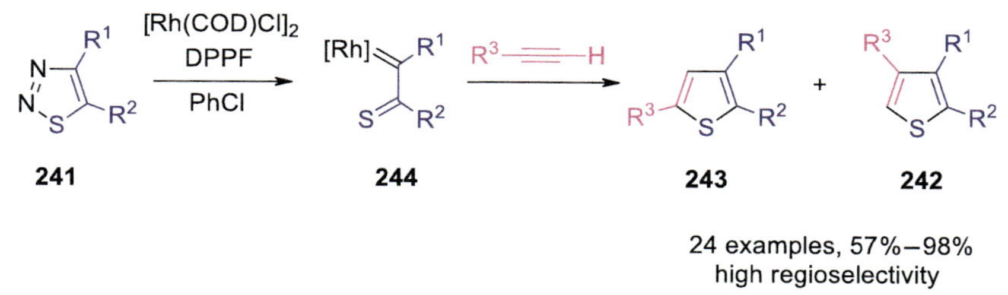

Scheme 82

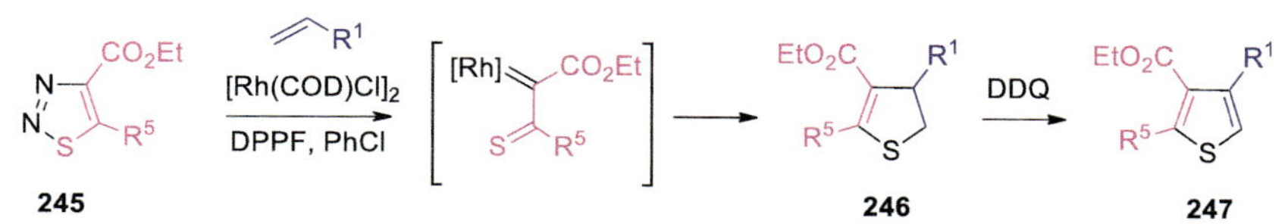

Scheme 83

1,2,3-Thiadiazoles **251** have been shown to react with nitriles in the presence of [Rh(COD)Cl]$_2$ as catalyst and DPPF as ligand to afford isothiazoles **252**, rather than the expected thiazoles (Scheme 85). The authors explained the phenomenon by stating that the α-thiavinyl carbene acts as an umpoled 1,3-dipole equivalent when reacting with a nitrile (2016OL5050).

Thus, application of rhodium catalysts in cleavage of the 1,2,3-thiadiazole ring allows the generation of intermediate rhodium α-thiovinylcarbenes and hence their further [3 + 2]-cycloaddition to different 1,3-dipolarophiles.

The data in this section allow us to conclude that ring transformation and ring cleavage of 1,2,3-thiadiazoles can serve as a valuable source of highly reactive intermediates and constitute original and effective methods for the synthesis of different classes of sulfur-containing organic compounds.

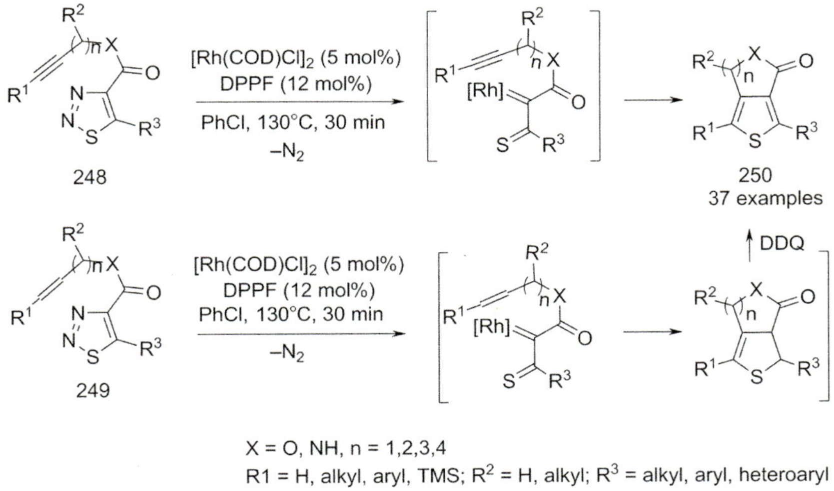

Scheme 84

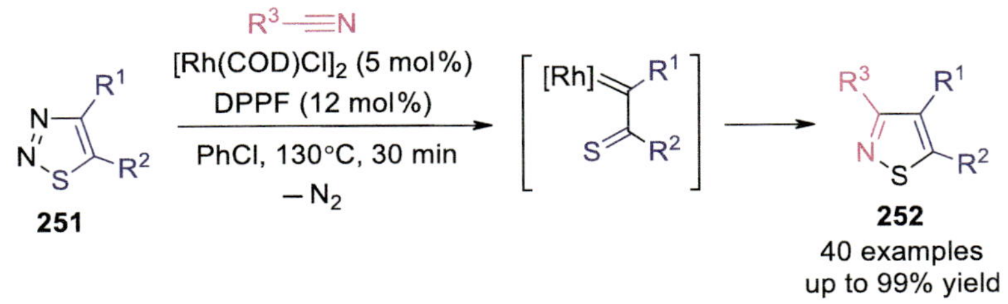

Scheme 85

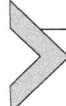

4. 1,2,3-THIADIAZOLES IN MEDICINE AND AGRICULTURE

Many 1,2,3-thiadiazole derivatives have been shown to exhibit antimicrobial, antithrombotic, and cardiovascular activity as well as various types of pesticidal properties (2004M[2]). Scheme 86 shows some derivatives that have either entered the market or display outstanding activity. Thus, cefunozam™, an antibiotic of the cephalosporine series, is used for chemotherapy of bacterial infections and contains a 1,2,3-thiadiazole ring in its structure. bion™, a systemic acquired resistance (SAR) inductor (elicitor) for plants, has an S-methyl benzo[d][1,2,3]thiadiazole-7-carbothioate structure and is used in plant cultivation for protection of dicotyledonous and monocotyledonous plants from fungi, bacterial, and viral diseases (2011SH(225)290, 2011MYB290). In plant biotechnology, thidiazuron™ (N-thiadiazol-5-yl-N'-phenylurea), possessing properties close to naturally occurring cytokinin phytohormones, has been applied for several decades. In recent years, this preparation is used for growth regulation and plant development, particularly for enhancing the growth of cellular cultures and tissues of new hybrid and transgenic plants (2011AJB8984).

Because of the success of thidiazuron™ and bion™ on the market, many researchers have paid attention to the elaboration of new, more efficient methods for the synthesis of new 1,2,3-thiadiazole derivatives and have carried out extensive screening of the compounds prepared. In the last decades, many investigations have aimed at studying the effect of derivatives of 1,2,3-thiadiazoles on the growth and development of plants and to their pesticidal activity, which has brought the discovery of new highly active derivatives.

Scheme 86

Among them is tiadinil™ (N-(3-chloro-4-methylphenyl)-4-methyl-1,2,3-thiadiazole-5-carboxamide), which was introduced into the market as an agent for controlling rice blast disease by Nihon Nohyaku Co., Ltd. in 2003. In 2004, Nakashita et al. showed that tiadinil™ does not possess any antimicrobial effect, but rather acts as an inductor of SAR (elicitor) (2004JPE46). A new inductor of SAR, methiadinil™, has been discovered (2011CNP101544633A, 2016CNP105230635). Recently, several other potential elicitors based on tiadinil™ and methiadinil™ (4-methyl-N-(5-methylthiazol-2-yl)-1,2,3-thiadiazole-5-carboxamide) have been found (2009JFA4279, 2010JFA2755), as well as a great number of various plant-protecting mixtures and compositions thereof (2017CNP107027804, 2014CNP105010357).

Testing of the phytohormonal properties of 1,2,3-thiadiazoles has revealed compounds with herbicidal activity (2010ARK330, 2012JHC732, 2010PS 2024, 2014PS379, 2015PS1884). Also several compounds possessing fungicidal (2009CPB561, 2012CPB25, 2014JHC690, 2010JFA2755, 2013BML 5821, 2011ASJ4064, 2010JFA2630, 2010JFA2755), antibacterial (2013JFA 11952, 2014OBC5911), insecticidal (2009CPB561, 2011JFA628, 2010WOP 127928, 2013CCL1134), and antiviral (2013CCL889, 2013CCL1134, 2010 JFA7846, 2010JFA2755, 2012CCL1233) activities have been found.

The therapeutic properties of 1,2,3-thiadiazoles have also been studied in the reviewed period. As a result, compounds with antiviral (2011WOP94823, 2010EJM(45)1919, 2004EP1137645, 2009BMC5920) as well as with antithrombotic (2009JME6588), antidiabetic (2009USP0036450), and antitumor (2011USP0046066) activities were discovered. For example, it was found that the 1,2,3-thiadiazole derivative **253** has anti-HIV activity higher than the known medicines nevirapine™ and delaviridine™ (2009BMC5920). Peptide-like compound **254** inhibits tumor growth when dosed orally in the MDA-MB-231 breast cancer xenograft model and successfully passed preclinic and clinic trials (2012JME4101).

Several 1,2,3-thiadiazoles were found to control the activity of enzymes (2014WOP100533, 2012USP0238561, 2014BML3278, 2014WOP075393, 2014WOP023754, 2009USP131431) and of the receptors of different biomolecules (2012WOP174199, 2009USP0093472, 2011USP0263612).

5. CONCLUDING REMARKS

The development of the well-known Hurd–Mori synthesis and new methods for the synthesis of 1,2,3-thiadiazoles based on oxidative heterocyclization of hydrazones has made 1,2,3-thiadiazoles available for wide investigations of their chemical and biological properties. This has favored the discovery of new thermal transformations and rearrangements of 1,2,3-thiadiazoles, and the synthesis of fused and 1,2,3-thiadiazoles linearly connected to other heterocycles were discovered in 2003–17. Data on new derivatives of 1,2,3-thiadiazoles exhibiting antiviral, anticancer, antithrombotic, and antidiabetic activities and new plant-activating chemicals against various diseases have appeared in the literature in this period. We believe that the discovery of metal-catalyzed transformation of 1,2,3-thiadiazoles will give new impetus to further development of the chemistry of 1,2,3-thiadiazoles.

ACKNOWLEDGMENT

The authors thank the Russian Scientific Foundation (Project 15-13-10031) for financial support.

REFERENCES

1904LAC333	L. Wolff, *Liebigs. Ann. Chem.*, **333**, 1 (1904).
1955JA5359	C.D. Hurd and R.I. Mori, *J. Am. Chem. Soc.*, **77**, 5359 (1955).
1975JPR317	K. Gewald and U. Hain, *J. Prakt. Chem.*, **317**, 329 (1975).
1982JCS(P1)1223	R.N. Butler and D.A. O'Donoghue, *J. Chem. Soc., Perkin Trans. 1*, 1223 (1982).
1989T7329	V.A. Bakulev, A.T. Lebedev, E.F. Dankova, V.S. Mokrushin, and V.S. Petrosyan, *Tetrahedron*, **45**, 7329 (1989).
1993CHE724	Yu.M. Shafran, V.A. Bakulev, V.A. Shevyrin, and M.Yu. Kolobov, *Chem. Heterocycl. Compd.*, **29**, 724 (1993).
1995JCS(P1)2079	G. L'abbé, P. Vossen, W. Dehaen, and S. Toppet, *J. Chem. Soc., Perkin Trans. 1*, 2079 (1995).
1995RCR99	V.A. Bakulev, *Russ. Chem. Rev.*, **64**, 99 (1995).
1996PS89	M.J. Gil, A. Reliquet, F.M. Reliquet, and C. Meslin, *Phosphorus Sulfur Silicon Relat. Elem.*, **117**, 89 (1996).
1997JHC605	M. Al-Smadi, N. Hanold, and H. Meier, *J. Heterocycl. Chem.*, **34**, 605 (1997).
1998H(48)259	P. Stanetty and M. Kremslehner, *Heterocycles*, **48**, 259 (1998).
1999THS(3)265	P. Stanetty, M. Turner, and M. Mihovilovich, In O.A. Attanasi and D. Spinelli, editors: *Targets in Heterocyclic Systems: Chemistry and Properties*, Società Chimica Italiana: Roma, 1999, **Vol. 3**, p 265.
2000ANH37	W. Dehaen, M. Voets, and V.A. Bakulev, *Adv. Nitrogen Heterocycl.*, **4**, 37 (2000).
2002RJO1510	M.L. Petrov, W. Dehaen, M.A. Abramov, I.P. Abramova, and D.A. Androsov, *Russ. J. Org. Chem.*, **38**, 1510 (2002).
2003BML3513	Y. Li, Y. Xu, X. Qian, and B. Qu, *Bioorg. Med. Chem. Lett.*, **13**, 3513 (2003).
2003BOC288	K. Pacaud, D. Tritsch, A. Burger, and J.-F. Biellmann, *Bioorg. Chem.*, **31**, 288 (2003).
2003CHE679	Yu.Yu. Morzherin, T.V. Glukhareva, and V.A. Bakulev, *Chem. Heterocycl. Compd*, **39**, 679 (2003).

2003FA63	A.R. Jalilian, S. Sattari, M. Bineshmarvasti, M. Daneshtalab, and A. Shafiee, *Farmacoterapia*, **58**, 63 (2003).
2003IJB189	M.S. Gaikwad, A.S. Mane, R.V. Hangarge, V.P. Chavan, and M.S. Shingare, *Indian J. Chem. B*, **42**, 189 (2003).
2003JCC1934	N. Siddiqui and Shamsuzzaman*J. Chil. Chem. Soc.*, **58**, 1934 (2013).
2003JHC149	V. Padmavathi, R. Ramana, V. Tippireddy, R. Audisesha, K. Audisesha Reddy, and D. Bhaskar Reddy, *J. Heterocycl. Chem.*, **40**, 149 (2003).
2003JHC427	M. Shekarchi, F. Ellahiyan, T. Akbarzadeh, and A. Shafiee, *J. Heterocycl. Chem.*, **40**, 427 (2003).
2003M[1]	D.J. Wilkins, P.A. Bradley, in "Science of Synthesis", George Thieme, Stuttgart, 2003, Vol. 13, 253.
2003RJO284	M.L. Petrov, D.A. Androsov, M.A. Abramov, and W. Dehaen, *Russ. J. Org. Chem.*, **39**, 284 (2003).
2003S2559	B. Zaleska, B. Trzewik, J. Grochowki, and P. Serda, *Synthesis*, **2559**, 2003.
2003WOP082862	R.A. Ancliff et al., WO Patent 82862 (2003).
2004EP1137645	J.D. Bloom, K.J. Curran, M.J. Digrandi, R.G. Dushin, S.A. Lang, E.B. Norton, A.A. Ross, and B.M. O'Hara, *European Patent*, **1137645**, 2004.
2004JPE46	M. Yasuda, H. Nakashita, and S. Yoshida, *J. Pest. Sci.*, **29**, 46 (2004).
2004M[2]	V.A. Bakulev and W. Dehaen, *The Chemistry of 1,2,3-Thiadiazoles*, John Wiley & Sons, Inc: USA, 2004.
2004RCB1311	T.V. Glukhareva, Yu.Yu. Morzherin, L.V. Dyudya, K.V. Malysheva, A.V. Tkachev, A. Padwa, and V.A. Bakulev, *Russ. Chem. Bull.*, **53**, 1311 (2004).
2004RJO818	M.L. Vasil'eva, M.V. Mukhacheva, N.P. Bel'skaya, V.A. Bakulev, R.J. Anderson, and P.V. Groundwater, *Russ. J. Org. Chem.*, **40**, 818 (2004).
2004RJO1092	N.I. Medvedeva, O.B. Flekhter, E.V. Tretyakova, F.Z. Galin, L.A. Baltina, L.V. Spirikhin, and G.A. Tolstikov, *Russ. J. Org. Chem.*, **40**, 1092 (2004).
2004RJO1338	P.E. Prokhorova, T.A. Kalinina, T.V. Glukhareva, and Yu.Yu. Morzherin, *Russ. J. Org. Chem.*, **48**, 1338 (2012).
2004RJO1691	M.L. Petrov, D.A. Androsov, and Yu.I. Lyakhovetskii, *Russ. J. Org. Chem.*, **40**, 1691 (2004).
2004RJO870	E.V. Tarasov, N.N. Volkova, Y.Y. Morzherin, and V.A. Bakulev, *Russ. J. Org. Chem.*, **40**(6), 870 (2004).
2004RJO90	O.B. Flekhter, E.V. Tret'yakova, N.I. Medvedeva, L.A. Baltinax, F.Z. Galin, G.A. Tolstikov, and V.A. Bakulev, *Russ. J. Org. Chem.*, **40**, 90 (2004).
2004USP002524	R. Chesworth, and L.D. Gegnas, US Patent 2,524 (2004).
2005BMC1615	Q. Yang, X. Qian, J. Xu, Y. Sun, and Y. Li, *Bioorg. Med. Chem.*, **13**, 1615 (2005).
2005CHE1186	I.B. Dzvinchuk, *Chem. Heterocycl. Compd.*, **41**, 1186 (2005).
2005CHE542	T.V. Glukhareva, L.V. Dyudya, T.A. Pospelova, V.A. Bakulev, A.V. Tkachev, and Yu.Yu. Morzherin, *Chem. Heterocycl. Compd.*, **41**, 542 (2005).
2005EJMC728	T. Balasankar, M. Gopalakrishnan, and S. Nagarajan, *Eur. J. Med. Chem.*, **40**, 728 (2005).
2005IJB783	P. Venkateswarlu and N.R. Vasireddy, *Indian J. Chem. B*, **44**, 783 (2005).

2005MOL367	P. Stanetty, M. Turner, and M.D. Mihovilovic, *Molecules*, **10**, 367 (2005).
2005MOL818	M. Mushfiq, M. Alam, M.S. Akhtar, and S. Mohd, *Molecules*, **10**, 818 (2005).
2005OBC1835	M. Kreis, C.F. Nising, M. Schroen, K. Knepper, and S. Bräse, *Org. Biomol. Chem.*, **3**, 1835–1837 (2005).
2005OBC932	S.L. Bogza, K.I. Kobrakov, A.P. Malienko, I.F. Perepichka, S.Y. Sujkov, M.R. Bryce, S.B. Lyubchik, A.S. Batsanov, and N.M. Bogdan, *Org. Biomol. Chem.*, **3**, 932 (2005).
2005PS1593	L. Karimi, L. Navidpour, M. Amini, and A. Shafiee, *Phosphorus Sulfur Silicon Relat. Elem.*, **180**, 1593 (2005).
2005T6634	Z. Li, Q. Yang, and X. Qian, *Tetrahedron*, **61**, 6634 (2005).
2005USP250783	J.A. Johnson, J. Lloyd, and A. Kover, *US Patent 250,783*, 2005.
2006CHE681	N.S. Joshi, B.K. Karale, and C.H. Gill, *Chem. Heterocycl. Compd.*, **42**, 681 (2006).
2006MC76	P.E. Kropotina, L.V. Dyudya, T.V. Glukhareva, Yu. Yu. Morzherin, V.A. Bakulev, K.V. Hecke, L.V. Meervelt, and W. Dehaen, *Mendeleev Commun.*, **16**, 76 (2006).
2006RJO1735	M.L. Petrov and V.A. Kuznetsov, *Russ. J. Org. Chem.*, **42**, 1735 (2006).
2006RUP2276146	A.M. Uskov, L.M. Nesterova, S.V. Yarovenko, and L.K. Fan, Russian Federation Patent Granted 2276146 (2006).
2006S2573	T. Kalai, B. Bognar, J. Jeko, and K. Hideg, *Synthesis*, 2573 (2006).
2006TL4129	H.T. Yang, G.W. Wang, Y. Xu, and J.C. Huang, *Tetrahedron Lett.*, **47**, 4129 (2006).
2007ASJC1783	M. Al-Smadi, *Asian. J. Chem.*, **19**, 1783 (2007).
2007BML4694	F.E. Boyer, et al., *Bioorg. Med. Chem. Lett.*, **17**, 4694 (2007).
2007BML869	M. Wu, Q. Sun, C. Yang, D. Chen, J. Ding, Y. Chen, L. Lin, and Y. Xie, *Bioorg. Med. Chem. Lett.*, **17**, 869 (2007).
2007EP1852428	K. Umetani, T. Shimaoka, M. Yamaguchi, M. Oda, N. Kyomura, T. Takemoto, and K. Kikutake, European Patent EP 1852428 (2007).
2007JHC1165	V. Padmavathi, K. Mahesh, P. Thriveni, and T.V.R. Reddy, *J. Heterocycl. Chem.*, **44**, 1165 (2007).
2007JOC5368	D.A. Androsov and D.C. Neckers, *J. Org. Chem.*, **72**, 5368 (2007).
2007MCR392	M. Gopalakrishnan, J. Thanusu, and V. Kanagarajan, *Med. Chem. Res.*, **16**, 392 (2007).
2007RJO1870	D.A. Androsov, M.L. Petrov, and A.A. Shchipalkin, *Russ. J. Org. Chem.*, **43**, 1870 (2007).
2007RJO630	M.L. Petrov, A.A. Shchipalkin, and V.A. Kuznetsov, *Russ. J. Org. Chem.*, **43**, 630 (2007).
2007SL2513	H. Ying, Y. Fan, T. Liu, and Y. Hu, *Synlett*, (16), 2513 (2007).
2007T3042	M.L. Kondratieva, A.V. Pepeleva, N.P. Belskaia, A.V. Koksharov, P.V. Groundwater, K. Robeyns, L. Van Meervelt, W. Dehaen, Z.-J. Fan, and V.A. Bakulev, *Tetrahedron*, **63**, 3042 (2007).
2007USP185165	A.P. Combs, A. Takvorian, W. Zhu, and R.B. Sparks, *US Patent 185,165*, 2007.
2007WOP017261	R.J.D. Hatley, A.M. Mason, and I.L. Pinto, WO Patent 17261 (2007).
2007WOP36733	C.D. Edlin, S. Holman, P.S. Jones, S.E. Keeling, M.K. Lindvall, C.J. Mitchell, and N. Trivedi, WO Patent 36733 (2007).

2008CHE233	P.E. Kropotina, T.V. Glukhareva, I.S. Isakova, E.A. Alekseeva, and Yu.Yu. Morzherin, *Chem. Heterocycl. Compd.*, **44**, 233 (2008).
2008CHECIII(5)468	D.J. Wilkins, In A.R. Katritzky, C.A. Ramsden, E.F.V. Scriven, and R.J.K. Taylor, editors: *Comprehensive Heterocyclic Chemistry III*, **5**, Elsevier: Oxford, 2008, **5**, p 468.
2008EP1970377	E. Umemura, Y. Wakiyama, K. Ueda, K. Kumura, S. Masaki, T. Watanabe, M. Yamamoto, C. Kaji, and K. Ajito, European Patent EP 1970377 (2008).
2008JME3005	S. Biswas, S. Hazeldine, B. Ghosh, I. Parrington, E. Kuzhikandathil, M.E.A. Reith, and A.K. Dutta, *J. Med. Chem.*, **51**, 3005 (2008).
2008M1067	W. Zhu, Z. Zhao, and Y. Xu, *Monatsh. Chem.*, **139**, 1067 (2008).
2008RJO1089	M.L. Petrov, A.A. Shchipalkin, and V.A. Kuznetsov, *Russ. J. Org. Chem.*, **44**, 1089 (2008).
2008RJO774	A.A. Shchipalkin, M.L. Petrov, and V.A. Kuznetsov, *Russ. J. Org. Chem.*, **44**, 774 (2008).
2008SC4407	Z. Liu, Y. Mu, J. Lin, and Y. Chen, *Synth. Commun.*, **38**, 4407 (2008).
2008WOP011045	S.X. Cai, J.A. Drewe, H.-Z. Zhang, S. Khasibhatla, N.S. Sirisoma, and W.E. Kemnitzer, WO Patent 11045 (2008).
2008WOP042571	R. Cadilla, A. Larkin, D.L. McDougald, A.S. Randawa, J.A. Ray, K. Stetson, E.L. Stewart, P.S. Turnbull, and H. Zhou, WO Patent 42571 (2008).
2009BMC5920	P. Zhan, X. Liu, Z. Li, Z. Fang, Z. Li, D. Wang, C. Pannecouque, and E. De Clercq, *Bioorg. Med. Chem.*, **17**, 5920 (2009).
2009BML1089	I. Cikotiene, E. Kazlauskas, J. Matuliene, V. Michailoviene, J. Torresan, J. Jachno, and D. Matulis, *Bioorg. Med. Chem. Lett.*, **19**, 1089 (2009).
2009CPB561	V. Padmavathi, K. Mahesh, A.V.N. Mohan, and A. Padmaja, *Chem. Pharm. Bull.*, **57**, 561 (2009).
2009EJI137	E. Cerrada, M. Laguna, and N. Lardies, *Eur. J. Inorg. Chem.*, 137 (2009).
2009JFA4279	Z. Fan, Z. Shi, H. Zhang, X. Liu, L. Bao, L. Ma, X. Zuo, Q. Zheng, and N. Mi, *J. Agric. Food Chem.*, **57**, 4279 (2009).
2009JIC259	E. Cerrada, M. Laguna, and N. Lardíes, *Eur. J. Inorg. Chem.*, **2009**, 137 (**2009**).
2009JME6588	M. Jones, et al., *J. Med. Chem.*, **52**, 6588 (2009).
2009RJO1727	M.L. Petrov, F.S. Teplyakov, D.A. Androsov, and M. Yekhlef, *Russ. J. Org. Chem.*, **45**, 1727 (2009).
2009S2539	M. Al-Smadi, N. Hanold, H. Kalbitz, and H. Meier, *Synthesis*, 2539 (2009).
2009T9989	J.K. Augustine, V. Vairaperumal, S. Narasimhan, P. Alagarsamy, and A. Radhakrishnan, *Tetrahedron*, **65**, 9989 (2009).
2009USP0036450	M. Takagi, T. Nakamura, I. Matsuda, T. Kiguchi, N. Ogawa, and H. Ozeki, *US Patent 36,450*, 2009.
2009USP0093472	V.M. Paradkar, M. Brescia, R. James, J. Liu, R. Liu, J.R. Merritt, M. Morris, M.J. Ohlmeyer, C. Zhang, and R. Zhang, *US Patent*, **93**, 472 (2009).
2009USP131431	C.D. Edlin, S. Holman, P.S. Jones, S.E. Keeling, M.K. Lindvall, C.J. Mitchell, and N. Trivedi, *US Patent 131,431*, 2009.

2009WOP134110	D. Matulis, I. Cikotiene, E. Kazlauskas, and J. Matuliene, WO Patent 134110 (2009).
2010ARK275	N.P. Belskaya, W. Dehaen, and V.A. Bakulev, *Arkivoc*, **i**, 275 (2010).
2010ARK330	T.-T. Wang, G.-F. Bing, X. Zhang, Z.-F. Qin, H.-B. Yu, X. Qin, H. Dai, and J.-X. Fang, *ARKIVOC*, **ii**, 330 (2010).
2010EJM(45)1919	W.-L. Dong, Z.-X. Liu, X.-H. Liu, Z.-M. Li, and W.-G. Zhao, *Eur. J. Med. Chem.*, **45**, 1919 (2010).
2010EJM4920	S.M.S. Atta, D.S. Farrag, A.M.K. Sweed, and A.H. Abdel-Rahman, *Eur. J. Med. Chem.*, **45**, 4920 (2010).
2010EJMC6127	F. Hayat, A. Salahuddin, J. Zargan, and A. Azam, *Eur. J. Med. Chem.*, **45**, 6127 (2010).
2010HAC131	A. Siwek, J. Stefańska, I. Wawrzycka-Gorczyca, and M. Wujec, *Heteroat. Chem.*, **21**, 131 (2010).
2010HAC521	A. Siwek, M. Wujec, M. Dobosz, and I. Wawrzycka-Gorczyca, *Heteroat. Chem.*, **21**, 521 (2010).
2010JFA2630	Z. Fan, Z. Yang, H. Zhang, N. Mi, H. Wang, F. Cai, X. Zuo, Q. Zheng, and H. Song, *J. Agric. Food Chem.*, **58**, 2630 (2010).
2010JFA2755	X. Zuo, N. Mi, Z. Fan, Q. Zheng, H. Zhang, H. Wang, and Z. Yang, *J. Agric. Food Chem.*, **58**, 2755 (2010).
2010JFA7846	Q. Zheng, N. Ma, Z. Fan, X. Zuo, H. Zhang, H. Wang, and Z. Yang, *J. Agric. Food Chem.*, **58**, 7846 (2010).
2010PS1594	A.D. Shinde, S.S. Sonar, B.B. Shingate, and M.S. Shingare, *Phosphorus Sulfur Silicon Relat. Elem.*, **185**, 1594 (2010).
2010PS2024	W. Tang, Z.-H. Yu, and D.-Q. Shi, *Phosphorus Sulfur Silicon Relat. Elem.*, **185**, 2024 (2010).
2010PS361	N. Guravaiah and V.R. Rao, *Phosphorus Sulfur Silicon Relat. Elem.*, **185**, 361 (2010).
2010RCB867	P.E. Prokhorova, T.V. Glukhareva, L.V. Dyudya, E.A. Alekseeva, and Yu.Yu. Morzherin, *Russ. Chem. Bull.*, **59**, 867 (2010).
2010RJO1214	M.L. Petrov, A.A. Shchipalkin, V.A. Kuznetsov, and B.N. Viktorov, *Russ. J. Org. Chem.*, **46**, 1214 (2010).
2010T1040	D.A. Androsov, A.Y. Solovyev, M.L. Petrov, R.J. Butcher, and J.P. Jasinski, *Tetrahedron*, **66**, 2474 (2010).
2010T5472	M. Turner, T. Linder, M. Schnürch, M.D. Mihovilovic, and P. Stanetty, *Tetrahedron*, **66**, 5472 (2010).
2010WOP043893	K. Barvian, G.S. Basarab, M.R. Gowravaram, S.I. Hauck, and F. Zhou, WO Patent 43893 (2010).
2010WOP127928	P. Maienfisch, C. Godfrey, C.R. Ayles, P.J.M. Jung, O.F. Hueter, and P. Renold, WO Patent, 127928 (2010).
2011AJB8984	B. Guo, B.H. Abbasi, A. Zeb, L.L. Xu, and Y.H. Wei, *Afr. J. Biotechnol.*, **10**, 8984 (2011).
2011ASJ4064	X.-H. Liu, J.-Q. Weng, and C.-X. Tan, *Asian J. Chem.*, **23**, 4064 (2011).
2011CNP101544633A	Z. Fan, Z. Yang, X. Zuo, Z. Fan, Q. Wu, Q. Zheng, and H. Zhang, *Chinese Patent*, 101544633 (2011A).
2011JFA628	H. Wang, Z. Yang, Z. Fan, Q. Wu, Y. Zhang, N. Mi, S. Wang, Z. Zhang, H. Song, and F. Liu, *J. Agric. Food Chem.*, **59**, 628 (2011).
2011JPC(A)14300	G. Burdzinski, M. Sliwa, Yu. Zhang, and S. Delbaere, *J. Phys. Chem. A.*, **115**, 14300 (2011).
2011MYB290	M.F. Abdel-Monaim, M.E. Ismail, and K.M. Morsy, *Mycrobiology*, **39**, 290 (2011).

2011PAC715	Yu.Yu. Morzherin, P.E. Prokhorova, D.A. Musikhin, T.V. Glukhareva, and Zh. Fan, *Pure Appl. Chem.*, **83**, 715 (2011).
2011RCB896	N.P. Bel'skaya, A.I. Bolgova, M.L. Kondrat'eva, O.S. El'tsov, and V.A. Bakulev, *Russ. Chem. Bull. Int. Ed.*, **60**, 896 (2011).
2011RCB981	T.A. Kalinina, P.E. Prokhorova, T.V. Glukhareva, and Yu. Yu. Morzherin, *Russ. Chem. Bull.*, **60**, 981 (2011).
2011RJO1191	A.A. Shchipalkin, M.L. Petrov, and V.A. Kuznetsov, *Russ. J. Org. Chem.*, **47**, 1191 (2011).
2011RJO1777	M.L. Petrov, A.A. Shchipalkin, and V.A. Kuznetsov, *Russ. J. Org. Chem.*, **47**, 1777 (2011).
2011RJO1878	A.A. Shchipalkin, M.L. Petrov, and V.A. Kuznetsov, *Russ. J. Org. Chem.*, **47**, 1878 (2011).
2011S2154	M.E. Trusova, E.A. Krasnokutskaya, P.S. Postnikov, Y. Choi, K.-W. Chi, and V.D. Filimonov, *Synthesis*, 2154 (2011).
2011SH(225)290	S. Farouk, B.E.A. Belal, and H.H.A. El-Sharkawy, *Sci. Hortic*, **225**, 646 (2017).
2011TMC490	M. Al-Noaimi, M. Al-Smadi, S.F. Haddad, S. Al-Omari, A. Haniyeh, and A.M. Rawashdeh, *Transit. Met. Chem.*, **36**, 409 (2011).
2011USP0046066	C. Ndubaku, J.A. Flygare, and F. Cohen, *US Patent*, **46**, 066 (2011).
2011USP0263612	J.P. Whitten, Y. Pei, Z. Wang, E. Rogers, B. Dyck, and J. Grey, *US Patent*, **263**, 612 (2011).
2011WOP079118	R.A. Galemmo et al., WO Patent 079118 (2011).
2011WOP25859	F. Pierre, M. Haddach, C.F. Reganm, and D.M. Ryckman, WO Patent 25859 (2011).
2011WOP29459	S. Hell, V.N. Belov, V.P. Boyarskiy, C. Wurm, S. Jakobs, and C. Geisler, WO Patent 29459 (2011).
2011WOP94823	J.P. Mitchell, G. Pitt, A.G. Draffan, P.A. Mayes, L. Andrau, and K. Anderson, *WO Patent*, 94823 (2011).
2012BML7237	H.-B. Hea, L.-X. Gaoc, Q.-F. Dengb, W.-P. Mac, C.-L. Tangc, W.-W. Qiub, J. Tanga, J.-Y. Lic, J. Lic, and F. Yanga, *Bioorg. Med. Chem. Lett.*, **22**, 7237 (2012).
2012CCL1233	W.T. Mao, H. Zhao, Z.J. Fan, X.T. Ji, X.W. Hua, T. Kalinina, Yu.Yu. Morzherin, and V.A. Bakulev, *Chin. Chem. Lett.*, **23**, 1233 (2012).
2012CPB25	D. Guo, Z. Wang, Z. Fan, H. Zhao, W. Zhang, J. Cheng, J. Yang, Q. Wu, Y. Zhang, and Q. Fan, *Chin. J. Chem.*, **30**, 2522 (2012).
2012JC1276	M.A. Hosny, T.H. El-Sayed, and E.A. El-Sawi, *E-J. Chem.*, **9**, 1276 (2012).
2012JFA346	D. Qingshan, Z. Weiping, Z. Zhenjiang, Q. Xuhong, and X. Yufang, *J. Agric. Food Chem.*, **60**, 346 (2012).
2012JHC732	F. Cheng and D.-Q. Shi, *J. Heterocycl. Chem.*, **49**, 732 (2012).
2012JME3122	J. Xu, Z. Li, J. Luo, F. Yang, T. Liu, M. Liu, W.-W. Qiu, and J. Tang, *J. Med. Chem.*, **55**, 3122 (2012).
2012JME4101	J.A. Flygare, et al., *J. Med. Chem.*, **55**, 4101 (2012).
2012JME8271	V. Bavetsias, et al., *J. Med. Chem.*, **55**, 8271 (2012).
2012JOC9391	A. Kumar, M.K. Muthyala, S. Choudhary, R.K. Tiwari, and K. Parang, *J. Org. Chem.*, **77**, 9391 (2012).
2012RCI1999	X.-H. Liu, W.-G. Zhao, B.-L. Wang, and Z.-M. Li, *Res. Chem. Intermed.*, **38**, 1999 (2012).
2012RJO1333	P.E. Prokhorova, T.A. Kalinina, T.V. Glukhareva, and Y.Y. Morzherin, *Russ. J. Org. Chem.*, **48**(10), 1333–1336 (2012).

2012RJO1338	P.E. Prokhorova, T.A. Kalinina, T.V. Glukhareva, and Yu. Yu. Morzherin, *Russ. J. Org. Chem.*, **48**, 1338 (2012).
2012RJO728	M.L. Petrov, M. Yekhlef, F.S. Teplyakov, and D.A. Androsov, *Russ. J. Org. Chem.*, **48**, 728 (2012).
2012USP0238561	G.J. MacDonald, B.C. Albert, G. De Boeck, and J.E. Leenaerts, *US Patent*, **238**, 561 (2012).
2012WOP174199	H.D. Fiji, M.J. Kelly III, III, J.C. Kern, M.E. Layton, J.E. Pero, A.J. Reif, and M.A. Ross, *WO Patent*, **174199**, 2012.
2012WOP47617	C.P. Miller, *WO Patent*, **47617**, 2012.
2013BML5821	W.-M. Xu, S.-Z. Li, M. He, S. Yang, X.-Y. Li, and P. Li, *Bioorg. Med. Chem. Lett.*, **23**, 5821 (2013).
2013CCL1134	Y.-D. Li, et al., *Chin. Chem. Lett.*, **24**, 1134 (2013).
2013CCL889	S.-X. Wang, Z. Fang, Z.-J. Fan, D. Wang, Y.-D. Li, X.-T. Ji, X.-W. Hua, Y. Huang, T.A. Kalinina, V.A. Bakulev, and Yu. Yu. Morzherin, *Chin. Chem. Lett.*, **24**, 889 (2013).
2013CHE1880	V.A. Bakulev, I.V. Efimov, N.A. Belyaev, S.S. Zhidovinov, Y.A. Rozin, N.N. Volkova, A.A. Khabarova, and O.S. El'tsov, *Chem. Heterocycl. Compd.*, **48**, 1880 (2013).
2013CHE350	T.A. Kalinina, D.V. Shatunova, T.V. Glukhareva, and Yu. Yu. Morzherin, *Chem. Heterocycl. Compd.*, **49**, 350 (2013).
2013CMC1295	D.E. Aubele, et al., *Chem. Med. Chem.*, **8**, 1295 (2013).
2013CNP103058956	L.V. Huan, M.A. Wenyuan, and Zhang Xinghuan, *Chinese Patent* CN103058956 (2013).
2013GC954	M.S. Singh, A. Nagaraju, G.K. Verma, G. Shukla, R.K. Verma, A. Srivastava, and K. Raghuvanshi, *Green Chem.*, **15**, 954–962 (2013).
2013IJC427	N. Ahmed, P.K. Bhattacharyya, and S.K. Bhattacharjee, *Int. J. Chem.*, **2**, 427 (2013).
2013JFA11952	P. Kodisundaram, S. Amirthaganesan, and T. Balasankar, *J. Agric. Food Chem.*, **61**, 2013. 11952.
2013JHC630	C. Rajitha, P.K. Dubey, V. Sunku, V.R. Veeramaneni, and M. Pal, *J. Heterocycl. Chem.*, **50**, 630 (2013).
2013JME2110	L. Feng, et al., *J. Med. Chem.*, **56**, 2110 (2013).
2013JPC(A)4551	G. Burdzinski, H.L. Luk, C.S. Reid, Yu. Zhang, C.M. Hadad, and M.S. Platz, *J. Phys. Chem. A.*, **117**, 4551 (2013).
2013LDDD2	A. Siwek, P. Stączek, A. Strzelczyk, and J. Stefańska, *Lett. Drug Des. Discovery*, **10**, 2 (2013).
2013MOL12725	N.-B. Sun, J.-Q. Fu, J.-Q. Weng, J.-Z. Jin1, C.-X. Tan, and X.-H. Liu, *Molecules*, **18**, 2013. 12725.
2013OL4038	F.S. Teplyakov, T.G. Vasileva, M.L. Petrov, and D.A. Androsov, *Org. Lett.*, **15**, 4038 (2013).
2013PPS895	G. Burdzinski, M. Sliwa, Yu. Zhang, S. Delbaere, T. Pedzinski, and J. Réhault, *Photochem. Photobiol. Sci.*, **12**, 895 (2013).
2013RJO479	M.L. Petrov and D.A. Androsov, *Russ. J. Org. Chem.*, **49**, 479 (2013).
2013T7423	A.I. Bolgova, K.I. Lugovik, J.O. Subbotina, P.A. Slepukhin, V.A. Bakulev, and N.P. Belskaya, *Tetrahedron*, **69**, 7423 (2013).
2014BML3278	B. Evranos-Aksoz, S. Yabanoglu-Ciftci, G. Ucar, K. Yelekci, and R. Ertan, *Bioorg. Med. Chem. Lett.*, **24**, 3278 (2014).
2014CBD223	C. Gao, F.-J. Dai, H.-W. Cui, S.-H. Peng, Y. He, X. Wang, Z.-F. Yi, and W.-W. Qiu, *Chem. Biol. Drug Des.*, **84**, 223 (2014).
2014CHE972	L.N. Dianova, V.S. Berseneva, O.S. El'tsov, Z.-J. Fan, and V.A. Bakulev, *Chem. Heterocycl. Compd.*, **50**, 972 (2014).

2014CNP105010357	J. Lv, Y. Gu, X. Chen, and X. Hu, *Chinese Patent*, **105010357**, 2014.
2014EJO3684	I. Efimov, et al., *Eur. J. Org. Chem.*, (17), 3684 (2014).
2014JHC690	C.-X. Tan, Y.-X. Shi, J.-Q. Weng, X.-H. Liu, W.-G. Zhao, and B.-J. Lic, *J. Heterocycl. Chem.*, **51**, 690 (2014).
2014OBC5911	K. Paulrasu, A. Duraikannu, M. Palrasu, A. Shanmugasundaram, M. Kuppusamy, and B. Thirunavukkarasu, *Org. Biomol. Chem.*, **12**, 5911 (2014).
2014PS1895	G.-X. Sun, M.-Y. Yang, Z.-H. Sun, H.-K. Wu, X.-H. Liu, and Y.-Y. Wei, *Phosphorus Sulfur Silicon Relat. Elem.*, **189**, 1895 (2014).
2014PS379	L.-J. Min, C.-X. Tan, J.-Q. Weng, and X.-H. Liu, *Phosphorus Sulfur Silicon Relat. Elem.*, **189**, 379 (2014).
2014PSS379	L.-J. Min, C.-X. Tan, J.-Q. Weng, and X.-H. Liu, *Phosphorus Sulfur Silicon Relat. Elem.*, **189**, 379 (2014).
2014RJC1863	M. Yekhlef, M.L. Petrov, and L.M. Pevzner, *Russ. J. Gen. Chem.*, **84**, 1863 (2014).
2014RJG2405	D.A. Androsov, E.A. Popova, M.L. Petrov, and A.I. Ponyaev, *Russ. J. Gen. Chem.*, **84**, 2405 (2014).
2014TL2430	A. Nagaraju, B.J. Ramulu, G. Shukla, A. Srivastava, G.K. Verma, K. Raghuvanshi, and M.S. Singh, *Tetrahedron Lett.*, **55**, 2430 (2014).
2014WOP023754	S. Altamura, et al., *WO Patent*, **23754**, 2014.
2014WOP075393	A. Abe Achab, et al., *WO Patent*, **75393**, 2014.
2014WOP100533	U. Velaparthi, C.P. Darne, D.S. Dodd, A.J. Sampognaro, M.D. Wittman, S. Kumaravel, D. Mullick, C. Reddy R, and P. Liu, *WO Patent*, **100533**, 2014.
2015CHE589	T.A. Kalinina, O.A. Bystrykh, V.A. Pozdina, T.V. Glukhareva, M.V. Ulitko, and Yu.Yu. Morzherin, *Chem. Heterocycl. Compd.*, **51**, 589 (2015).
2015PS1884	L.-J. Min, M.-Y. Yang, J.-X. Mu, Z.-X. Sun, C.-X. Tan, J.-Q. Weng, X.-H. Liu, and Y.-G. Zhang, *Phosphorus Sulfur Silicon Relat. Elem.*, **190**, 1884 (2015).
2015RJC600	K.V. Kuticheva, L.M. Pevzner, and M.L. Petrov, *Russ. J. Gen. Chem.*, **85**, 600 (2015).
2015RJC866	L.M. Pevzner, R. Maadadi, and M.L. Petrov, *Russ. J. Gen. Chem.*, **85**, 866 (2015).
2015RJO1040	M.L. Petrov, E.A. Popova, and D.A. Androsov, *Russ. J. Org. Chem.*, **51**, 1040 (2015).
2015TL1545	Y. Rozin, et al., *Tetrahedron Lett.*, **56**, 1545 (2015).
2015WOP022073	F. Voss, S. Ritter, S. Nordhoff, S. Wachten, S. Oberborsch, and A. Kless, WO Patent 22073 (2015).
2016CHE206	N.A. Belyaev, T.V. Beryozkina, and V.A. Bakulev, *J. Heterocycl. Comp.*, **52**, 206 (2016).
2016CHE743	I.V. Efimov, Y.M. Shafran, N.N. Volkova, N.A. Beliaev, P.A. Slepukhin, and V.A. Bakulev, *Chem. Heterocycl. Compd.*, **52**, 743 (2016).
2016CM6390	Z. Chen, et al., *Chem. Mater.*, **28**, 6390 (2016).
2016CNP105230635	Y. Huang, F. Qiu, and Y. Yin, *Chinese Patent*, **105230635**, 2017.
2016CNP105294598	Y. Yang, Z. Ding, S. Wang, X. Xu, H. Xu, J. Yang, J. Rui, X. Cao, Y. Wang, Z. Wang, L. Yang, Q. Zhang, N. Sun, Y. Zhang, H. Quang, and W. Gu, Chineese Patent CN105294598 (2016).

2016EJM245	N.P. Belskaya, K.I. Lugovik, V.A. Bakulev, J. Bauer, I. Kitanovic, P. Holenya, M. Zakhartsev, and S. Wölfl, *Eur. J. Med. Chem.*, **108**, 245 (2016).
2016JEIMC481	T. Fralczek, A. Paneth, R. Kaminski, A. Krakowiak, and P. Paneth, *J. Enzyme Inhib. Med. Chem.*, **31**, 481 (2016).
2016JHC206	N.A. Belyaev, T.V. Beryozkina, and V.A. Bakulev, *J. Heterocycl. Comp.*, **52**, 206 (2016).
2016JME8549	F. Micheli, et al., *J. Med. Chem.*, **59**, 8549 (2016).
2016JOC271	J. Chen, Y. Jiang, J.-T. Yu, and J. Cheng, *J. Org. Chem.*, **81**, 271 (2016).
2016OL1258	W. Fan, Q. Li, Y. Li, H. Sun, B. Jiang, and G. Li, *Org. Lett.*, **18**, 1258 (2016).
2016OL1804	D. Kurandina and V. Gevorgyan, *Org. Lett.*, **18**, 1804 (2016).
2016OL5050	B. Seo, Y.G. Kim, and P.H. Lee, *Org. Lett.*, **18**, 5050 (2016).
2016OL5408	J.-Y. Son, J. Kim, S.H. Han, S.H. Kim, and P.H. Lee, *Org. Lett.*, **18**, 5408 (2016).
2016RGC1762	M. Yekhlef, M.L. Petrov, and L.M. Pevzner, *Russ. J. Gen. Chem.*, **86**, 1762 (2016).
2016WOP67043	S. Cremonesi, F. Micheli, T. Semeraro, and L. Tarsi, WO Patent 67043 (2016).
2017ACSC4986	T. Ishikawa, M. Kimura, T. Kumoi, and H. Iida, *ACS Catal.*, **7**, 4986 (2017).
2017BMC2336	K. Ono, H. Banno, M. Okaniwa, T. Hirayama, N. Iwamura, Y. Hikichi, S. Murai, M. Hasegawa, Y. Hasegawa, K. Yonemori, A. Hata, K. Aoyama, and D.R. Cary, *Bioorg. Med. Chem.*, **25**, 2336 (2017).
2017CCL372	J.-P. Zhang, Y.-G. Qin, Y.-W. Dong, D.-L. Song, H.-X. Duan, and X.-L. Yang, *Chin. Chem. Lett.*, **28**, 372 (2017).
2017CHE236	O.A. Vysokova, T.A. Kalinina, M.A. Tokareva, T.A. Pospelova, T.V. Glukhareva, and Y.Y. Morzherin, *Chem. Heterocycl. Compd.*, **53**, 236 (2017).
2017CNP107027804	Y. Song, Ch. Wang, R. Dai, N. Wei, Zh. Qiao, F. Duan, Yu. Zhang, and T. Pang, *Chinese Patent*, **107027804**, 2017.
2017EJM48	G. Wei, Y. Wu, X.-L. He, T. Liu, M. Liu, J. Luo, and W.-W. Qiu, *Eur. J. Med. Chem.*, **131**, 48 (2017).
2017EP3133075	M. Okaniwa, H. Banno, T. Hirayama, D.R. Cary, K. Ono, and N. Iwamura, European Patent EP3133075 (2017).
2017JCP224302	X.-Y. Liu, Y.-G. Fang, B.-B. Xie, W.-H. Fang, and G.L. Cui, *J. Chem. Phys.*, **146**, 224302 (2017).
2017JHC137	T.A. Kalinina, O.A. Bystrykh, T.V. Glukhareva, and Yu. Yu. Morzherin, *J. Heterocycl. Chem.*, **54**, 137 (2017).
2017JOC1437	J.E. Kim, J. Lee, H. Yun, Y. Baek, and P.H. Lee, *J. Org. Chem.*, **82**, 1437 (2017).
2017JOC4056	V.O. Filimonov, L.N. Dianova, K.A. Galata, T.V. Beryozkina, M.S. Novikov, V.S. Berseneva, O.S. Eltsov, A.T. Lebedev, P.A. Slepukhin, and V.A. Bakulev, *J. Org. Chem.*, **82**, 4056 (2017).
2017JPB(145)315	D.M. Mazur, M.E. Zimens, V.A. Bakulev, and A.T. Lebedev, *J. Pharm. Biomed. Anal.*, **145**, 315 (2017).
2017RJG259	R. Maadadi, L.M. Pevzner, and M.L. Petrov, *Russ. J. Gen. Chem.*, **87**, 259 (2017).
2017USP110666	A. Facchetti, Z. Chen, and J.E. Brown, *US Patent 110,666*, 2017.

CHAPTER FOUR

The Literature of Heterocyclic Chemistry, Part XVI, 2016

Leonid I. Belen'kii[1], Yulia B. Evdokimenkova

Russian Academy of Sciences, Moscow, Russian Federation
[1]Corresponding author: e-mail address: libel31@mail.ru

Contents

1. Introduction	174
2. General Sources and Topics	175
2.1 General Books and Reviews	175
2.2 General Topics by Reaction Type	176
2.3 Specialized Heterocycles	194
2.4 Natural and Synthetic Biologically Active Heterocycles	195
3. Three-Membered Rings	207
3.1 One Heteroatom	207
4. Four-Membered Rings	208
4.1 One Nitrogen Atom	208
4.2 One Oxygen or Sulfur Atom	208
5. Five-Membered Rings	208
5.1 One Heteroatom	208
5.2 Two Heteroatoms	212
5.3 Three Heteroatoms	214
5.4 Four Heteroatoms	215
6. Six-Membered Rings	215
6.1 One Heteroatom	215
6.2 Two Heteroatoms	217
6.3 Three Heteroatoms	220
7. Rings With More Than Six Members	220
7.1 Seven-Membered Rings	220
7.2 Large Rings	221
8. Heterocycles Containing Unusual Heteroatoms	222
8.1 Phosphorus Heterocycles	222
8.2 Boron Heterocycles	222
8.3 Silicon Heterocycles	223
8.4 Metallacycles	223
Appendix	223
References	224

Abstract

A systematized survey of reviews and monographs published in 2016 on all aspects of heterocyclic chemistry is given.

Keywords: Five-membered rings, Four-membered rings, Heterocycles containing unusual heteroatoms, Rings with more than six members, Six-membered rings, Three-membered rings

1. INTRODUCTION

This survey is a sequel to 15 already published [(66AHC(7)225), (79AHC(25)303), (88AHC(44)269), (92AHC(55)31), (98AHC(71)291), (99AHC(73)295), (01AHC(79)199), (04AHC(87)1), (06AHC(92)145), (11AHC(102)1), (13AHC(108)195), (14AHC(111)147), (15AHC(116)193), (17AHC(122)245), (18AHC(124)121)]. It includes monographs and reviews published during 2016, as well as some published earlier but omitted in Part XV.

As in Parts III–XV, sources not only in English but also in Russian, Japanese, Chinese, and other languages are surveyed and classified. This feature of the survey should cause no problem because some of the sources are available in English translations (references to the latter are given for reviews published in Russian after those to original publications) and practically all others have informative English abstracts as well as quite understandable and useful schemes and lists of references. As before, carbohydrates are not covered. Such compounds are mentioned only in general cases (e.g., anomeric effect) as well as when carbohydrates serve as starting compounds for the synthesis of other heterocycles or they are present as fragments of a complex system including another heterocyclic moiety, such as nucleosides.

An avalanche-like growth of the number of reviews makes it difficult to publish a full version of the survey. Therefore, the volume was diminished by a partial change of subdivisions, elimination of some doubling, and in some cases low-informative references as well as those having purely medicinal or commercial contents. All excluded data might not be lost for readers who can get them by e-mail as "Supplementary materials" from the authors libel31@mail.ru or u_ksenoff@mail.ru and/or https://doi.org/10.1016/bs.aihch.2018.02.003.

2. GENERAL SOURCES AND TOPICS
2.1 General Books and Reviews
2.1.1 Textbooks and Handbooks
Arene chemistry (general monograph): (16MI1).
Chemistry of heteroaromatic compounds (textbook): (15MI1).
Organic nanoreactors: (16MI2).
NMR data interpretation: (16MI3).
Organic redox systems: (16MI4).
Cross conjugation: (16MI5).
Applications of supramolecular chemistry: (16MI6).
Encapsulation phenomenon: (16MI7).
Milestones in microwave chemistry: (16MI8).
Cycloadditions in bioorthogonal chemistry: (16MI9).
Electrochemistry of N_4 macrocyclic metal complexes: (16MI10).

2.1.2 Annual Reports
Three-membered ring systems: (16PHC57).
Four-membered ring systems: (16PHC95).
Thiophenes and Se/Te derivatives: (16PHC121).
Pyrroles and benzo analogs: (16PHC165).
Furans and benzofurans: (16PHC219).
Five-membered ring systems with more than one N atom: (16PHC275).
Five-membered ring systems with N and S (Se) atoms: (16PHC317).
Five-membered ring systems with O and S (Se, Te) atoms: (16PHC341).
Five-membered ring systems with O and N atoms: (16PHC361).
Pyridines and benzo derivatives: (16PHC391).
Diazines and benzo derivatives: (16PHC439).
Triazines and tetrazines: (16PHC493).
Six-membered ring systems with O and/or S atoms: (16PHC523).
Seven-membered rings: (16PHC579).
Eight-membered and larger rings: (16PHC623).

2.1.3 Specialized Reports Devoted to One or Several Recent Years
Carbazole scaffold in medicinal chemistry and natural products (2010–2015): (16CTMC1290).

Cyclic hypervalent iodine reagents for atom-transfer reactions (since 2011): Beyond trifluoromethylation: (16AG(E)4436).
The Diels–Alder reaction in total synthesis (2009–2014): (16COC2284).
Indole molecules as inhibitors of tubulin polymerization: Potential new anticancer agents (2013–2015): (16FMC1291).
Inverse electron demand Diels–Alder reactions: Synthesis of heterocycles and natural products along with bioorthogonal and material sciences applications (2013–2014): (16COC2136).
Organocatalytic asymmetric nitro-Michael reactions (2014–March 2015): (16COS687).
Synthesis of cyclopenta-fused heterocycles by metal catalysis (2010–2015): (16CAJ642).
Progress in 1,2,3-triazole synthesis under transition-metal-free conditions (2010–2015): (16CJOC1779)

2.1.4 History of Heterocyclic Chemistry, Biographies

History and new developments in synthesis of the rare-earth tetrapyrrolic complexes: (16CCR(319)110).
Ionic liquids as novel media for electrophilic/onium ion chemistry and metal-mediated reactions: A progress from the author's laboratory during the past 15 years: (16ARK(1)150).
New marine natural products and application to drug development by J. Kobayashi group: (16CPB1079).
Tomás Torres' research in a nutshell: (16JPP966).
What's in a name? The MacDonald condensation: (16JPP855).

2.1.5 Bibliography of Monographs and Reviews

The literature of heterocyclic chemistry, Part XV, 2015: (18AHC(124)121).

2.2 General Topics by Reaction Type
2.2.1 General Sources and Topics

Alkynes as a versatile platform for construction of chemical molecular complexity and realization of molecular 3D printing: (16UK226).
Bowl-shaped conjugated polycycles: (16CCL1166).
Calculating the aromaticity of heterocycles: (16AHC(120)301).
Halogen bonding: A powerful, emerging tool for constructing high-dimensional metal-containing supramolecular networks: (16CCR(308)1).

Light-driven molecular machines based on ruthenium(II) polypyridine complexes: Strategies and recent advances: (16CCR(325)125).
Neutral tris(azolyl)phosphanes: An intriguing class of molecules in chemistry: (16CCR(329)16).
New developments in heterocyclic tautomerism: Desmotropes, carbenes and betaines: (16AHC(119)209).
Recent advances in the chemistry of Se-containing 5-membered heterocycles: (16CCR(312)149).
Self-replicating systems (and the origin of life on Earth): (16OBC4170).
Semi-conjugated heteroaromatic rings: A missing link in heterocyclic chemistry: (16PHC1).
Controlled folding, motional, and constitutional dynamic processes of polyheterocyclic molecular strands: (16AG(E)4130).
Integrated assessment of physico-chemical properties of new energy-rich materials: (16IAN2315).
Recent advances in molecular recognition with tetrathiafulvalene-based receptors: (16TL5416).
Targeted design of complex molecular structures from easily available precursors: (16IAN1418).

2.2.2 Structure and Stereochemistry
2.2.2.1 Stereochemical Aspects
Amino acid-derived bifunctional phosphines for enantioselective transformations: (16ACR1369).
Asymmetric copper-catalyzed azide-alkyne cycloadditions: (16ACSC3629).
Asymmetric dearomatization reactions: (16MI11).
Asymmetric hydrogenation of nonaromatic heterocycles and carbocycles: (16CRV14769).
Asymmetric reactions employing 1,3-dipoles: (16KGS616).
Asymmetric ring-opening reactions of donor-acceptor cyclopropanes and cyclobutanes: (16IJC463).
Carbohydrates as enantioinduction components in stereoselective catalysis: (16OBC4008).
Catalytic kinetic resolution of saturated N-heterocycles by enantioselective amidation with chiral hydroxamic acids: (16ACR2807).
Effect of additives on asymmetric catalysis: (16CRV4006).
N-Heterocyclic carbenes in asymmetric hydrogenation: (16ACSC5978).
Memory of chirality in intramolecular sp^3 C–H amination: (16SL486).
Memory of chirality in reactions involving monoradicals: (16FRR(S)102).

A new generation of chiral phase-transfer catalysts: (16OBC5367).
Organocatalytic enantioselective desymmetrization of *meso*-aziridines and prochiral azetidines: (16COC1851).
Stereo- and regioselectivity of the hetero-Diels–Alder reaction of nitroso derivatives with conjugated dienes: (16BJOC1949).

2.2.2.2 Betaines and Other Unusual Structures
Activation of small molecules (CO, NO, N_2O, H_2, and NH_3) at N-heterocyclic carbene centers: (16SL477).
Anionic N-heterocyclic carbenes: Synthesis, coordination chemistry and applications in homogeneous catalysis: (16CCR(316)68).
Cooperative catalysis and activation with N-heterocyclic carbenes: (16AG(E)14912).
Hypervalent heterocycles: (16AHC(119)57).
N-Heterocyclic carbenes: (16AHC(119)143).
Azafullerene $C_{59}N$ in donor–acceptor dyads: Synthetic approaches and properties: (16CEJ1206).
Cationic rhenium complexes ligated with N-heterocyclic carbenes—An overview: (16DT15).
Chemistry of cyclic imides: An overview on the past, present and future: (16COC1955).
Polyaromatic N-heterocyclic carbene ligands and π-stacking. Catalytic consequences: (16CC5777).
Recent progress of supported N-heterocyclic carbene catalyst in organic reactions: (16CJOC1484).
Stable N- and N/S-rich heterocyclic radicals: Synthesis and applications: (16AHC(119)173).
SF_5-Substituted aromatic heterocycles: (16AHC(120)1).
25 years of N-heterocyclic carbenes: Activation of both main-group element–element bonds and NHCs themselves: (16DT5880).

2.2.3 Reactivity
2.2.3.1 General Topics
Alkenylation of arenes and heteroarenes with alkynes: (16CRV5894).
Catalytic control in cyclizations: From computational mechanistic understanding to selectivity prediction: (16ACR1042).
1,1-Carboboration through activation of silicon–carbon and tin–carbon bonds: (16EJIC300).

Computational approach to diarylprolinol-silyl ethers in aminocatalysis: (16ACR974).
Cooperative capture synthesis: Yet another playground for copper-free click chemistry: (16CSR3766).
Dearomatization through halofunctionalization reactions: (16CEJ11918).
Deep eutectic solvents: The organic reaction medium of the century: (16EJOC612).
The developing concept of bifunctional catalysis with transition metal N-heterocyclic carbene complexes: (16EJIC1448).
Developments in heterocyclic microwave chemistry: (16AHC(120)275).
N,N'-Diamidocarbenes: Isolable divalent carbons with *bona fide* carbene reactivity: (16ACR1458).
Electrocatalytic conversion of furanic compounds: (16ACSC6704).
Elucidation of mechanisms and selectivities of metal-catalyzed reactions using quantum chemical methodology: (16ACR1006).
Enantioconvergent catalysis: (16BJOC2038).
Gold-catalyzed oxidative cyclization: (16CSR4448).
Gold-catalyzed reactions of diynes: (16CSR4471).
1H-Imidazol-4(5H)-ones and thiazol-4(5H)-ones as emerging pronucleophiles in asymmetric catalysis: (16BJOC918).
Mechanistic perspectives on organic photoredox catalysis for aromatic substitutions: (16ACR2316).
Metal-free oxidative biaryl coupling by hypervalent iodine reagents: (16COC580).
Multicomponent and multicatalytic reactions—A synthetic strategy inspired by nature: (16EJIC1306).
Pentamethylcyclopentadienyl Ir(III) metallacycles applied to homogeneous catalysis for fine chemical synthesis: (16CCC1755).
Pyrolysis of energetic ionic salts based on nitrogen-rich heterocycles: (16COC2514).
QSAR of heterocyclic compounds in large descriptor spaces: (16AHC(120)237).
Reactivity of donor-acceptor cyclopropanes with saturated and unsaturated heterocycles: (16IJC512).
Recent advances in organometallic chemistry and their role in catalysis: (16AOC175).
Regioselectivity in Pd-catalyzed direct arylation of 5-membered ring heteroaromatics: (16CST2005).

Research progress in the cycloaddition reactions of nitroso compounds: (16CJOC1994).
Ru-catalyzed azide alkyne cycloaddition reaction: Scope, mechanism, and applications: (16CRV14726).
Silver-catalyzed carboxylation: (16CSR4524).
Site-selective catalysis: (16TCC(372)1).
Sustainable catalysis: (15MI2, 15MI3, 15MI4, 16MI13).
Synthesis and catalytic applications of C_3-symmetric tris(triazolyl)methanol ligands and derivatives: (16CC1997).
The kinetics and spectroscopy of photoredox catalysis and transition-metal-free alternatives: (16ACR1320).
Visible-light-promoted and photoredox-catalyzed radical addition to triple bonds resulting in formation of heterocycles: (16SL2659).
When CuAAC "Click Chemistry" goes heterogeneous: (16CST923).

2.2.3.2 Reactions With Electrophiles and Oxidants

Alternative mechanisms of electrophilic substitution in azoles series: (16IAN1441).
Catalytic enantioselective Friedel–Crafts reactions of naphthols and electron-rich phenols: (16S2151).
Copper-catalyzed oxidation of alkenes and heterocycles: (16S2323).
Domino electrophilic activation of amides: (16JOC4421).
Pd-catalyzed construction of quaternary carbon centers with propargylic electrophiles: (16TL3586).
Recent advances in bromination of aromatic and heteroaromatic compounds: (16S615).
Recent advances in transition-metal-catalyzed iodination of arenes: (16S2969).

2.2.3.3 Reactions With Nucleophiles and Reducing Agents

Asymmetric hydrogenation of heteroarenes with multiple heteroatoms: (16S1769).
Recent advances in the field of direct C–H functionalization of aromatics and heteroaromatics through nucleophilic aromatic substitution of hydrogen: (16TL2665).
Substrate induced diastereoselective hydrogenation/reduction of arenes and heteroarenes: (16RSCA18419).

2.2.3.4 Reactions Toward Free Radicals, Carbenes, etc.

Advances in annulation reactions initiated by phosphorus ylides generated in situ (to give carbo- and heterocycles as final products): (16EJOC1937).
Copper-catalyzed atom transfer radical cyclization reactions: (16EJOC2231).
Free radical chemistry enabled by visible light-induced electron transfer leading to trifluoromethylated (hetero)arenes: (16ACR2295).
From regulation of elementary step to controlled radical process leading to macromolecule: (16ZOK1551).
Imidoyl radical-involved reactions and their use for synthesis of functionalized heterocycles: (16OBC2593).
New amination strategies based on nitrogen-centered radical chemistry: (16CSR3069).
Photo degradation for remediation of cyclic nitramine and nitroaromatic explosives: (16RSCA77603).
Radical cation salts: From single-electron oxidation to C–H activation: (16S18).
Radical C–H functionalization to construct heterocyclic compounds: (16CC2220).
Radical-mediated fluoroalkylations: (16CR47).
Radicals: Reactive intermediates with translational potential: (16JA12692).
Visible light photoredox-controlled reactions of N-radicals and radical ions: (16CSR2044).

2.2.3.5 Cross-Coupling and Related Reactions

Applications of palladium-catalyzed C–N cross-coupling reactions (2008–2015): (16CRV12564).
Au-Catalyzed synthesis and functionalization of heterocycles: (16THC(46)1).
C–H Bond activation and catalytic functionalization: (16TOC(55)1, 16TOC(56)1).
Copper-catalyzed electrophilic amination of sp^2 and sp^3 C–H bonds: (16MI14).
Homo- and heterobimetallic complexes in catalysis. Cooperative catalysis: (16TOC(59)1).
Homogeneous rhodium(I)-catalysis in de novo heterocycle syntheses: (16OBC4986).

The privileged pincer-metal platform: Coordination chemistry and applications: (16TOC(54)1).
Rhodium-catalyzed C–C coupling reactions via double C–H activation: (16OBC4554).
Synthesis and modification of heterocycles by metal-catalyzed cross-coupling reactions: (16THC(45)1).
Transition-metal-catalyzed acyloxylation: Activation of $C(sp^2)$–H and $C(sp^3)$–H bonds: (16EJOC3282).
Transition metal catalyzed carbonylative synthesis of heterocycles: (16THC(42)1).

2.2.3.6 Heterocycles as Intermediates in Organic Synthesis

2-Acetylphenothiazines as synthon in heterocyclic synthesis: (16RCI6143).
Applications of N-heterocyclic imines in main group chemistry: (16CSR6327).
5-Chloropyrazole-4-carboxaldehydes as synthons in heterocyclic synthesis: (16RCI2119).
Heterocycle-mediated *ortho*-functionalization of aromatic compounds containing a small-ring heterocycle as a directing metalating group: The DoM methodology and synthetic utility: (16S1993).
One-pot access to a privileged library of six-membered nitrogenous heterocycles through multi-component cascade approach: (16RCI5147).
Recent advances of organocatalytic enantioselective Michael-addition to chalcone: (16RCI2731).
Recent developments in the heterocyclic ketene aminal-based synthesis of heterocycles: (16RCI5617).
Recent synthetic scenario on imidazo[1,2-*a*]pyridines chemical intermediate: (16RCI2749).

2.2.3.7 Organocatalysts

The aminoindanol core as a key scaffold in bifunctional organocatalysts: (16BJOC505).
Asymmetric organocatalysis: The utility of α,β-unsaturated acylammonium salts: (16AG(E)13934).
Catalytic efficiency of primary β-amino alcohols and their derivatives in organocatalysis: (16EJOC4124).
Computational studies on *Cinchona* alkaloid-catalyzed asymmetric organic reactions: (16ACR1250).

Computational insights into the central role of nonbonding interactions in modern covalent organocatalysis including N-heterocyclic carbene catalysis: (16ACR1279).
Cupreines and cupreidines: An established class of bifunctional *Cinchona* organocatalysts: (16BJOC429).
Emerging roles of in situ generated quinone methides in metal-free catalysis: (16JOC10145).
Enantioselective catalysis by chiral isothioureas: (16EJOC5589).
Enantioselective organocatalyzed transformations of β-ketoesters: (16CRV9375).
Hydrazones as singular reagents in asymmetric organocatalysis: (16CEJ13430).
Meldrum's acid: A useful platform in asymmetric organocatalysis: (16CCC1882).
New approaches to organocatalysis based on C–H and C–X bonding for electrophilic substrate activation: (16BJOC2834).
Organocatalytic enantioselective desymmetrisation: (16CSR5474).
Recent advances in N-heterocyclic carbene catalyzed benzoin reactions: (16BJOC444).
Recent developments in organocatalytic dynamic kinetic resolution: (16T3133).

2.2.4 Synthesis
2.2.4.1 General Topics and Nonconventional Synthetic Methodologies
Strategies and tactics in organic synthesis: (16MI15).
Click reactions in organic synthesis: (16MI16).
Carbocation chemistry: Applications in organic synthesis: (16MI17).
Organic chemistry from retrosynthesis to asymmetric synthesis: (16MI18).
Asymmetric Brønsted acid catalysis: (16MI19).
Practical methods for biocatalysis and biotransformations: (16MI20).
The modern face of synthetic heterocyclic chemistry: (16JOC10109).
Flow synthesis of heterocycles: (16AHC(119)25).
Brønsted acid-promoted synthesis of common heterocycles and related bio-active and functional molecules: (16RSCA37784).
Fermentation and organic synthesis for the production of complex heterocycles: (16JOC10136).
Freones in catalytic olefination reaction and synthesis of fluorinated compounds (including heterocycles) based on the reaction products: (16ZOK1087).

Hypervalent iodine(III) fluorinations of alkenes and diazo compounds: New opportunities in fluorination chemistry, in particular, the subsequent heterocyclization: (16CSR6270).
Synthesis of 4- to 7-membered heterocycles by ring expansion. Aza-, oxa- and thiaheterocyclic small-ring systems: (16THC(41)1).
Ketenes as privileged synthons in the synthesis of six-membered heterocycles: (16AHC(118)195).
Progress in heterocyclic metallosupramolecular construction: (16AHC(120)195).
Recent progress in the use of functionalized β-lactams as building blocks in heterocyclic chemistry: (16PHC27).
Synthesis of aromatic heterocycles using ring-closing metathesis: (16AHC(120)43).
Visible light mediated photoredox catalytic arylation and heteroarylation reactions: (16ACR1566).
Employing arynes in Diels-Alder reactions and transition-metal-free multicomponent coupling and arylation reactions: (16ACR1658).
Exploration of visible-light photocatalysis in heterocycle synthesis and functionalization: (16ACR1911)
Organic synthesis enabled by light-irradiation of electron donor–acceptor complexes: Theoretical background and synthetic applications: (16ACSC1389).
π-Conjugated carbocycles and heterocycles via annulation through C–H and X–Y activation across C–C triple bonds: (16ARK(2)9).
Intramolecular Baylis-Hillman reaction: Synthesis of heterocyclic molecules: (16ARK(2)172).
Controlled fluoroalkylation reactions of heteroarenes by visible-light photoredox catalysis: (16ACR2284).
Divergent synthesis of carbo- and heterocycles via gold-catalyzed reactions: (16ACSC2515).
Synthesis of unsaturated N-heterocycles by cycloadditions of aziridines and alkynes: (16ACSC6651).
Advances in synthetic applications of hypervalent iodine compounds: (16CRV3328).
Triazolinediones as highly enabling synthetic tools: (16CRV3919).
Fifty years of π-conjugated triazenes (derived from coupling of NHCs with azides): (16EJOC860).
Rh-catalyzed enantioselective C–H functionalization in asymmetric synthesis: (16EJOC1459).

Recent advances in target synthesis using SmI$_2$: (15OMC(40)1).
Chemistry of ketene N,S-acetals: An overview including their role in heterocyclic synthesis: (16CRV287).
1,3-Dien-5-ynes: Versatile building blocks for the synthesis of carbo- and heterocycles: (16CRV8256).
Use of bromine and bromo-organic compounds in organic synthesis: (16CRV6837).

2.2.4.2 Synthetic Strategies and Individual Methods
2.2.4.2.1 General Problems
The catalytic, formal homo-Nazarov cyclization as a template for diversity-oriented synthesis: (16IJC499).
Developments of Corey-Chaykovsky reaction in synthetic organic chemistry and total synthesis of natural products: (16COS308).
Iodine-mediated synthesis of heterocycles via electrophilic cyclization of alkynes: (16OBC7639).
Synthetic transformations to give benzofurans and dibenzopyrans through alkynoxy palladium interactions and C–H activation: (16ACR67).
Recent developments in synthetic methods for benzo[b]heteroles containing heteroatoms other than nitrogen and oxygen: (16OBC5402).
Synthesis of benzo-fused cyclic compounds via intramolecular cyclization of aryltriazenes: (16SL1318).
Synthesis of heterobiaryls featuring 1,4 N···S inter-ring interactions: (16S184).
Synthesis of N-, O-, S, and Se-heterocycles via aryne intermediates: (16YGK326).

2.2.4.2.2 Synthetic Applications of Photoreactions and Alternative Energy Input
Carbon-carbon bond forming reactions via photogenerated intermediates: (16CRV9850).
Continuous-flow photochemistry in organic synthesis, material science, and water treatment: (16CRV10276).
Dual catalysis strategies in photochemical synthesis (cooperative interactions of photocatalysts with redox mediators, Lewis and Brønsted acids, organocatalysts, enzymes, or transition metal complexes): (16CRV10035).
Illustrations of efficient solar driven organic reactions: (16COS372).
Organic photoredox catalysis, including oxidative transformations to give heterocycles: (16CRV10075).

Photochemical electron and hydrogen transfer in organic synthesis: The control of selectivity: (16S1782).
Photochemical synthesis of azaheterocycles: (16S2009).
Photogenerated amine radical cations in cascade reactions: From carbocycles to heterocycles: (16ACR1957).
Photoinduced rearrangements of dihetarylethenes: (16KGS658).
Supramolecular photochemistry as a potential synthetic tool: Photocycloaddition: (16CRV9914).

2.2.4.2.3 Synthetic Application of Metal-Catalyzed Reactions
Transition metal-catalyzed heterocycle synthesis via C–H activation: (16MI21).
Ligand design in metal chemistry: Reactivity and catalysis: (16MI22).
Mild metal-catalyzed C–H activation: Examples and concepts: (16CSR2900).
Recent developments in microwave-assisted metal-catalyzed C–*H functionalization of heteroarenes for medicinal chemistry and material applications: (16S3879).
Transition-metal catalyzed C–N bond activation: (16CSR1257).
Ytterbium triflate catalyzed synthesis of heterocycles: (16S4305).

2.2.4.2.4 Synthesis of Heterocycles via Cycloadditions and Multicomponent Reactions
Anionic cascade routes to sulfur and nitrogen heterocycles originating from thio- and aminophosphate precursors: (16EJOC4249).
Arene-alkene cycloaddition: (16CRV9816).
2 + 2 Cycloaddition reactions promoted by group 11 metal-based catalysts: (16T355).
Diels–Alder approaches for the synthesis of bridged bicyclic systems: Synthetic applications of (7-hetero)norbornadienes: (16COC2393).
Diels-Alder "click" reactions: (16COC2211).
Isocyanide based [4 + 1] cycloaddition as an indispensable tool in multi-component reactions: (16CC6958).
Metal promoted Diels–Alder reactions: (16COC2222).
Nitrile N-oxides and nitrile imines as new fuels for the discovery of novel isocyanide-based multicomponent reactions: (16S2721).
Lewis acid catalyzed domino-ring opening-cyclization of donor-acceptor cyclopropanes to give, particularly, tetrahydrocarbazoles: (16IJC445).

Multicomponent reactions: A simple and efficient route to heterocyclic phosphonates: (16BJOC1269).
Multicomponent syntheses of heterocycles initiated by catalytic generation of ynones and ynediones: (16AHC(120)67).
Multicomponent syntheses of fluorophores initiated by metal catalysis: (16EJOC2907).
Multicomponent syntheses of functional chromophores: (16CSR2825).
Progress in 1,3-dipolar cycloadditions in the recent decade: (16T1603).
External-oxidant-based multicomponent reactions: (16S4050).
Recent contributions to hetero Diels–Alder reactions: (16COC2161).
Recent progress in dehydro(genative) Diels-Alder reaction: (16CEJ1558).
Alkenes in [2 + 2 + 2] cycloadditions: (16CEJ6720).
Multicomponent and domino reactions of 3-aroylacrylic acids in the synthesis of heterocycles: (16KGS651).
Synthesis and (3 + 2) cycloaddition reactions of N,N E-1 and C,N-cyclic azomethine imines: (16KGS627).
Application of Ugi/de-Boc/cyclization strategy in the synthesis of N-heterocycles: (16CJOC1241).
Strain-promoted cycloadditions for development of copper-free click reactions: (16COC1902).
6-Aminouracil-involved multicomponent reactions in synthesis of heterocyclic scaffolds: (16RSCA38827).
Current advances of organocatalytic Michael–Michael cascade reaction in the synthesis of highly functionalized cyclic molecules: (16RSCA96154).

2.2.4.2.5 Miscellaneous Methods
Applications of dimedone in the synthesis of heterocycles: (16COC1676).
Applications of indan-1,3-dione in heterocyclic synthesis: (16COS385).
Diverse applications of nitrones for the synthesis of heterocyclic compounds: (16AJOC9).
Hydroxamic acids, recent breakthroughs in stereoselective synthesis and biological evaluations: (16COS659).
Recent advances in the synthesis of azonia aromatic heterocycles: (16JOC10126).
Synthetic routes to benzosuberone-based fused- and spiro-heterocyclic ring systems: (16RSCA17955).
Unsaturated polyfluoroalkyl ketones in the synthesis of nitrogen-bearing heterocycles: (16RSCA1984).

2.2.4.3 Versatile Synthons and Specific Reagents

5-Alkoxyoxazole—A versatile building block in (bio)organic synthesis: (16EJOC3264).

Cyclobutanones as useful synthons for the preparation of bioactive molecules and fused-ring chemicals: (16COC1878).

Fluorinated diazoalkanes—A versatile class of reagents for the synthesis of fluorinated compounds: (16OBC10547).

Generation, reactivity and uses of sulfines (thioketone S-oxides) in organic synthesis: (16EJOC1630).

Octa-1,7-diene-4,5-diamine derivatives: Useful intermediates for stereoselective synthesis of nitrogen heterocycles and ligands for asymmetric catalysis: (16EJOC3143).

Recent advances in the chemistry of organic thiocyanates, particularly, as precursors of heterocycles: (16CSR494).

2.2.4.4 Ring Synthesis From Nonheterocyclic Compounds

Borylative cyclization reactions to form boryl-substituted carbo- and heterocycles: (16EJOC5446).

Diazonaphthoquinones: Synthesis, reactions and applications: (16H(92) 1761).

(Hetero)aromatics from dienynes, enediynes and enyne–allenes: (16CSR4364).

Solid-phase synthesis of heterocycles with α-haloketones as the key building blocks: (16S3684).

Substrate-assisted carbon dioxide activation as a versatile approach for heterocyclic synthesis: (16S3863).

Synthesis and reactivity (mostly, formation of heterocycles possessing the N–N–C=O motif) of N-substituted isocyanates, -isothiocyanates and their masked/blocked derivatives: (16S3625).

Transition metal-catalyzed couplings of alkynes to 1,3-enynes: Modern methods and applications in carbocycle, heterocycle, and natural product syntheses: (16CSR2212).

2.2.5 Properties and Applications (Except Drugs and Pesticides)
2.2.5.1 Dyes and Intermediates

Azadipyrromethenes: From traditional dye chemistry to leading edge applications: (16CSR3846).

Development of D-π-A dye sensitizers with azine ring and their photovoltaic performances of dye-sensitized solar cells: (16YGK760).

Dipole-dipole interaction driven self-assembly of merocyanine dyes: From dimers to nanoscale objects and supramolecular materials: (16ACR868).
Dye-sensitized solar hydrogen production from water: The emerging role of metal-free organic sensitizers: (16EJOC5194).
Light-emitting donor-acceptor dyes in water: Creation of light-emitting system based on aggregation of donor-acceptor dyes: (16YGK781).
Nile red and Nile blue: Applications and syntheses of structural analogues: (16CEJ13764).
Perylene bisimide dye assemblies as archetype functional supramolecular materials: (16CRV962).
Phenothiazine-based dyes for efficient dye-sensitized solar cells: (16JMCC2404).
Recent advances of the near-infrared fluorescent aza-BODIPY dyes: (16COC1736).

2.2.5.2 Substances With Luminescent and Related Properties

CuAAC click reactions for the design of multifunctional luminescent ruthenium complexes: (16DT2338).
Cyclometalated Ir(III) chelates—A new exceptional class of the electrochemiluminescent luminophores: (16ABC7013).
Development of fluorescent zinc chemosensors based on various fluorophores and their applications in zinc recognition: (16DP100).
Electrochemiluminescence of acridines: (16EA2672).
Electrofluorochromic systems: Molecules and materials exhibiting redox-switchable fluorescence: (16EPJ478).
Evolution of 2, 3′-bipyridine class of cyclometallating ligands as efficient phosphorescent Ir(III) emitters for applications in organic light emitting diodes: (16JPC(29)29).
Fluorescent probes for glutathione detection: (16COC2718).
Luminescent bioactive NHC-metal complexes to bring light into cells: (16DT440).
Multicolor bioluminescence obtained using firefly luciferin: (16CTMC2648).
Optically active BODIPYs: (16CCR(318)1).
1,3,4-Oxadiazoles as luminescent materials for organic light emitting diodes via cross-coupling reactions: (16JMCC8596).
Phosphorescent platinum(II) complexes with C^C* cyclometalated NHC ligands: (16ACR2680).

Photochromism of diarylpyrans, spectrokinetic properties and functional materials: (16JPC(29)73).

Recent progress of benzodithiophene-based efficient small molecule organic solar cells: (16CJOC2284).

Room-temperature phosphorescence from purely organic materials, particularly, heterocycle/heteroatom containing compounds: (16CCL1231).

The structure-property relationships of D-π-A BODIPY dyes for dye-sensitized solar cells: (16CR719).

A tale of two luciferins: Fungal and earthworm new bioluminescent systems: (16ACR2372).

2.2.5.3 Organic Conductors and Photovoltaics

Benzothiadiazole and its π-extended, heteroannulated derivatives: Useful acceptor building blocks for high-performance donor-acceptor polymers in organic electronics: (16JMCC6200).

Benzothiadiazole building units in solution-processable small molecules for organic photovoltaics: (16JMCA15771).

Carbazole based conjugated molecules for highly efficient organic solar cell application: (16TL243).

10,15-Dihydro-5H-diindolo[3,2-a: 3′,2′-c]carbazole (triazatruxene or triindole) based materials for organic electronics and optoelectronics: (16JMCC10574).

Influence of ancillary ligands in dye-sensitized solar cells: (16CRV9485).

Molecular design of benzodithiophene-based organic photovoltaic materials: (16CRV7397).

New thiophene-based functional units towards single-molecule electronics: (16YGK676).

Oligomer molecules based on triphenylamine, benzodithiophene, and indacenodithiophene units for efficient organic photovoltaics: (16ACR175).

Phenazines as chemosensors of solution analytes and as sensitizers in organic photovoltaics: (16ARK(1)82).

Photoactive compounds based on the thiazolo[5,4-d]thiazole core and their application in organic and hybrid photovoltaics: (16EJOC233).

Polymethine dyes in hybrid photovoltaics: Structure-properties relationships: (16EJOC2244).

Research progress of benzo[1,2-b:4,5-b′]difuran organic photovoltaic materials: (16CJOC687).

2.2.5.4 Coordination Compounds

Application of 1,2,4-triazolo[1,5-*a*]pyrimidines for the design of coordination compounds with interesting structures and new biological properties: (16CCR(327)221).
Carbon-based two electron σ-donor ligands beyond classical N-heterocyclic carbenes: (16DT16081).
Coordination chemistry and applications of versatile 4,5-diazafluorene derivatives: (16DT32).
Coordination chemistry of thiazoles, isothiazoles and thiadiazoles: (16CCR(308)32).
Coordination complexes of pnictogen(V) cations: (16CCR(324)1).
Coordination polymers built from 1,4-bis(imidazol-1-ylmethyl)benzene: (16DT11233).
Cu(I) complexes—Thermally activated delayed fluorescence. Photophysical approach and material design: (16CCR(325)2).
Cyclic alkyl(amino) carbene stabilized complexes with low coordinate metals of enduring nature: (16ACR357).
Excited state decay of cyclometalated polypyridine ruthenium complexes: Insight from theory and experiment: (16DT13631).
Metal complexes of multitopic, third generation poly(pyrazolyl)methane ligands: Multiple coordination arrangements: (16EJIC2253).
Quantum chemical modelling of extraction separation of minor actinides and lanthanides: The state of the art including evaluation of efficiency of N-heterocyclic ligands: (16UK917).
Recent advances in coordination chemistry of metal complexes based on nitrogen heteroaromatic alcohols. Synthesis, structures and potential applications: (16CCR(327)242).
Recent advances in 1,10-phenanthroline ligands for chemosensing of cations and anions: (16RSCA23169).
Ruthenium(II) complexes with dipyrido[3,2-*a*:2′,3′-*c*]phenazine: From molecular photoswitch to biological applications: (16DT13261).
Single-molecule magnets based on pyridine alcohol ligands: (16JCC3131).
Transition metal mediated C–H activation and functionalization: The role of poly(pyrazolyl)borate and poly(pyrazolyl)alkane ligands: (16EJIC2296).

2.2.5.5 Polymers

Carbazole derivatives synthesis and their electropolymerization: (16JSSE2599).

Click-chemistry approaches to π-conjugated polymers for organic electronics applications: (16CS6298).
Conjugated polymers: Catalysts for photocatalytic hydrogen evolution: (16AG(E)15712).
Coordination polymers and metal-organic frameworks based on poly(pyrazole)-containing ligands: (16CCR(307)1).
Cyclic polymers by ring-closure strategies: (16AG(E)13944).
Diketopyrrolopyrrole polymers for organic solar cells: (16ACR78).
Direct (hetero)arylation polymerization: Simplicity for conjugated polymer synthesis: (16CRV14225).
Dynamic ordering and phase segregation in hydrogen-bonded polymers: (16ACR1409).
Low bandgap semiconducting polymers for polymeric photovoltaics: (16CSR4825).
Membrane materials based on polyheteroarylenes and their application for pervaporation: (16UK81).
Metal complexes containing nitrogen-heterocycle based aryloxide or arylamido derivatives as discrete catalysts for ring-opening polymerization of cyclic esters: (16DT17557).
Polymerization of ethylene oxide, propylene oxide, and other alkylene oxides: Synthesis, novel polymer architectures, and bioconjugation: (16CRV2170).
Poly(1,2,3-triazolium)s: A new class of functional polymer electrolytes: (16CC2433).
Production of bio-based 2,5-furan dicarboxylate polyesters: Recent progress and critical aspects in their synthesis and thermal properties: (16EPJ202).
Rational design of benzodithiophene based conjugated polymers for better solar cell performance: (16RSCA23760).
The removal of heavy metal ions from wastewater/aqueous solution using polypyrrole-based adsorbents: (16RSCA14778).
Semiconducting polymers—A novel trend in organic electronic chemistry: (16COS861).
Spin effects in ethylene polymerization with bis(imino) pyridine iron(II) complexes: (16JOMC(811)48).
Ring-opening copolymerization of epoxides and cyclic anhydrides with discrete metal complexes: Structure-property relationships: (16CRV15167).

Supramolecular helical systems: Helical assemblies of small molecules, foldamers, and polymers with chiral amplification and their functions: (16CRV13752).
Trends and perspectives of direct (hetero)arylation polymerization: (16JA10056).
Triazine-based covalent organic polymers: Design, synthesis and applications in heterogeneous catalysis: (16JMCA16288).

2.2.5.6 Ionic Liquids
Acidic ionic liquids: (16CRV6133).
Active chemisorption sites in functionalized ionic liquids for carbon capture: (16CSR4307).
The applications of ionic liquid as functional material: (16COC2109).
Covalently supported ionic liquid phases: An advanced class of recyclable catalytic systems: (16CCC664).
Extractive denitrogenation of fuel oils using ionic liquids: (16RSCA93932).
Green chemical conversion of biomass into 5-hydroxymethylfurfural in ionic liquids: (16RSCA63991).
Ionic liquids completely UnCOILed: (16MI23).
Ionic liquid crystals: Versatile materials (2005–2015): (16CRV4643).
Ionic liquid properties. From molten salts to room temperature ionic liquids: (16MI24).
Ionic liquid transported into metal-organic frameworks: (16CCR (307)382).
Ionic liquids derived from organosuperbases: En route to superionic liquids: (16RSCA9194).
Latest development on the methods of synthesizing ionic liquids: (16H (92)1171).
Magnetic ionic liquids in analytical chemistry: (16ANCA9).
Meta-analysis of the ionic liquid literature and toxicity: (16CHS266).
Microwave-assisted organic synthesis in ionic liquids: (16JHC1697).

2.2.5.7 Miscellaneous
Advances in click chemistry for silica-based material construction: (16RSCA21979).
Alphabet-inspired design of (hetero)aromatic push-pull chromophores: (16CR1886).

Fe N-Heterocyclic carbene complexes as promising photosensitizers: (16ACR1477).
Five-membered products of cyclometallation reactions as materials in sensing devices: (16JOMC(823)50).
N-Functionalization strategies for N-heterocyclic frameworks in high energy density materials: (16ACR4).
F-18-Labeling of arenes and heteroarenes for applications in positron emission tomography: (16CRV719).
Heavy main group elements with π-conjugated materials for optoelectronic applications: (16CC9485).
Merging thiophene with boron: New building blocks for conjugated materials: (16DT13996).
Pyridylborates as a new type of robust scorpionate ligand: From metal complexes to polymeric materials: (16EJIC2227).
Quaterpyridines as scaffolds for functional metallosupramolecular materials: (16CRV14620).
Recent development of chemosensors based on cyanine platforms: (16CRV7768).
Scorpionate hydrotris(indazolyl)borate ligands as tripodal platforms for surface-mounted molecular gears and motors: (16EJIC2214).
"Sticky"-ends-guided creation of oligopyridine-based functional hollow nanopores for guest encapsulation and water transport: (16ACR922).

2.3 Specialized Heterocycles
2.3.1 Nitrogen Heterocycles (Except Alkaloids)
Advances in the synthesis and applications of cationic N-heterocycles through transition metal-catalyzed C–H activation: (16CAJ448).
Assembly of nitrogen heterocycles initiated by aza-Michael reaction: (16IAN1687).
Deployment of small-ring azaheterocycles as building blocks for the synthesis of organofluorine compounds: (16SL1486).
Direct imine acylation: A versatile method for the synthesis of nitrogen-containing heterocycles, spirocycles and natural products: (16SL2051).
Developments in preparation of N-heterocycles using Pd-catalyzed C–H activation: (16JOMC(801)139).
Recent developments in the stereoselective synthesis of nitrogen-containing heterocycles using *N*-acylimines as reactive substrates: (16ASC3657).

2.3.2 Oxygen Heterocycles

Tandem bond-forming reactions of 1-alkynyl ethers to give complex α-functionalized β-, γ-, or δ-lactones: (16ACR1168).

Recent developments in the lithiation reactions of oxygen heterocycles: (16AHC(118)91).

Stereo- and regioselective synthesis of cyclic ethers by means of organoselenium-mediated cyclization of unsaturated alcohols: (16COC777).

2.3.3 Sulfur Heterocycles

Thiahelicenes: From basic knowledge to applications: (16AHC(118)1).

2.4 Natural and Synthetic Biologically Active Heterocycles
2.4.1 General Sources and Topics

Studies in natural products chemistry: (16MI25, 16MI26, 16MI27, 16MI28).

Cyclohepta[b]indoles: A privileged structure motif in natural products and drug design: (16ACR2390).

Metal complexes as fluorescent probes for sensing biologically relevant gas molecules: (16CCR(318)16).

Microbial resistance to natural compounds: Challenges for developing novel alternatives to antibiotics: (16COC2983).

Natural products containing nitrogen heterocycles—Some highlights 1990–2015: (16AHC(119)81).

Natural products containing oxygen heterocycles—Synthetic advances 1990–2015: (16AHC(119)107).

The novelty of phytofurans, isofurans, dihomo-isofurans and neurofurans: Discovery, synthesis and potential application: (16BCH49).

Plant natural products targeting bacterial virulence factors: (16CRV9162).

Trifluoromethyl nitrogen heterocycles: Synthetic aspects and potential biological targets: (16CC3077).

Web resources for prognosis of biological activity of organic compounds: (16IAN384).

Water-soluble perylenediimides: Design concepts and biological applications: (16CSR1513).

2.4.1.1 Biological Functions of Natural and Synthetic Bioactive Heterocycles

Azaacenes as active elements for sensing and bio applications: (16JMCB7060).

Biological potential of pyrimidine derivatives in a new era: (16RCI6777).
Biofunctionalization of polydioxythiophene derivatives for biomedical applications: (16JMCB4952).
Biologically active γ-lactams: Synthesis and natural sources: (16OBC10134).
Biosynthetic approaches to creating bioactive fungal metabolites: Pathway engineering and activation of secondary metabolism: (16BMCL5843).
Capturing biological activity in natural product fragments by chemical synthesis: (16AG(E)3882).
Catalytic repertoire of bacterial bisindole formation: (16COCB74).
Chemical transformations of bacteriochlorophyll *a* and their applications: (16IAN333).
Furan fatty acids: Their role in plant systems: (16PCR121).
Luciferins, luciferases, their mechanisms of action and applications in chemical analysis, biology and medicine: (16CSR6048).
Phosphono- and phosphinolactones in the life sciences: (16AHC(118)129).
Methoxylated flavones: Occurrence, importance, biosynthesis: (16PCR363).
Naturally occurring thiophenes: Isolation, purification, structural elucidation, and evaluation of bioactivities: (16PCR197).
Nitropyrrole natural products: Isolation, biosynthesis and total synthesis: (16OBC5390).
Recent advances of Diels–Alderases involved in natural product biosynthesis: (16JAB500).
Synthetic entries to and biological activity of pyrrolopyrimidines (2000–2015): (16CRV80).
Tetrahydroselenophene. A unique structural motif for biochemical applications: (16COC155).

2.4.1.2 General Approaches to Syntheses of Biologically Active Heterocycles

The Achmatowicz reaction and its application in the syntheses of bioactive molecules: (16RSCA111564).
Biomimetic assembly lines producing natural product analogs: Strategies from a versatile manifold to skeletally diverse scaffolds: (16CR652).
Cascade polycyclizations in natural product synthesis: (16CSR1557).
Catalytic conversion of furans and pyrroles to natural products and analogues utilizing donor–acceptor substituted cyclopropanes as key intermediates: (16IJC531).

Direct (hetero)arylation reactions of (hetero)arenes as tools for the step- and atom-economical synthesis of biologically active unnatural compounds including pharmaceutical targets: (16S3821).
The direct oxidative diene cyclization to tetrahydrofurans and related reactions in natural product synthesis: (16BJOC2104).
Diverse natural products including ring-expanded γ-butyrolactam, γ-butyrolactone, and cyclopentanone derivatives from α,α-dichlorocyclobutanones: (16ACS252).
Efficient synthesis of polyaromatic hydrocarbon via a formal [2+2] cycloaddition to give benzo-fused polycyclic cyclobutanols and (after acid-promoted rearrangement)—Heterocycles: (16YZ1517).
Electrophilic selenocyclofunctionalization in the synthesis of biologically relevant molecules: (16COC2606).
The Grandberg reaction in synthesis of bioactive compounds: (16IAN1709).
A journey in the chemistry of ynamides: From synthesis to applications in medicinal chemistry and natural product synthesis: (16CL574).
Late-stage annulative convergency in natural product synthesis: (16T3345).
Methods for the synthesis of aza(deaza)xanthines as a basis of biologically active compounds: (16UK308).
Multicomponent hetero-[4+2] cycloaddition/allylboration reaction: From natural product synthesis to drug discovery: (16ACR2489).
Natural product synthesis using multicomponent reaction and structure-activity relationships: (16YGK426).
Photochemical approaches to complex chemotypes: Applications in natural product synthesis: (16CRV9683).
Preparation of reduced pyrazino[2,1-a]isoquinoline derivatives: Important heterocycles in the field of bioactive compounds: (16S3646).
Recent advances in the synthesis of flavaglines from genus *Aglaia*, a family of potent bioactive natural compounds originating from traditional Chinese medicine: (16EJOC5908).
Recent advances in the synthesis of biologically relevant selenium-containing 5-membered heterocycles: (16COC166).
Recent advances and perspectives in the synthesis of bioactive coumarins: (16RSCA46394).
Recent application of oxa-Michael reaction in complex natural product synthesis: (16TL5519).
Recent applications in natural product synthesis of dihydrofuran and -pyran formation by ring-closing alkene metathesis: (16OBC5875).

Recent synthetic approaches to pyrazole-based fipronil, a super-effective and safe pesticide: (16RCI6805).
Stereoselective desymmetrization methods in the assembly of complex natural molecules: (16MI12).
Stereoselective halogenation in natural product synthesis: (16AG(E)4396).
Stereoselective reactions of nitro compounds in the synthesis of natural compound analogs and active pharmaceutical ingredients: (16T6191).
Syntheses and reactions of chalcogen-containing heterocycles: (16YZ841).
Synthetic and biophysical studies on the tridachiahydropyrone family of natural products: (16MI29).
Synthetic strategies employed for the construction of α,β-unsaturated lactone fostriecin and related natural products: (16CRV15035).
Synthetic studies of bioactive heterocyclic natural products and fused heterocyclic compounds based on the thermal electrocyclic or azaelectocyclic reaction of 6π-electron or aza-6π-electron systems: (16YZ607).
Transition-metal-free synthesis of oxazoles as structural fragments in drug discovery: (16RSCA93016).
Using singlet oxygen to synthesize natural products and drugs: (16CRV9994).
Wittig-Horner reagents: Powerful tools in the synthesis of 5-and 6-membered heterocycles useful in pharmaceutical chemistry: (16TJC225).

2.4.1.3 Total Syntheses of Natural Products

Cutting-edge and time-honored strategies for stereoselective construction of C–N bonds in total synthesis: (16CRV4441).
Gold catalysis in total synthesis—Recent achievements: (16CSR1331).
From biosynthesis to total synthesis: Strategies and tactics for natural products: (16MI30).
Nucleophilic addition to N-alkoxyamides: Development and application to the total synthesis of gephyrotoxin: (16YGK599).
Oxazolidinones as chiral auxiliaries in the asymmetric alkylation in total synthesis: (16RSCA30498).
Total synthesis of marine natural products: Cephalosporolides: (16AJOC839).
Trends in applying C–H oxidation to the total synthesis of natural products: (16NPR562).

2.4.2 Alkaloids

2.4.2.1 General

Alkaloids and isoprenoids modification by copper(I)-catalyzed Huisgen 1,3-dipolar cycloaddition (click chemistry): Toward new functions and molecular architectures: (16CRV5689).
Biocatalysts from alkaloid producing plants: (16COCB22).
Exploiting plant alkaloids for medicine: (16COBT155).
Total synthesis, biosynthesis and biological profiles of clavine alkaloids: (16OBC5894).

2.4.2.2 Synthesis

Alkaloid synthesis using chiral secondary amine organocatalysts: (16OBC409).
Alkaloid synthesis via carbenoid intermediates: (16COC82).
Asymmetric synthesis of isoquinoline alkaloids: 2004–2015: (16CRV12369).
Cycloaddition of carbonyl ylides in synthesis of alkaloids: (16IAN2183).
Development of an artificial assembly line generating skeletally diverse indole alkaloids inspired by biogenetic strategy: (16YGK854).
Occurrence of enantioselectivity in nature: The case of (S)-norcoclaurine alkaloid: (16CHIR169).
Palladium-catalyzed cross-coupling reaction of indolylborates and its application for the syntheses of indole alkaloids: (16YGK117).
Recent advances of 1,3-dipolar cycloaddition chemistry for alkaloid synthesis: (16AHC(119)241).
Strategies for the total synthesis of an ergot alkaloid clavicipitic acid: (16CEJ5468).
Synthesis of quinolizidine-containing lycopodium alkaloids and related natural products: (16T4989).
Synthetic strategies towards the azabicyclo[3.3.0]octane core of natural pyrrolizidine alkaloids: (16OPPI223).
Unified total synthesis of tryptophan-based dimeric diketopiperazine alkaloids: (16YGK104).

2.4.2.3 Individual Groups of Alkaloids

Amaryllidaceae and *Sceletium* alkaloids: (16NPR1318).
Apoptosis-inducing effects of *Amaryllidaceae* alkaloids: (16CMC161).
Brazilian Amaryllidaceae as a source of acetylcholinesterase inhibitory alkaloids: (16PCR147).

The biosynthesis and genetic engineering of bioactive indole alkaloids in plants: (16JPB203).
The chemistry of the akuammiline alkaloids: (16ACB171).
The diaza[5.5.6.6]fenestrane skeleton—Synthesis of leuconoxine alkaloids: (16CEJ3600).
The 3,4-dioxygenated 5-hydroxy-4-aryl-quinolin-2(1*H*)-one alkaloids. Results of 20 years of research, uncovering a new family of natural products: (16NPR1425).
Discovery and metabolic engineering of iridoid/secoiridoid and monoterpenoid indole alkaloid biosynthesis: (16PCR339).
Emergence of diversity and stereochemical outcomes in the biosynthetic pathways of cyclobutane-centered marine alkaloid dimers: (16NPR820).
Marine pyridoacridine alkaloids: Biosynthesis and biological activities: (16CBD37).
Matrine, a quinolizidine alkaloid: Bioactivities and structural modifications: (16CTMC3365).
Mechanism of creaming down based on chemical characterization of a complex of caffeine and tea catechins: (16CPB676).
Monoterpenoid indole alkaloids biosynthesis and its regulation in *Catharanthus roseus*: A literature review from genes to metabolites: (16PCR221).
Monoterpenoid bisindole alkaloids: (16ACB259).
Muscarine, imidazole, oxazole and thiazole alkaloids: (16NPR1268).
The radical cation mediated cleavage of catharanthine leading to the vinblastine type alkaloids: Implications for total synthesis and drug design: (16RSCA18002).
A recent perspective on discovery and development of diverse therapeutic agents inspired from isatin alkaloids: (16CTMC1262).
Simple analogues of marine asmarine alkaloids potent against some cancer cell lines: (16SL1145).
Structure, chemical synthesis, and biosynthesis of prodiginine alkaloids: (16CRV7818).
Structures and biological activities of indole diketopiperazine alkaloids from fungi: (16JAFC6659).

2.4.3 Antibiotics

Biosynthesis of tetrahydroisoquinoline antibiotics: (16CTMC1717).
Chemistry and bioactivities of aristeromycins: (16CTMC3258).
Chemistry of antibiotics and related drugs: (16MI31).

Enduracididine, a rare amino acid component of peptide antibiotics: Natural products and synthesis: (16BJOC2325).
Semi-synthetic zwitterionic rifamycins: A promising class of antibiotics; survey of their chemistry and biological activities: (16RSCA114758).
Synthesis of antibiotic ciprofloxacin-based compounds: (16SC1849).
Targeting antibiotic resistance: (16AG(E)6599).

2.4.4 Vitamins
Biotechnology of riboflavin (vitamin B_2): (16AMB2107).
Diverse and important contributions by medicinal chemists to the development of pharmaceuticals: An example of active vitamin D_3 analog, eldecalcitol: (16H(92)1013).
Photolytic properties of cobalamins: A theoretical perspective: (16DT4457).
Total synthesis of vitamin B_{12}—A fellowship of the ring: (16JPP1).

2.4.5 Drugs
2.4.5.1 General
Innovative drug synthesis: (16MI32).
Synthetic methods in drug discovery: (15MI5, 16MI33).
Privileged scaffolds in medicinal chemistry: (15MI6).
Synthesis of best-seller drugs: (16MI34).
Profiles of drug substances, excipients and related methodology: (16MI35).
Scaffold diversity synthesis and its application in probe and drug discovery: (16AG(E)7586).
The medicinal chemist's toolbox for late stage functionalization of drug-like molecules: (16CSR546).
Synthesis of heterocycles in contemporary medicinal chemistry: (16THC(44)1).
Modern advances in heterocyclic chemistry in drug discovery: (16OBC6611).
Dihydropyrimidin-2(1H)-one and its analogues as a platform for the design and synthesis of new biologically-active compounds: (16UK1056).
Mechanisms and pharmaceutical consequences of stereoisomerizations: (16EJPS(88)101).
Carbon monoxide-releasing molecules with a wide variety of medical applications: (15OMC(40)140).

Quinazolines and quinazolinones as ubiquitous structural fragments in medicinal chemistry: An update on the development of synthetic methods and pharmacological diversification: (16BMC2361).
Computational exploration of molecular scaffolds in medicinal chemistry: (16JMC4062).
The medicinal chemistry of therapeutic oligonucleotides: (16JMC9645).
Nucleoside derived antibiotics and probable new utilities for an established class of drugs to fight microbial drug resistance: (16JMC10343).
Natural endoperoxides as drug lead compounds: (16CMC383).
Isoxazolidine: A privileged scaffold for organic and medicinal chemistry: (16CRV15235).
Metabolism, biochemical actions, and chemical synthesis of anticancer nucleosides, nucleotides, and base analogs: (16CRV14379).
Silver coordination compounds: A new horizon in medicine: (16CCR(327)349).
Hybrid molecules: The privileged scaffolds for various pharmaceuticals: (16EJMC(124)500).
Sulfur-containing scaffolds in drugs: Synthesis and application in medicinal chemistry: (16CTMC1200).
Synthetic approaches, structure–activity relationship and biological applications for pharmacologically attractive pyrazole/pyrazoline-thiazolidine-based hybrids: (16EJMC(113)145).
Thiazolo[4,5-d]pyrimidines as a privileged scaffold in drug discovery: (16EJMC(113)198).
A real-world perspective on molecular design: (16JMC4087).
Use of photosensitizers in semisolid formulations for microbial photodynamic inactivation: (16JMC4428).
Quadruplex nucleic acids as novel therapeutic targets: (16JMC5987).
Chemical space of DNA-encoded libraries: (16JMC6629).
Challenges and opportunities for the application of boron clusters in drug design: (16JMC7738).
Underexplored opportunities for natural products in drug discovery: (16JMC9295).

2.4.5.2 Definite Types of Activity
2.4.5.2.1 Antibacterial Activity
The oxadiazole antibacterials: (16COMB13).
Recent development of benzimidazole-containing antibacterial agents: (16CHMC646).

2.4.5.2.2 Antitumor Activity
Anti-cancer N-heterocyclic carbene complexes of Au(III), Au(I) and Pt(II): (16MI36).
AXL Inhibitors in cancer: A medicinal chemistry perspective: (16JMC3593).
Challenges and perspectives on the development of small-molecule EGFR inhibitors against T790M-mediated resistance in non-small-cell lung cancer: (16JMC6580).
Flavonoids with therapeutic potential in prostate cancer: (16ACA1205).
Metal N-heterocyclic carbene complexes as potential anti-tumor metallodrugs: (16CCR(329)191).
Toxicogenomics in predictive carcinogenicity: (16MI37).
Recent developments of 4-substituted coumarin derivatives as anticancer agents: (16EJMC(119)141).
Recent developments on 1,2,4-triazole nucleus in anticancer compounds: (16ACA465).
1,3,4-Thiadiazole based anticancer agents: (16ACA1301).
Quinoxaline nucleus: A promising scaffold in anti-cancer drug discovery (2010–2015): (16ACA1339).
Perspectives of benzimidazole derivatives as anticancer agents in the new era: (16ACA1403).
Recent advances in cancer treatment by iron chelators: (16BMCL251).

2.4.5.2.3 CNS-Targeted Drugs
The evolution of P2X7 antagonists with a focus on CNS indications: (16BMCL3838).
Emerging targets and new small molecule therapies in Parkinson's disease treatment: (16BMC1419).
Heterocyclic scaffolds for the treatment of Alzheimer's disease: (16CPD3971).
Pivotal role of glycogen synthase kinase-3: A therapeutic target for Alzheimer's disease: (16EJMC(107)63).
Progress in multifunctional metal chelators as potential drugs for Alzheimer's disease: (16CCR(327)287).
Recent progress in repositioning Alzheimer's disease drugs based on a multitarget strategy: (16FMC2113).

2.4.5.2.4 Anti-HIV Activity
Anti-HIV drug discovery and development: Current innovations and future trends: (16JMC2849).

Discovery of non-peptide small molecular CXCR4 antagonists as anti-HIV agents: Recent advances and future opportunities: (16EJMC(114)65).
Efavirenz (drug used in the treatment of HIV-AIDS) a nonnucleoside reverse transcriptase inhibitor of first-generation: Approaches based on its medicinal chemistry: (16EJMC(108)455).
Small molecule CXCR3 antagonists: (16JMC2894).
Synthesis routes to anti-HIV drugs: (16T3389).

2.4.5.2.5 Antimalarial and Related Activity

Antimalarial chemotherapy: Natural product inspired development of preclinical and clinical candidates with diverse mechanisms of action: (16JMC5587).
Artemisinin-derived dimers: Potent antimalarial and anticancer agents: (16JMC7360).
π-Delocalized lipophilic cations as new candidates for antimalarial, anti-trypanosomal and anti-leishmanial agents: Synthesis, evaluation of anti-protozoal potency, and insight into action mechanisms: (16CPB656).
The medicinal chemistry of dengue virus: (16JMC5622).
Recent advances in novel heterocyclic scaffolds for the treatment of drug-resistant malaria: (16JEIMC173).
Recent updates on development of drug molecules for human African trypanosomiasis: (16CTMC2245).
Spiroindolone NITD609 is a novel antimalarial drug that targets the P-type ATPase PfATP4: (16FMC227).

2.4.5.2.6 Receptor Antagonisting and Relative Activities

Endothelin-receptor antagonists beyond pulmonary arterial hypertension: Cancer and fibrosis: (16JMC8168).
G-Protein-coupled receptor kinase 2 inhibitors: Current trends and perspectives: (16JMC9277).
Drug-like antagonists of P2Y receptors—From lead identification to drug development: (16JMC9981).
Transient receptor potential melastatin 8 channel (TRPM8) modulation: Cool entryway for treating pain and cancer: (16JMC10006).
Disrupting acetyl-lysine recognition: Progress in development of bromodomain inhibitors: (16JMC1271).
Nociceptin opioid receptor (NOP) as a therapeutic target: Progress in translation from preclinical research to clinical utility: (16JMC7011).
Orexin receptor antagonists: New therapeutic agents for the treatment of insomnia: (16JMC504).

G protein-coupled receptor 119 (GPR119) agonists for the treatment of diabetes: Recent progress and prevailing challenges: (16JMC3579).
Recent progress on bile acid receptor modulators for treatment of metabolic diseases: (16JMC6553).
Recent progress on third generation covalent epidermal growth factor receptor inhibitors: (16BMCL1861).

2.4.5.2.7 Enzyme Inhibitors and Activators
Various enzyme inhibitors and activators: their therapeutic evaluation: ((16AP507), (16BMC5007), (16BMCL5139), (16CTMC1362), (16CTMC1392), (16CTMC2125), (16BMC5017), (16BMCL2383), (16EJMC(112)347), (16EJMC(121)671), (16JMC1308), (16JMC2269), (16JMC3609), (16JMC7029), (16JMC8667), (16JMC9305), (16JMC9575), (16JMC10030), (16OBC4970), (16RSCA2119), (16RSCA7575), (16RSCA42660), (16ZOK159)).

2.4.5.3 Individual Substances and Groups of Compounds
N-Heterocycles with potential anticancer activities: (16CMC4338, 16COS544).
Bioactivity of flavonoids and analogs: (16CHMC2102, 16CMC3078, 16CMC4151, 16EJMC(121)47).
Coumarins and isocoumarins in medicinal chemistry: (16EJMC(116) 290, 16EJMC(123)236, 16IJPS482).
Indoles in drug development: (16ACSID382, 16EJMC(122)366, 16EJMC(123)858, 16EJPS(91)1).
Isoxazole and isoxazoline as promising drugs: (16CHMC270, 16CTMC2863).
Pyridines and imidazopyridines with medicinal significance: (16ARJC1443, 16CMC1100, 16CTMC3274, 16CTMC3590).
Pyrrole, hydropyrroles, and fused pyrroles as probable therapeutic agents: (16EJMC(110)13, 16EJMC(114)257, 16HL817).
Bioactivity and pharmacology of xanthone derivatives: (16CMC3654, 16DDT1814, 16EJMC(116)267).
Thiazole derivatives in medicinal chemistry: (16AJPT62, 16EJMC(109) 89, 16IJPS1375).

2.4.6 Miscellaneous
2.4.6.1 Enzymes, Coenzymes, and Their Models
The *Amaryllidaceae* alkaloids: Biosynthesis and methods for enzyme discovery: (16PCR317).

Enzymology of pyrimidine metabolism and neurodegeneration: (16CMC1408).
Pyrazinamide and pyrazinoic acid derivatives directed to mycobacterial enzymes against tuberculosis: (16CPPS213).

2.4.6.2 Amino Acids and Peptides

Antimycin-type depsipeptides: Discovery, biosynthesis, chemical synthesis, and bioactivity: (16NPR1146).
Cyclic opioid peptides: (16CMC1288).
Heterocyclic scaffolds in the design of peptidomimetic integrin ligands: Synthetic strategies, structural aspects, and biological activity: (16CTMC343).
Self-assembly of cyclic peptides and peptidomimetic macrocycles: Linking structure with function: (16T3379).
Side-chain cyclized aromatic amino acids: Great tools as local constraints in peptide and peptidomimetic design: (16JMC10865).
Targeting "undruggable" proteins: Design of synthetic cyclopeptides: (16CMC748).
Unusual amino acids in medicinal chemistry: (16JMC10807).

2.4.6.3 Plant Metabolites

Constituents from *Chloranthaceae* plants and their biological activities: (16HC175).
Dibenzofurans and derivatives from lichens and ascomycetes: (16NPR801).
Synthesis of *Amaryllidaceae* constituents and unnatural derivatives: (16AG(E)5642).
Synthesis and mode of action of oligomeric sesquiterpene lactones: (16NPR602).

2.4.6.4 Heterocycles Produced by Marine Organisms

Bromopyrrole alkaloids from Okinawan marine sponges *Agelas* spp.: (16CPB691).
Marine natural products: (16NPR382).
Structural diversity and chemical synthesis of peroxide and peroxide-derived polyketide metabolites from marine sponges: (16NPR861).
Synthesis of oxazole, oxazoline and isoxazoline derived marine natural products: (16COC898).

2.4.6.5 Other Natural Products

Chemical diversity, origin, and analysis of phycotoxins: (16JNP662).
Curing hepatitis C virus infection with direct-acting antiviral agents: (16JMC7311).
Hepatitis C virus inhibitors targeting NS4B: From target to preclinical drug candidates: (16JMC16).
Heterocyclic aromatic amines in cooked meat products: Causes, formation, occurrence, and risk assessment: (16CRF269).
Narcotic substances of natural origin and methods of their determination: (16ZAK3).
Recent progress of on-resin cyclization for the synthesis of glycopeptidomimetics: (16CCL1731).

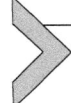

3. THREE-MEMBERED RINGS

3.1 One Heteroatom

3.1.1 One Nitrogen Atom

Bicyclic aziridinium ions in azaheterocyclic chemistry—Preparation and synthetic application of 1-azoniabicyclo[n.1.0]alkanes: (16ASC3485).
Evolution of the aza-Diels-Alder reaction of 2H-azirines: (16SL2171).
Oxidative aminoaziridination: Past, present, and future: (16TL3575).
Advances in catalytic selective synthesis of epoxides and aziridines via diazocarbonyl compounds: (16COC19).
Research progress in the cycloaddition reactions of aziridines: (16CJOC939).
The synthesis of cyclophellitol-aziridine and its configurational and functional isomers: (16EJOC3671).
Synthetic applications and methodological developments of donor-acceptor cyclopropanes, aziridines, and related compounds: (16IJC431).

3.1.2 One Oxygen Atom

Aniline-terephthalaldehyde resin p-toluenesulfonate (ATRT) as a highly efficient and reusable catalyst for alcoholysis, hydrolysis, and acetolysis of epoxides: (16SC1781).
Epoxidation of olefins catalyzed by Mn(II, III, V) in different valence states: (16CJOC1765).
Molecular epoxidation reactions catalyzed by rhenium, molybdenum, and iron complexes: (16CR349).

Metal-catalyzed directed regio- and enantioselective ring-opening of epoxides: (16ACR193).

Organocatalytic asymmetric reactions of epoxides: Recent progress: (16CEJ3632).

Recent advances and perspectives in the manganese-catalyzed epoxidation reactions: (16T1).

Recent developments and perspectives in the ruthenium-catalyzed olefin epoxidation: (16T6175).

Recent trends in ring opening of epoxides by amines as nucleophiles (2011–2015): (16SC831).

Synthesis of fluorinated heterocycles based on polyfluorooxiranes and O, N, S-dinucleophiles: (16IAN2163).

Utility of unsaturated terminal epoxides in cyclopropane synthesis: (16COC4).

4. FOUR-MEMBERED RINGS
4.1 One Nitrogen Atom

A comprehensive review on structural and biological activity relationship of 3-acetoxyazetidin-2-ones: (16AJPT96).

Recent developments in the synthesis of condensed β-lactams: (16RSCA99220).

Selective synthesis of β-lactams via catalytic metal carbene C–H insertion reactions: (16COC29).

4.2 One Oxygen or Sulfur Atom

Oxetanes: Recent advances (2011–2015) in synthesis, reactivity, and medicinal chemistry: (16CRV12150).

Regioselectivity in the ring opening and expansion of unsymmetric thietanes: (16COS73).

Ring-opening reactions of oxetanes: Methodology development and synthetic applications: (16SC1397).

5. FIVE-MEMBERED RINGS
5.1 One Heteroatom
5.1.1 One Nitrogen Atom
5.1.1.1 Monocyclic Pyrroles

Chemistry of pyrroles (general monograph): (16MI38).

Anion-responsive π-electronic systems based on a 1,3-dipyrrolylpropane-1,3-dione framework and providing ion-pairing assemblies: (16YGK243).
Developments in the synthesis of pyrroles catalyzed by group-1B-metals (Cu, Ag and Au): (16OBC7136).
The oxidation of pyrrole: (16CAJ155).
A new avenue to the synthesis of highly substituted pyrroles: Synthesis from N-propargylamines: (16RSCA18619).
Active methylenes in the synthesis of a pyrrole motif: An imperative structural unit of pharmaceuticals, natural products and optoelectronic materials: (16RSCA37039).

5.1.1.2 Hydropyrroles

Advances toward the synthesis of functionalized γ-lactams: (16OPPI254).
Nitrogen-containing acetals and ketals in the synthesis of pyrrolidine derivatives: (16KGS753).
Synthesis and biological activities of spiro oxindole–pyrrolidine compounds: (16SC1643).

5.1.1.3 Porphyrins and Related Systems

The breaking and mending of *meso*-tetraarylporphyrins: Transmuting the pyrrolic building blocks: (16ACR1080).
Breakdown of chlorophyll in higher plants phyllobilins as abundant, yet hardly visible signs of ripening, senescence, and cell death: (16AG(E)4882).
"Click"-reaction: An alternative tool for new architectures of porphyrin based derivatives: (16CCR(306)1).
Cooperative self-assembly of porphyrins with polymers possessing bioactive functions: (16CC13543).
Porphyrins in complex with DNA: Modes of interaction and oxidation reactions: (16CCR(308)460).
Recent advances in the template-directed synthesis of porphyrin nanorings: (16CC10205).
Single-molecule magnetism of tetrapyrrole lanthanide compounds with sandwich multiple-decker structures: (16CCR(306)195).
Structure modification and spectroscopic properties of artificial porphyrinoids: (16IJC119).

Synthesis of chlorins, bacteriochlorins and their tetraaza analogues: (16UK700).
Synthesis, properties and self-organization of *meso*-arylporphyrins with higher alkyl substituents: (16UK477).
Visible light-driven photophysics and photochemistry of water-soluble metalloporphyrins: (16CCR(325)59).
The evolution of corrole synthesis—From simple one-pot strategies to sophisticated ABC-corroles: (16JPP96).
Structurally characterized bimetallic porphyrin complexes of Pb, Bi, Hg and Tl based on unusual coordination modes: (16JPP117).
A molecule for all seasons: The heme: (16JPP134).
Phosphorus(V) complexes of porphyrinoid macrocycles: (16JPP895).
Synthesis and anion binding properties of porphyrins and related compounds: (16JPP950).
Synthesis of low melting point porphyrins: A quest for new materials: (16JPP843).

5.1.1.4 Indoles, Carbazoles, Related Systems, and Hydrogenated Derivatives

Access to spiro and fused indole derivatives from α,β-unsaturated aldehydes enabled by N-heterocyclic carbene catalysis: (16CR1489).
Cascade reactions of nitrones and allenes for the synthesis of indole derivatives: (16JOC9521).
Catalytic asymmetric synthesis of biologically important 3-hydroxyoxindoles: (16BJOC1000).
1-/2-/3-Fluoroalkyl-substituted indoles, promising medicinally and biologically beneficial compounds: Synthetic routes, significance and potential applications: (16JFC(185)118).
C-3 Functionalization of indole derivatives with isoquinolines: (16COC2038).
Indole ring synthesis: From natural products to drug discovery: (16MI39).
Indoloquinolines: Possible biogenesis from common indole precursors and their synthesis using domino strategies: (16COS58).
New syntheses of cycloalka[*b*]indoles: (16ZOK1239).
Reactivity of indolizines in organic synthesis: (16SC719).
Recent advance in the synthesis of indolizines and their π-expanded analogues: (16OBC7804).
Recent advances of the access to indolinones via C–H bond functionalization/cyclization cascade strategy: (16CJOC875).

Synthesis and some biological properties of pyrrolo[1,2-a]indoles: (16JHC685).
Synthesis of spirocyclic indolenines: (16CEJ2856).
Various difunctionalizations of acrylamide: An efficient approach to synthesize oxindoles: (16OBC4365).

5.1.1.5 Isoindoles (Including Phthalocyanins and Porphyrazines)
Liquid crystalline metal phthalocyanines: Structural organization on the substrate surface: (16CCR(310)131).
Multisensor systems based on phthalocyanines for monitoring the quality of grapes: (16JPP889).
Optically active porphyrin and phthalocyanine systems: (16CRV6184).
Phthalocyanines as molecular scaffolds to block disease-associated protein aggregation: (16ACR801).
Synthesis and application of trifluoroethoxy-substituted phthalocyanines and subphthalocyanines: (16YGK154).
Synthesis, characterization and electrochemical properties of tetra 7-oxy-3-biphenylcoumarin substituted metal-free, zinc(II), cobalt(II) and indium(III) phthalocyanines: (16DP311).
Tetrapyrazinoporphyrazines and their metal derivatives: Synthesis and basic structural information: (16CCR(309)107).

5.1.2 One Oxygen Atom
5.1.2.1 Furans
The Butin reaction: (16KGS973).
Debenzylative cycloetherification: A key strategy for complex tetrahydrofuran synthesis: (16CEJ9456).
One-pot catalytic conversion of carbohydrates into furfural and 5-hydroxymethylfurfural: (16CST3694).
One-pot conversion of carbohydrates into furan derivatives via furfural and 5-hydroxylmethylfurfural as intermediates: (16CSS2015).
Rearrangements vs fragmentations in the dye-sensitized photo-oxygenation of N-aryl α-furanamides: (16COC2117).
Recent advances in the synthesis of tetrahydrofurans and applications in total synthesis: (16T5003).
Recent trends of ionic liquids for the synthesis of 5-hydroxymethylfurfural: (16COC736).
Solid acid catalyzed synthesis of furans from carbohydrates: (16CRSE36).
Zeolite and zeotype-catalyzed transformations of biofuranic compounds: (16GC5701).

5.1.2.2 Annulated Furans
Annulation of furan rings to arenes (in Russian): (16UK817).
Reactivity of benzo[b]furans: A full perspective: (16COC1069).
Silver- and gold-catalyzed routes to furans and benzofurans: (16OBC9184).
Synthesis of biologically active compounds containing benzo[b]furans as a framework: (16COS780).
Synthesis of highly condensed aromatic compounds by using isobenzofurans: (16YGK316).

5.1.2.3 Five-Membered Lactones
Homogeneous catalyzed reactions of levulinic acid: To γ-valerolactone and beyond: (16CSS2037).
Nonracemic γ-lactones from the Sharpless asymmetric dihydroxylation of β,γ-unsaturated carboxylic esters: (16EJOC5060).
Synthesis of γ-valerolactone from carbohydrates and its applications: (16CSS156).

5.1.3 One Sulfur Atom
Aromatic metamorphosis of dibenzothiophenes: (16SL1765).
Recent progress in synthesis and application of thiophene oligomers based on bithiophene dicarbanions: (16CR797).
3-Sulfolenes and their derivatives: Synthesis and applications: (16S1).

5.2 Two Heteroatoms
Recent advances in the synthesis of 1,3-azoles: (16CTMC3617).

5.2.1 Two Nitrogen Atoms
5.2.1.1 Pyrazoles and Annulated Pyrazoles
Chemistry of antipyrine: (16SC1567).
Chemistry of 4,4-(arylmethylene)-bis(3-methyl-1-phenyl-1H-pyrazol-5-ol)s: (16JHC356).
Dicopper(II) pyrazolenophanes: Ligand effects on their structures and magnetic properties: (16CCR(315)135).
Fluorine-containing indazoles: Synthesis and biological activity: (16JFC(192)1).
Functionalization of indazoles by means of transition metal-catalyzed cross-coupling reactions: (16T6711).
Synthetic routes to 3(5)-phosphonylated pyrazoles: (16UK667).
Synthetic routes to pyrazole-3(5)-carboxylates: (16JHC13).

5.2.1.2 Imidazoles and Annulated Imidazoles

Aza-Michael addition of imidazole analogues: (16S2681).
Chemistry of 2-aminoimidazoles: (16JHC345).
Design, synthesis, and functionalization of imidazoheterocycles: (16CR1868).
Diversity oriented synthesis of benzimidazole-based biheterocyclic molecules by combinatorial approach: (16RSCA50384).
Excited-state intramolecular proton transfer in 2′-(2-hydroxyphenyl) imidazo[1,2-*a*]pyridines: (16JPC(28)116).
Imidazolium ionic liquids, imidazolylidene heterocyclic carbenes, and zeolitic imidazolate frameworks for CO_2 capture and photochemical reduction: (16AG(E)2308).
Recent advances in the synthesis of benzimidazol(on)es via rearrangements of quinoxalin(on)es: (16RSCA42132).
Recent progress in the chemistry of 2-unsubstituted 1*H*-imidazole 3-oxides: (16COC1359).
Synthetic design of functional boron imidazolate frameworks: (16CCR(307)255).
Zwitterionic imidazolium salt: Recent advances in organocatalysis: (16S1269).

5.2.2 One Nitrogen and One Oxygen Atom
5.2.2.1 1,2-Heterocycles

Functionalized isoxazole and isothiazole ligands: Design, synthesis, palladium complexes, homogeneous and heterogeneous catalysis in aqueous media: (16IAN321).
Isoxazolidines as biologically active compounds: (16COS726).
Synthesis of 3-substituted 1,2-benzisoxazole derivatives: (16CSTR8).

5.2.2.2 1,3-Heterocycles

Azlactone (oxazolone) reaction developments: (16CEJ10294).
Progress in the synthesis of 2-aminobenzoxazole derivatives: (16CJOC2634).
Recent development of synthesis of functionalized polysubstituted oxazoles: (16CJOC1000).
Recent progresses in the chemistry of münchnones: (16CJOC913).

5.2.3 One Nitrogen and One Sulfur Atom

Exploration of different methodologies for synthesizing biologically important benzothiazoles: (16COS41).

Progress in the chemistry of 5-acetylthiazoles: (16SC1081).

Recent advances in chemistry of condensed 4-thiazolidinones: (16JHC1687).

Synthetic routes to thiazoloquinazolines: (16KGS766).

Thiazole: Chemistry, synthesis and therapeutic importance of its derivatives: (16CTMC2841).

5.3 Three Heteroatoms
5.3.1 Three Nitrogen Atoms

Advances of azide-alkyne cycloaddition-click chemistry over the recent decade: (16T5257).

Advances on the metallocarbene formation reactions based on triazole derivatives: (16CJOC1555).

The benzotriazole story: (16AHC(119)1).

Chemistry of benzotriazole derivatives: (16THC(43)1).

Copper nanoparticle-catalyzed synthesis of 1,4-disubstituted 1,2,3-triazoles in water: (16ARK(1)307).

Geminal bis- and tristriazoles: Long-known but still uncommon heterocyclic entities: (16S1437).

Hypervalent-iodine(III)-mediated oxidative methodology for the synthesis of fused triazoles: (16CAJ1988).

Recent advances in the synthesis of heterocycles and related substances based on α-imino rhodium carbene complexes derived from N-sulfonyl-1,2,3-triazoles: (16CEJ17910).

Recent advances in the tandem reaction of azides with alkynes or alkynols: (16OBC11317).

Regioselective synthesis of multisubstituted 1,2,3-triazoles: Moving beyond the copper-catalyzed azide-alkyne cycloaddition: (16CC14188).

Syntheses of triazoloquinoxalines: (16H(92)1931).

5.3.2 Two Nitrogen Atoms and One Oxygen, Sulfur, or Selenium Atom

Advances in the synthesis of non-annulated polynuclear heterocyclic systems comprising the 1,2,5-oxadiazole ring: (16UK1097).

Advances in the synthesis of poly-nitro furazans: (16CJOC1528).

Fluorinated 1,3,4-oxadiazoles and other azoles in cycloaddition reactions: (16KGS675).

Synthetic and biological aspects of thiadiazoles and their condensed derivatives: (16CTMC2884).

Tandem intramolecular Diels-Alder/1,3-dipolar cycloaddition cascade of 1,3,4-oxadiazoles: (16ACS241).

5.4 Four Heteroatoms

Energetic salts based on tetrazole N-oxide: (16CEJ7670).
Mesoionic tetrazoles—Progress since 1980: (16H(92)185).
Microwave activation in tetrazole chemistry: (16KGS887).
Tetrazoles with oxygen-, sulfur-, and selenium-containing substituents: (16IAN923).

6. SIX-MEMBERED RINGS
6.1 One Heteroatom
6.1.1 One Nitrogen Atom
6.1.1.1 Pyridines

New page to access pyridine derivatives: Synthesis from N-propargylamines: (16RSCA71662).
Recent uses of Kröhnke methodology: A short survey: (16S1974).
Synthesis of nitropyridines and nitroanilines using three-component ring transformation: (16YGK130).
Transition metal-catalyzed pyridine synthesis: (16MI40).

6.1.1.2 Pyridinium Compounds, Ylides, and Pyridine N-Oxides

Cyclization reactions of Kröhnke-Mukaiyama salts: (16KGS666).
Metal-free or nonnoble-metal-catalyzed oxidation of alkynes by pyridine N-oxides as oxidants: (16SL493).
Recent advances in application of pyridinium chlorochromate in organic synthesis: (16COS220).

6.1.1.3 Applications of Pyridines

Mononuclear ruthenium polypyridine complexes that catalyze water oxidation: (16CS6591).
Pyridinyl trifluoromethyl sulfonates: Synthesis and application in organic synthesis: (16IAN2559).

6.1.1.4 Hydropyridines

N-Methyltetrahydropyridines and pyridinium cations as toxins and comparison with naturally-occurring alkaloids: (16FCT23).
Progress in synthesis of 1,4-dihydropyridines: (16CJOC2858).
Recent developments in multicomponent synthesis of structurally diversified tetrahydropyridines: (16RSCA42045).

6.1.1.5 Pyridines Annulated With Carbocycles

Asymmetric hydrogenation of quinoline derivatives catalyzed by cationic transition metal complexes of chiral diamine ligands: Scope, mechanism and catalyst recycling: (16CR2697).

Novel routes to quinoline derivatives from N-propargylamines: (16RSCA49730).

The Povarov reaction: A versatile strategy leading to 1,2,3,4-tetrahydroquinoline derivatives: (16COS157).

Synthetic approaches to 4-(het)aryl-3,4-dihydroquinolin-2(1H)-ones: (16KGS509).

6.1.1.6 Pyridines Annulated With Heterocycles

Functional naphthalene diimides: Synthesis, properties, and applications: (16CRV11685).

Recent developments in the chemistry of pyrido[1,2-c]pyrimidines: (16SC1477).

Recent developments in the chemistry of pyrido[4,3-d]pyrimidines: (16RSCA71827).

6.1.2 One Oxygen Atom
6.1.2.1 Pyrans and Hydropyrans

Biosynthesis of α-pyrones: (16BJOC571).

6.1.2.2 Annulated Pyrans

Biflavanoids. Chemical and pharmacological aspects: (16MI42).

Chemistry of 3-carbonyl-2-methyl-4-oxo-4H-1-benzopyrans (1980–2015): (16ARK(1)111).

Chemistry of 2-amino-4-oxo-4H-1-benzopyran-3-carboxaldehydes: (16ARK(1)375).

Convenient methods for the synthesis of pentacyclic fused heterocycles with coumarin moiety: (16SC569).

Dimeric pyranonaphthoquinones: Isolation, bioactivity, and synthetic approaches: (16EJOC5778).

Flavonoids and related compounds: Bioavailability and function: (16MI41).

Fungi as a source of natural coumarins production: (16AMB6571).

Recent advances in the synthesis of coumarin derivatives via Pechmann condensation: (16COC798).

Recent advances in transition-metal-catalyzed synthesis of coumarins: (16S2303).
Recent developments in the enzymatic O-glycosylation of flavonoids: (16AMB4269).
Recent progress on the asymmetric synthesis of chiral flavanones: (16SL656).
Synthesis and application of fluorescent α-amino acids, in particular, those bearing coumarin and flavone derived chromophores: (16OBC8911).
Synthesis and chemical properties of chromone-3-carboxylic acid: (16KGS71).
Synthesis of (iso)quinoline, (iso)coumarin and (iso)chromene derivatives from acetylene compounds: (16UK637).
Synthetic protocols on 6H-benzo[c]chromen-6-ones: (16TJC1).

6.1.3 One Sulfur Atom
6.1.3.1 Thiopyran Compounds
Synthesis and chemical properties of thiochromone and its 3-substituted derivatives: (16KGS427).

6.2 Two Heteroatoms

Transition metal catalyzed pyrimidine, pyrazine, pyridazine and triazine synthesis: (16MI43).
Quinoxalines. Synthesis, reactions, mechanisms and structure: (16MI44).

6.2.1 Two Nitrogen Atoms
6.2.1.1 1,2-Heterocycles: Pyridazines
Synthesis of spiropyridazines, partially-saturated pyridazines and their fused derivatives: (16COC1512).
The use of 3,6-pyridazinediones in organic synthesis and chemical biology: (16JCR1).

6.2.1.2 1,3-Heterocycles: Monocyclic Pyrimidines and Hydropyrimidines (Except Pyrimidine Nucleoside Bases and Nucleosides, and Annulated Pyrimidines (Except Purines))
Acid-base and metal ion-binding properties of thiopyrimidine derivatives: (16CCR(327)200).

Advanced synthetic strategies for constructing quinazolinone scaffolds: (16S1253).
Biginelli reaction: (16TL5135).
The chemistry and bio-medicinal significance of pyrimidines and condensed pyrimidines: (16CTMC3133).
Microwave-assisted Biginelli reaction: An old reaction, a new perspective: (16COS569).
New routes for creation of fluoroalkyl-substituted heteroannulated pyrimidines: (16IAN1700).
Playing around with the size and shape of quinolizinium derivatives: Versatile ligands for duplex, triplex, quadruplex and abasic site-containing DNA: (16SL1775).
Progress in synthesis of enantiomerically pure 3,4-dihydropyrimidin-2-thione derivatives: (16CJOC283).
Pyrimidine-fused derivatives: Synthetic strategies and medicinal attributes: (16CTMC3175).
Recent developments in transition metal catalysis for quinazolinone synthesis: (16OBC8014).
Synthesis and antitumour activity of 4-aminoquinazoline derivatives: (16UK759).
Synthetic approaches toward the benzo[a]quinolizidine system: (16OPPI425).
Synthetic chemistry of pyrimidines and fused pyrimidines: (16SC645).
Utilization of isatoic anhydride in the syntheses of quinazoline and quinazolinone derivatives: (16SC993).

6.2.1.3 Pyrimidine Nucleoside Bases and Purines

The renaissance of metal pyrimidine nucleobase coordination chemistry: (16ACR1537).

6.2.1.4 Nucleotides and Nucleosides

Base-modified nucleosides: Etheno derivatives: (16FC19).
C-Linked 8-aryl guanine nucleobase adducts: Biological and fluorescent properties: (16CS3482).
Cyclic dinucleotide signalings have come of age to be inhibited by small molecules: (16CC9327).
Double-headed nucleotides: Building blocks for new nucleic acid architectures: (16AJC194).

Function-oriented synthesis: How to design simplified analogues of antibacterial nucleoside natural products: (16CR1106).
Recent trends in nucleotide synthesis (since 2000): (16CRV7854).
Supramolecular gels made from nucleobase, nucleoside and nucleotide analogs: (16CSR3188).

6.2.1.5 Nucleic Acids
Furan oxidation based cross-linking: A new approach for the study and targeting of nucleic acid and protein interactions: (16CC1539).
The application of fluorescence-conjugated pyrrole/imidazole polyamides in the characterization of protein-DNA complex formation: (16BS391).
DNA stains as surrogate nucleobases in fluorogenic hybridization probes: (16ACR714).
DNA minor groove binders-inspired by nature: (16ACSL689).
Nature-inspired design of smart biomaterials using the chemical biology of nucleic acids: (16BCJ843).
Properties and reactivity of nucleic acids relevant to epigenomics, transcriptomics, and therapeutics: (16CSR2637).
Quantum mechanical studies on the photophysics and the photochemistry of nucleic acids and nucleobases: (16CRV3540).

6.2.1.6 1,4-Heterocycles: Pyrazines and Hydropyrazines
Opportunities and challenges for direct C–H functionalization of piperazines: (16BJOC702).

6.2.2 One Nitrogen and One Oxygen Atoms
Metal-catalyzed cyclizations to pyran and oxazine derivatives: (16S3470).

6.2.3 One Nitrogen and One Sulfur Atoms
6.2.3.1 1,3-Thiazine Derivatives
Synthesis of 1,3-benzothiazinone derivatives: (16CJOC2024).

6.2.3.2 1,4-Thiazine Derivatives
N-Alkylphenothiazines—Synthesis, structure and application as ligands in metal complexes: (16HI461).

6.2.4 Two Oxygen Atoms
6.2.4.1 1,3-Dioxane Derivatives

1,3-Dioxane motif—A useful tool in monitoring molecular and supramolecular architectures: (16TL2683).

6.3 Three Heteroatoms

A systematic review on synthetic methods of symmetric triazines: (16AJPT57).

Advances in the domain of 4-amino-3-mercapto-1,2,4-triazine-5-ones: (16RSCA24010).

Ancient evolution and recent evolution converge for the biodegradation of cyanuric acid and related triazines: (16AEM1638).

Recent advances in synthesis and antifungal activity of 1,3,5-triazines: (16COS484).

Ring contraction in 1,2,4-triazine derivatives to give imidazoles: (16IAN2172).

Synthesis and chemical behavior of 1,2,4-triazine derivatives bearing phosphorus amides as donor–acceptors: (16COS408).

Use of 2,4,6-trichloro-1,3,5-triazine (TCT) as organic catalyst in organic synthesis: (16SC1155).

7. RINGS WITH MORE THAN SIX MEMBERS
7.1 Seven-Membered Rings

Advances in 1,3,5-triazepines chemistry: (16RSCA37286).

Axial chirality originating in amide and sulfonamide structures: Flexible stereochemistry of benzo-fused seven-membered-ring heterocycles: (16YGK56).

Current development in multicomponent catalytic synthesis of 1,5-benzodiazepines: (16IJCT1).

Developments in 1,2,5-triazepine chemistry: Reactions and synthetic applications: (16SC93).

Mild radical oxidative sp^3-carbon-hydrogen functionalization: Innovative construction of isoxazoline and dibenz[b,f]oxepine/azepine derivatives: (16SL526).

New synthesis of 1,4-oxazepane and 1,4-diazepane derivatives from N-propargylamines: (16RSCA99781).

Progress in synthesis and application of benzo[b][1,4]diazepine derivatives: (16CJOC711).

Synthesis and biological screening of oxazepinedione derivatives: (16BRAC1137).
Synthesis of pyrido-annulated diazepines, oxazepines and thiazepines: (16CMC4784).

7.2 Large Rings
7.2.1 Structure, Stereochemistry, Reactivity, and Design
Calixarenes and beyond: (16MI46).
Covalent or supramolecular combinations of resorcinarenes and porphyrinoids: (16JPP571).
Halogen derivatives of benzo- and dibenzocrown ethers: Synthesis, structure, properties, and application: (16UK172).
Laser spectroscopic study of cold gas-phase host-guest complexes of crown ethers: (16CR1034).
Macrocyclic and supramolecular chemistry: (16MI45).
Multicavity macrocyclic hosts: (16CC12130).
Recent advances in discovery, biosynthesis and genome mining of medicinally relevant polycyclic tetramate macrolactams: (16CTMC1727).
Supramolecular explorations: Exhibiting the extent of extended cationic cyclophanes: (16ACR262).
Thiacalixarenes: Emergent supramolecules in crystal engineering and molecular recognition: (16JIPMC(85)1).
Unrecognized reactivity of N-alkyl unsaturated imines to produce 8-membered saturated heterocycles: Synthetic application and biological functions: (16YGK700).
Versatile ruthenium(II/III) tetraazamacrocycle complexes and their nitrosyl derivatives: (16CCR(306)652).

7.2.2 Synthesis
Creating molecular macrocycles for anion recognition: (16BJOC611).
Diversity-oriented synthesis of macrocycle libraries for drug discovery and chemical biology: (16S1457).
Macrolactam analogues of macrolide natural products: (16OBC11301).
Synthesis and separation of cucurbit[*n*]urils and their derivatives: (16OBC4335).
Synthesis of fluvirucins (bioactive macrolactam glycosides isolated from actinomycetes) and their aglycons, the fluvirucinins: (16S2705).
Synthesis of heterocyclic [8]circulenes and related structures: (16SL498).

7.2.3 Applications

Catalytic reduction of proton, oxygen and carbon dioxide with cobalt macrocyclic complexes: (16JPP935).

Macrocyclic cavitands cucurbit[n]urils: Prospects for application in biochemistry, medicine and nanotechnology: (16UK795).

Metallacrown-based compounds: Applications in catalysis, luminescence, molecular magnetism, and adsorption: (16CCR(327)304).

Molecular recognition and activation by polyaza macrocyclic compounds based on host-guest interactions: (16CC10322).

Sustainable metal alkynyl chemistry: 3d Metals and polyaza macrocyclic ligands: (16CC3271).

8. HETEROCYCLES CONTAINING UNUSUAL HETEROATOMS

8.1 Phosphorus Heterocycles

Analogy of phosphaalkenes and azaphospholes with their respective non-phosphorus analogues: (16COC2099).

Cyclodiphosphazanes: Options are endless: (16DT12252).

Electron-rich aromatic 1,3-heterophospholes—Recent syntheses and impact of high electron density at σ P(2) on the reactivity: (16EJIC575).

Organophosphorus chemistry: (16MI47).

Organophosphorus compounds, particularly, (hetero-)phospholes in organic electronics: (16CEJ10718).

Phosphaphenalenes: An evolution of the phosphorus heterocycles: (16SL2293).

Recent developments in the chemistry of $3H$-1,2,3,4-triazaphosphole derivatives: (16EJIC595).

Resolution of P-stereogenic P-heterocycles via the formation of diastereomeric molecular and coordination complexes: (16DT1823).

Specific functionalization of cyclotriphosphazene for synthesis of smart dendrimers: (16DT1810).

A very peculiar family of N-heterocyclic phosphines: Unusual structures and the unique reactivity of 1,3,2-diazaphospholenes: (16DT5896).

8.2 Boron Heterocycles

Boratabenzene rare-earth metal complexes: (16CCR(314)2).

Electrochemistry and photoluminescence of icosahedral carboranes, boranes, metallacarboranes, and their derivatives: (16CRV14307).

Functionalization of carborane via carboryne intermediates: (16MI48).

Half-sandwich late transition metal complexes based on functionalized carborane ligands: (16CCR(309)21).

Icosahedral boron clusters: A perfect tool for the enhancement of polymer features: (16CSR5147).

The organometallic chemistry of boron-containing pincer ligands based on diazaboroles and carboranes: (16BCJ269).

Recent advances in the chemistry of 1,2-dehydro-*o*-carboranes (carborynes): (16CCR(314)14).

Recent developments in the chemistry of boron heterocycles: (16AHC(118)47).

Ring expansion reactions of anti-aromatic boroles to give unsaturated boracycles: (16CC9985).

8.3 Silicon Heterocycles

The chemistry of benzodisilacyclobutenes and benzobis(disilacyclobutene)s: New development of transition-metal-catalyzed reactions, stereochemistry and theoretical studies: (16DT3210).

Efficient methods for preparing silicon compounds: (16MI49).

Reactions of hydrosilanes with transition metal complexes (2009–2013): (16CRV11291).

Structure and conformational analysis of silacyclohexanes and 1,3-silaheterocyclohexanes: (16T5027).

8.4 Metallacycles

Metallacycle-mediated cross-coupling in natural product synthesis: (16T7093).

APPENDIX

Supplementary material related to this chapter can be found on the accompanying CD or online at https://doi.org/10.1016/bs.aihch.2018.02.003.

REFERENCES

66AHC(7)225	A.R. Katritzky and S.M. Weeds, *Adv. Heterocycl. Chem.*, **7**, 225 (1966).
79AHC(25)303	A.R. Katritzky and P.M. Jones, *Adv. Heterocycl. Chem.*, **25**, 303 (1979).
88AHC(44)269	L.I. Belen'kii, *Adv. Heterocycl. Chem.*, **44**, 269 (1988).
92AHC(55)31	L.I. Belen'kii and N.D. Kruchkovskaya, *Adv. Heterocycl. Chem.*, **55**, 31 (1992).
98AHC(71)291	L.I. Belen'kii and N.D. Kruchkovskaya, *Adv. Heterocycl. Chem.*, **71**, 291 (1998).
99AHC(73)295	L.I. Belen'kii, N.D. Kruchkovskaya, and V.N. Gramenitskaya, *Adv. Heterocycl. Chem.*, **73**, 295 (1999).
01AHC(79)199	L.I. Belen'kii, N.D. Kruchkovskaya, and V.N. Gramenitskaya, *Adv. Heterocycl. Chem.*, **79**, 199 (2001).
04AHC(87)1	L.I. Belen'kii and V.N. Gramenitskaya, *Adv. Heterocycl. Chem.*, **87**, 1 (2004).
06AHC(92)145	L.I. Belen'kii, V.N. Gramenitskaya, and Yu.B. Evdokimenkova, *Adv. Heterocycl. Chem.*, **92**, 145 (2006).
11AHC(102)1	L.I. Belen'kii, V.N. Gramenitskaya, and Yu.B. Evdokimenkova, *Adv. Heterocycl. Chem.*, **102**, 1 (2011).
13AHC(108)195	L.I. Belen'kii and Yu.B. Evdokimenkova, *Adv. Heterocycl. Chem.*, **108**, 195 (2013).
14AHC(111)147	L.I. Belen'kii and Yu.B. Evdokimenkova, *Adv. Heterocycl. Chem.*, **111**, 147 (2014).
15AHC(116)193	L.I. Belen'kii and Yu.B. Evdokimenkova, *Adv. Heterocycl. Chem.*, **116**, 193 (2015).
15MI1	M.A. Yurovskaya, *Khimiya Aromaticheskikh Geterotsiklicheskikh Soedinenii (Chemistry of Heteroaromatic Compounds, in Russian)*, BINOM: Moscow: (2015).
15MI2	M. North, editor: *Sustainable Catalysis: Without Metals or Other Endangered Elements*, Royal Society of Chemistry: (2015), Pt. 1.
15MI3	M. North, editor: *Sustainable Catalysis: With Non-endangered Metals*, Royal Society of Chemistry: (2015), Pt. 1.
15MI4	M. North, editor: *Sustainable Catalysis: With Non-endangered Metals*, Royal Society of Chemistry: (2015), Pt. 2.
15MI5	D.C. Blakemore, P.M. Doyle, and Y.M. Fobian, editors: *Synthetic Methods in Drug Discovery*, Royal Society of Chemistry: (2015), **Vol. 1**.
15MI6	S. Bräse, editor: *Privileged Scaffolds in Medicinal Chemistry: Design, Synthesis, Evaluation*, Royal Society of Chemistry: (2015).
15OMC(40)1	X. Just-Baringo, I. Yalavac, and D. Procter, In I. Fairlamb and J. Lynam, editors: *Organometallic Chemistry*, Royal Society of Chemistry: (2015), **Vol. 40**, p 1.
15OMC(40)140	J.S. Ward, In I. Fairlamb and J. Lynam, editors: *Organometallic Chemistry*, Royal Society of Chemistry: (2015), **Vol. 40**, p 107.
16ACA465	R. Kaur, A.R. Dwivedi, B. Kumar, and V. Kumar, *Anticancer Agents Med. Chem.*, **16**, 465 (2016).
16ACA1205	B. Vue, S. Zhang, and Q.H. Chen, *Anticancer Agents Med. Chem.*, **16**, 1205 (2016).
16ACA1301	A. Aliabadi, *Anticancer Agents Med. Chem.*, **16**, 1301 (2016).
16ACA1339	A.C. Pinheiro, T.C.M. Nogueira, and M.V.N. de Souza, *Anticancer Agents Med. Chem.*, **16**, 1339 (2016).
16ACA1403	S. Yadav, B. Narasimhan, and H. Kaur, *Anticancer Agents Med. Chem.*, **16**, 1403 (2016).

16ABC7013	A. Kapturkiewicz, *Anal. Bioanal. Chem.*, **408**, 7013 (2016).
16ACB171	G.L. Adams and A.B. Smith III, *Alkaloids Chem. Biol.*, **76**, 171 (2016).
16ACB259	M. Kitajima and H. Takayama, *Alkaloids Chem. Biol.*, **76**, 259 (2016).
16ACR4	P. Yin, Q.H. Zhang, and J.M. Shreeve, *Acc. Chem. Res.*, **49**, 4 (2016).
16ACR67	Y. Minami and T. Hiyama, *Acc. Chem. Res.*, **49**, 67 (2016).
16ACR78	W.W. Li, K.H. Hendriks, M.M. Wienk, and R.A.J. Janssen, *Acc. Chem. Res.*, **49**, 78 (2016).
16ACR175	Y.Z. Lin and X.W. Zhan, *Acc. Chem. Res.*, **49**, 175 (2016).
16ACR193	C. Wang, L. Luo, and H. Yamamoto, *Acc. Chem. Res.*, **49**, 193 (2016).
16ACS241	J.E. Sears and D.L. Boger, *Acc. Chem. Res.*, **49**, 241 (2016).
16ACS252	J.P. Depres, P. Delair, J.F. Poisson, A. Kanazawa, and A.E. Greene, *Acc. Chem. Res.*, **49**, 252 (2016).
16ACR262	E.J. Dale, N.A. Vermeulen, M. Juricek, J.C. Barnes, R.M. Young, M.R. Wasielewski, and J.F. Stoddart, *Acc. Chem. Res.*, **49**, 262 (2016).
16ACR357	S. Roy, K.C. Mondal, and H.W. Roesky, *Acc. Chem. Res.*, **49**, 357 (2016).
16ACR714	F. Hovelmann and O. Seitz, *Acc. Chem. Res.*, **49**, 714 (2016).
16ACR801	A.A. Valiente-Gabioud, M.C. Miotto, M.E. Chesta, V. Lombardo, A. Binolfi, and C.O. Fernandez, *Acc. Chem. Res.*, **49**, 801 (2016).
16ACR868	F. Wurthner, *Acc. Chem. Res.*, **49**, 868 (2016).
16ACR922	Y.P. Huo and H.Q. Zeng, *Acc. Chem. Res.*, **49**, 922 (2016).
16ACR974	K.S. Halskov, B.S. Donslund, B.M. Paz, and K.A. Jorgensen, *Acc. Chem. Res.*, **49**, 974 (2016).
16ACR1006	S. Santoro, M. Kalek, G.P. Huang, and F. Himo, *Acc. Chem. Res.*, **49**, 1006 (2016).
16ACR1042	Q. Peng and R.S. Paton, *Acc. Chem. Res.*, **49**, 1042 (2016).
16ACR1080	C. Bruckner, *Acc. Chem. Res.*, **49**, 1080 (2016).
16ACR1168	T.G. Minehan, *Acc. Chem. Res.*, **49**, 1168 (2016).
16ACR1250	G. Tanriver, B. Dedeoglu, S. Catak, and V. Aviyente, *Acc. Chem. Res.*, **49**, 1250 (2016).
16ACR1279	D.M. Walden, O.M. Ogba, R.C. Johnston, and P.H.Y. Cheong, *Acc. Chem. Res.*, **49**, 1279 (2016).
16ACR1320	S.P. Pitre, C.D. McTiernan, and J.C. Scaiano, *Acc. Chem. Res.*, **49**, 1320 (2016).
16ACR1369	T.L. Wang, X.Y. Han, F.R. Zhong, W.J. Yao, and Y.X. Lu, *Acc. Chem. Res.*, **49**, 1369 (2016).
16ACR1409	S.B. Chen and W.H. Binder, *Acc. Chem. Res.*, **49**, 1409 (2016).
16ACR1458	J.P. Moerdyk, D. Schilter, and C.W. Bielawski, *Acc. Chem. Res.*, **49**, 1458 (2016).
16ACR1477	Y.Z. Liu, P. Persson, V. Sundstrom, and K. Warnmark, *Acc. Chem. Res.*, **49**, 1477 (2016).
16ACR1537	B. Lippert and P.J.S. Miguel, *Acc. Chem. Res.*, **49**, 1537 (2016).
16ACR1566	I. Ghosh, L. Marzo, A. Das, R. Shaikh, and B. Konig, *Acc. Chem. Res.*, **49**, 1566 (2016).
16ACR1658	S.S. Bhojgude, A. Bhunia, and A.T. Biju, *Acc. Chem. Res.*, **49**, 1658 (2016).
16ACR1911	J.R. Chen, X.Q. Hu, L.Q. Lu, and W.J. Xiao, *Acc. Chem. Res.*, **49**, 1911 (2016).
16ACR1957	S.A. Morris, J. Wang, and N. Zheng, *Acc. Chem. Res.*, **49**, 1957 (2016).
16ACR2284	T. Chatterjee, N. Iqbal, Y. You, and E.J. Cho, *Acc. Chem. Res.*, **49**, 2284 (2016).

16ACR2295	D. Staveness, I. Bosque, and C.R.J. Stephenson, *Acc. Chem. Res.*, **49**, 2295 (2016).
16ACR2316	M. Majek and A.J. von Wangelin, *Acc. Chem. Res.*, **49**, 2316 (2016).
16ACR2372	A.S. Tsarkova, Z.M. Kaskova, and I.V. Yampolsky, *Acc. Chem. Res.*, **49**, 2372 (2016).
16ACR2390	E. Stempel and T. Gaich, *Acc. Chem. Res.*, **49**, 2390 (2016).
16ACR2489	D.G. Hall, T. Rybak, and T. Verdelet, *Acc. Chem. Res.*, **49**, 2489 (2016).
16ACR2680	T. Strassner, *Acc. Chem. Res.*, **49**, 2680 (2016).
16ACR2807	I. Kreituss and J.W. Bode, *Acc. Chem. Res.*, **49**, 2807 (2016).
16ACSC1389	C.G.S. Lima, T.D. Lima, M. Duarte, I.D. Jurberg, and M.W. Paixao, *ACS Catal.*, **6**, 1389 (2016).
16ACSC2515	Y. Wei and M. Shi, *ACS Catal.*, **6**, 2515 (2016).
16ACSC3629	W.D.G. Brittain, B.R. Buckley, and J.S. Fossey, *ACS Catal.*, **6**, 3629 (2016).
16ACSC5978	D.B. Zhao, L. Candish, D. Paul, and F. Glorius, *ACS Catal.*, **6**, 5978 (2016).
16ACSC6651	J.J. Feng and J.L. Zhang, *ACS Catal.*, **6**, 6651 (2016).
16ACSC6704	Y. Kwon, K.J.P. Schouten, J.C. van der Waal, E. de Jong, and M.T.M. Koper, *ACS Catal.*, **6**, 6704 (2016).
16ACSID382	N. Ye, H.Y. Chen, E.A. Wold, P.Y. Shi, and J. Zhou, *ACS Infect. Dis.*, **2**, 382 (2016).
16ACSL689	A.I. Khalaf, A.A.H. Al-Kadhimi, and J.H. Ali, *Acta Chim. Slov.*, **63**, 689 (2016).
16AEM1638	J.L. Seffernick and L.P. Wackett, *Appl. Environ. Microbiol.*, **82**, 1638 (2016).
16AG(E)2308	S.B. Wang and X.C. Wang, *Angew. Chem. Int. Ed.*, **55**, 2308 (2016).
16AG(E)3882	E.A. Crane and K. Gademann, *Angew. Chem. Int. Ed.*, **55**, 3882 (2016).
16AG(E)4130	M. Barboiu, A.M. Stadler, and J.M. Lehn, *Angew. Chem. Int. Ed.*, **55**, 4130 (2016).
16AG(E)4396	W.J. Chung and C.D. Vanderwal, *Angew. Chem. Int. Ed.*, **55**, 4396 (2016).
16AG(E)4436	Y.F. Li, D.P. Hari, M.V. Vita, and J. Waser, *Angew. Chem. Int. Ed.*, **55**, 4436 (2016).
16AG(E)4882	B. Krautler, *Angew. Chem. Int. Ed.*, **55**, 4882 (2016).
16AG(E)5642	M. Ghavre, J. Froese, M. Pour, and T. Hudlicky, *Angew. Chem. Int. Ed.*, **55**, 5642 (2016).
16AG(E)6599	M.F. Chellat, L. Raguz, and R. Riedl, *Angew. Chem. Int. Ed.*, **55**, 6599 (2016).
16AG(E)7586	M. Garcia-Castro, S. Zimmermann, M.G. Sankar, and K. Kumar, *Angew. Chem. Int. Ed.*, **55**, 7586 (2016).
16AG(E)13934	S. Vellalath and D. Romo, *Angew. Chem. Int. Ed.*, **55**, 13934 (2016).
16AG(E)13944	T. Josse, J. De Winter, P. Gerbaux, and O. Coulembier, *Angew. Chem. Int. Ed.*, **55**, 13944 (2016).
16AG(E)14912	M.H. Wang and K.A. Scheidt, *Angew. Chem. Int. Ed.*, **55**, 14912 (2016).
16AG(E)15712	G.G. Zhang, Z.A. Lan, and X.C. Wang, *Angew. Chem. Int. Ed.*, **55**, 15712 (2016).
16AHC(118)1	E. Licandro, S. Cauteruccio, and D. Dova, *Adv. Heterocycl. Chem.*, **118**, 1 (2016).
16AHC(118)47	B.J. Wang and M.P. Groziak, *Adv. Heterocycl. Chem.*, **118**, 47 (2016).

16AHC(118)91	F.M. Perna, A. Salomone, and V. Capriati, *Adv. Heterocycl. Chem.*, **118**, 91 (2016).
16AHC(118)129	J.-N. Volle, R. Guillon, F. Bancel, Y.-A. Bekro, J.-L. Pirat, and D. Virieux, *Adv. Heterocycl. Chem.*, **118**, 129 (2016).
16AHC(118)195	M.M. Heravi and B. Talaei, *Adv. Heterocycl. Chem.*, **118**, 195 (2016).
16AHC(119)1	C.D. Hall and S.S. Panda, *Adv. Heterocycl. Chem.*, **119**, 1 (2016).
16AHC(119)25	M. Movsisyan, M.M.A. Moens, and C.V. Stevens, *Adv. Heterocycl. Chem.*, **119**, 25 (2016).
16AHC(119)57	V.V. Zhdankin, *Adv. Heterocycl. Chem.*, **119**, 57 (2016).
16AHC(119)81	J.A. Joule, *Adv. Heterocycl. Chem.*, **119**, 81 (2016).
16AHC(119)107	J. Cossy and A. Guérinot, *Adv. Heterocycl. Chem.*, **119**, 107 (2016).
16AHC(119)143	A. Schmidt, S. Wiechmann, and C.F. Otto, *Adv. Heterocycl. Chem.*, **119**, 143 (2016).
16AHC(119)173	C.P. Constantinides and P.A. Koutentis, *Adv. Heterocycl. Chem.*, **119**, 173 (2016).
16AHC(119)209	B. Stanovnik, *Adv. Heterocycl. Chem.*, **119**, 209 (2016).
16AHC(119)241	A. Padwa and S. Bur, *Adv. Heterocycl. Chem.*, **119**, 241 (2016).
16AHC(120)1	O.S. Kanishchev and W.R. Dolbier Jr, *Adv. Heterocycl. Chem.*, **120**, 1 (2016).
16AHC(120)43	H.K. Potukuchi, I. Colomer, and T.J. Donohoe, *Adv. Heterocycl. Chem.*, **120**, 43 (2016).
16AHC(120)67	C.F. Gers-Panther and T.J.J. Müller, *Adv. Heterocycl. Chem.*, **120**, 67 (2016).
16AHC(120)195	J.M. Ludlow III and G.R. Newkome, *Adv. Heterocycl. Chem.*, **120**, 195 (2016).
16AHC(120)237	M. Karelson and D.A. Dobchev, *Adv. Heterocycl. Chem.*, **120**, 237 (2016).
16AHC(120)275	J. Bariwal, R. Kaur, and E.V. Van der Eycken, *Adv. Heterocycl. Chem.*, **120**, 275 (2016).
16AHC(120)301	H. Szatylowicz, O.A. Stasyuk, and T.M. Krygowski, *Adv. Heterocycl. Chem.*, **120**, 301 (2016).
16AJOC9	L.L. Anderson, *Asian J. Org. Chem.*, **5**, 9 (2016).
16AJOC839	M.B. Halle and R.A. Fernandes, *Asian J. Org. Chem.*, **5**, 839 (2016).
16AJPT57	D.S. Rao, G.V.P. Kumar, B. Pooja, and G. Harika, *Am. J. PharmTech Res.*, **6**(4), 57 (2016).
16AJPT62	A. Bhalla, S. Berry, and S.S. Bari, *Am. J. PharmTech Res.*, **6**(6), 62 (2016).
16AJPT96	A. Bhalla, D. Narula, and S.S. Bari, *Am. J. PharmTech Res.*, **6**(6), 96 (2016).
16AMB2107	S.K. Schwechheimer, E.Y. Park, J.L. Revuelta, J. Becker, and C. Wittmann, *Appl. Microbiol. Biotechnol.*, **100**, 2107 (2016).
16AMB4269	B. Hofer, *Appl. Microbiol. Biotechnol.*, **100**, 4269 (2016).
16AMB6571	T.M. Costa, L.B.B. Tavares, and D. de Oliveira, *Appl. Microbiol. Biotechnol.*, **100**, 6571 (2016).
16ANCA9	K.D. Clark, O. Nacham, J.A. Purslow, S.A. Pierson, and J.L. Anderson, *Anal. Chim. Acta*, **934**, 9 (2016).
16AOC175	C. Pettinari, R. Pettinari, and F. Marchetti, *Adv. Organomet. Chem.*, **65**, 175 (2016).
16AP507	S.T.S. Hassan and M. Zemlicka, *Arch. Pharm.*, **349**, 507 (2016).
16ARJC1443	M. Malhotra, P. Ghai, B. Narasimhan, and A. Deep, *Arab. J. Chem.*, **9**, S1443 (2016).
16ARK(1)82	S. Banerjee, *ARKIVOC*, (1), 82 (2016).

16ARK(1)111	C.K. Ghosh and A. Chakraborty, *ARKIVOC*, (1), 111 (2016).
16ARK(1)150	K.K. Laali, *ARKIVOC*, (1), 150 (2016).
16ARK(1)307	K. Lal and P. Rani, *ARKIVOC*, (1), 307 (2016).
16ARK(1)375	C.K. Ghosh, A. Chakraborty, and C. Bandyopadhyay, *ARKIVOC*, (1), 375 (2016).
16ARK(2)9	Y. Yamamoto, A.I. Almansour, N. Arumugam, and R.S. Kumar, *ARKIVOC*, (2), 9 (2016).
16ARK(2)172	D. Basavaiah and G.C. Reddy, *ARKIVOC*, (2), 172 (2016).
16ASC3485	J. Dolfen, N.N. Yadav, N. De Kimpe, M. D'Hooghe, and H.J. Ha, *Adv. Synth. Catal.*, **358**, 3485 (2016).
16ASC3657	E. Marcantoni and M. Petrini, *Adv. Synth. Catal.*, **358**, 3657 (2016).
16BCJ269	M. Yamashita, *Bull. Chem. Soc. Jpn.*, **89**, 269 (2016).
16BCJ843	G.N. Pandian and H. Sugiyama, *Bull. Chem. Soc. Jpn.*, **89**, 843 (2016).
16BCH49	C. Cuyamendous, A. de la Torre, Y.Y. Lee, K.S. Leung, A. Guy, V. Bultel-Ponce, J.M. Galano, J.C.Y. Lee, C. Oger, and T. Durand, *Biochimie*, **130**, 49 (2016).
16BJOC429	L.A. Bryant, R. Fanelli, and A.J.A. Cobb, *Beilstein J. Org. Chem.*, **12**, 429 (2016).
16BJOC444	R.S. Menon, A.T. Biju, and V. Nair, *Beilstein J. Org. Chem.*, **12**, 444 (2016).
16BJOC505	I.G. Sonsona, E. Marques-Lopez, and R.P. Herrera, *Beilstein J. Org. Chem.*, **12**, 505 (2016).
16BJOC571	T.F. Schaberle, *Beilstein J. Org. Chem.*, **12**, 571 (2016).
16BJOC611	A.H. Flood, *Beilstein J. Org. Chem.*, **12**, 611 (2016).
16BJOC702	Z.S. Ye, K.E. Gettys, and M.J. Dai, *Beilstein J. Org. Chem.*, **12**, 702 (2016).
16BJOC918	A. Mielgo and C. Palomo, *Beilstein J. Org. Chem.*, **12**, 918 (2016).
16BJOC1000	B. Yu, H. Xing, D.Q. Yu, and H.M. Liu, *Beilstein J. Org. Chem.*, **12**, 1000 (2016).
16BJOC1269	M. Haji, *Beilstein J. Org. Chem.*, **12**, 1269 (2016).
16BJOC1949	L. Brulikova, A. Harrison, M.J. Miller, and J. Hlavac, *Beilstein J. Org. Chem.*, **12**, 1949 (2016).
16BJOC2038	J.T. Mohr, J.T. Moore, and B.M. Stoltz, *Beilstein J. Org. Chem.*, **12**, 2038 (2016).
16BJOC2104	J. Adrian, L.J. Gross, and C.B.W. Stark, *Beilstein J. Org. Chem.*, **12**, 2104 (2016).
16BJOC2325	D.J. Atkinson, B.J. Naysmith, D.P. Furkert, and M.A. Brimble, *Beilstein J. Org. Chem.*, **12**, 2325 (2016).
16BJOC2834	P. Nagorny and Z.K. Sun, *Beilstein J. Org. Chem.*, **12**, 2834 (2016).
16BMC1419	H.J. Zhang, R.S. Tong, L. Bai, J.Y. Shi, and L. Ouyang, *Bioorg. Med. Chem.*, **24**, 1419 (2016).
16BMC2361	I. Khan, S. Zaib, S. Batool, N. Abbas, Z. Ashraf, J. Iqbal, and A. Saeed, *Bioorg. Med. Chem.*, **24**, 2361 (2016).
16BMC5007	S.X. Gu, P. Xue, X.L. Ju, and Y.Y. Zhu, *Bioorg. Med. Chem.*, **24**, 5007 (2016).
16BMC5017	S.S. Laev, N.F. Salakhutdinov, and O.I. Lavrik, *Bioorg. Med. Chem.*, **24**, 5017 (2016).
16BMCL251	V. Corce, S.G. Gouin, S. Renaud, F. Gaboriau, and D. Deniaud, *Bioorg. Med. Chem. Lett.*, **26**, 251 (2016).
16BMCL1861	H.M. Cheng, S.K. Nair, and B.W. Murray, *Bioorg. Med. Chem. Lett.*, **26**, 1861 (2016).
16BMCL2383	S. Shah and J. Savjani, *Bioorg. Med. Chem. Lett.*, **26**, 2383 (2016).

16BMCL3838	J.C. Rech, A. Bhattacharya, M.A. Letavic, and B.M. Savall, *Bioorg. Med. Chem. Lett.*, **26**, 3838 (2016).
16BMCL5139	K.O. Cameron and R.G. Kurumbail, *Bioorg. Med. Chem. Lett.*, **26**, 5139 (2016).
16BMCL5843	T. Motoyama and H. Osada, *Bioorg. Med. Chem. Lett.*, **26**, 5843 (2016).
16BRAC1137	L. Joseph, M. George, and A.K. Alexander, *Biointerface Res. Appl. Chem.*, **6**, 1137 (2016).
16BS391	Y.W. Han, H. Sugiyama, and Y. Harada, *Biomater. Sci.*, **4**, 391 (2016).
16CAJ155	J.K. Howard, K.J. Rihak, A.C. Bissember, and J.A. Smith, *Chem. Asian J.*, **11**, 155 (2016).
16CAJ448	P. Gandeepan and C.H. Cheng, *Chem. Asian J.*, **11**, 448 (2016).
16CAJ642	M. Petrovic and E.G. Occhiato, *Chem. Asian J.*, **11**, 642 (2016).
16CAJ1988	R. Kamal, V. Kumar, and R. Kumar, *Chem. Asian J.*, **11**, 1988 (2016).
16CBD37	S.R.M. Ibrahim and G.A. Mohamed, *Chem. Biodivers.*, **13**, 37 (2016).
16CC1539	L.L.G. Carrette, E. Gyssels, N. De Laet, and A. Madder, *Chem. Commun.*, **52**, 1539 (2016).
16CC1997	P. Etayo, C. Ayats, and M.A. Pericas, *Chem. Commun.*, **52**, 1997 (2016).
16CC2220	J.T. Yu and C.D. Pan, *Chem. Commun.*, **52**, 2220 (2016).
16CC2433	M.M. Obadia and E. Drockenmuller, *Chem. Commun.*, **52**, 2433 (2016).
16CC3077	F. Meyer, *Chem. Commun.*, **52**, 3077 (2016).
16CC3271	T. Ren, *Chem. Commun.*, **52**, 3271 (2016).
16CC5777	E. Peris, *Chem. Commun.*, **52**, 5777 (2016).
16CC6958	T. Kaur, P. Wadhwa, S. Bagchi, and A. Sharma, *Chem. Commun.*, **52**, 6958 (2016).
16CC9327	C. Opoku-Temeng, J. Zhou, Y. Zheng, J.M. Su, and H.O. Sintim, *Chem. Commun.*, **52**, 9327 (2016).
16CC9485	S.M. Parke, M.P. Boone, and E. Rivard, *Chem. Commun.*, **52**, 9485 (2016).
16CC9985	J.H. Barnard, S. Yruegas, K.X. Huang, and C.D. Martin, *Chem. Commun.*, **52**, 9985 (2016).
16CC10205	S.P. Wang, Y.F. Shen, B.Y. Zhu, J. Wu, and S.J. Li, *Chem. Commun.*, **52**, 10205 (2016).
16CC10322	D.C. Zhong and T.B. Lu, *Chem. Commun.*, **52**, 10322 (2016).
16CC12130	W.B. Hu, W.J. Hu, Y.H.A. Liu, J.S. Li, B. Jiang, and K. Wen, *Chem. Commun.*, **52**, 12130 (2016).
16CC13543	L.Z. Zhao, R. Qu, A. Li, R.J. Ma, and L.Q. Shi, *Chem. Commun.*, **52**, 13543 (2016).
16CC14188	F. Wei, W.G. Wang, Y.D. Ma, C.H. Tung, and Z.H. Xu, *Chem. Commun.*, **52**, 14188 (2016).
16CCC664	F. Giacalone and M. Gruttadauria, *ChemCatChem*, **8**, 664 (2016).
16CCC1755	C. Michon, K. MacIntyre, Y. Corre, and F. Agbossou-Niedercorn, *ChemCatChem*, **8**, 1755 (2016).
16CCC1882	E. Pair, T. Cadart, V. Levacher, and J.F. Briere, *ChemCatChem*, **8**, 1882 (2016).
16CCL1166	X.Q. Hou, Y.T. Sun, L. Liu, S.T. Wang, R.L. Geng, and X.F. Shao, *Chin. Chem. Lett.*, **27**, 1166 (2016).
16CCL1231	Y. Liu, G. Zhan, Z.W. Liu, Z.Q. Bian, and C.H. Huang, *Chin. Chem. Lett.*, **27**, 1231 (2016).

16CCL1731	Z.M. Wu, S.Z. Liu, X.Z. Cheng, W.Z. Ding, T. Zhu, and B. Chen, *Chin. Chem. Lett.*, **27**, 1731 (2016).
16CCR(306)1	K. Ladomenou, V. Nikolaou, G. Charalambidis, and A.G. Coutsolelos, *Coord. Chem. Rev.*, **306**, 1 (2016).
16CCR(306)195	H.L. Wang, B.W. Wang, Y.Z. Bian, S. Gao, and J.Z. Jiang, *Coord. Chem. Rev.*, **306**, 195 (2016).
16CCR(306)652	F.G. Doro, K.Q. Ferreira, Z.N. da Rocha, G.F. Caramori, A.J. Gomes, and E. Tfouni, *Coord. Chem. Rev.*, **306**, 652 (2016).
16CCR(307)1	C. Pettinari, A. Tabacaru, and S. Galli, *Coord. Chem. Rev.*, **307**, 1 (2016).
16CCR(307)255	H.X. Zhang, M. Liu, T. Wen, and J. Zhang, *Coord. Chem. Rev.*, **307**, 255 (2016).
16CCR(307)382	K. Fujie and H. Kitagawa, *Coord. Chem. Rev.*, **307**, 382 (2016).
16CCR(308)1	B. Li, S.Q. Zang, L.Y. Wang, and T.C.W. Mak, *Coord. Chem. Rev.*, **308**, 1 (2016).
16CCR(308)32	L.M.T. Frija, A.J.L. Pombeiro, and M.N. Kopylovich, *Coord. Chem. Rev.*, **308**, 32 (2016).
16CCR(308)460	G. Pratviel, *Coord. Chem. Rev.*, **308**, 460 (2016).
16CCR(309)21	Z.J. Yao and W. Deng, *Coord. Chem. Rev.*, **309**, 21 (2016).
16CCR(309)107	M.P. Donzello, C. Ercolani, V. Novakova, P. Zimcik, and P.A. Stuzhin, *Coord. Chem. Rev.*, **309**, 107 (2016).
16CCR(310)131	T. Basova, A. Hassan, M. Durmus, A.G. Gurek, and V. Ahsen, *Coord. Chem. Rev.*, **310**, 131 (2016).
16CCR(312)149	M. Elsherbini, W.S. Hamama, and H.H. Zoorob, *Coord. Chem. Rev.*, **312**, 149 (2016).
16CCR(314)2	P. Cui and Y.F. Chen, *Coord. Chem. Rev.*, **314**, 2 (2016).
16CCR(314)14	D. Zhao and Z.W. Xie, *Coord. Chem. Rev.*, **314**, 14 (2016).
16CCR(315)135	I. Castro, W.P. Barros, M.L. Calatayud, F. Lloret, N. Marino, G. De Munno, H.O. Stumpf, R. Ruiz-Garcia, and M. Julve, *Coord. Chem. Rev.*, **315**, 135 (2016).
16CCR(316)68	A. Nasr, A. Winkler, and M. Tamm, *Coord. Chem. Rev.*, **316**, 68 (2016).
16CCR(318)1	H. Lu, J. Mack, T. Nyokong, N. Kobayashi, and Z. Shen, *Coord. Chem. Rev.*, **318**, 1 (2016).
16CCR(318)16	M. Strianese and C. Pellecchia, *Coord. Chem. Rev.*, **318**, 16 (2016).
16CCR(319)110	V.E. Pushkarev, L.G. Tomilova, and V.N. Nemykin, *Coord. Chem. Rev.*, **319**, 110 (2016).
16CCR(324)1	P.A. Gray and N. Burford, *Coord. Chem. Rev.*, **324**, 1 (2016).
16CCR(325)2	R. Czerwieniec, M.J. Leitl, H.H.H. Homeier, and H. Yersin, *Coord. Chem. Rev.*, **325**, 2 (2016).
16CCR(325)59	O. Horvath, Z. Valicsek, M.A. Fodor, M.M. Majora, M. Imran, G. Grampp, and A. Wankmuller, *Coord. Chem. Rev.*, **325**, 59 (2016).
16CCR(325)125	B. Colasson, A. Credi, and G. Ragazzon, *Coord. Chem. Rev.*, **325**, 125 (2016).
16CCR(327)200	A. Sigel, B.P. Operschall, A. Matera-Witkiewicz, J. Swiatek-Kozlowska, and H. Sigel, *Coord. Chem. Rev.*, **327**, 200 (2016).
16CCR(327)221	I. Lakomska and M. Fandzloch, *Coord. Chem. Rev.*, **327**, 221 (2016).
16CCR(327)242	J. Masternak, M. Zienkiewicz-Machnik, M. Kowalik, A. Jablonska-Wawrzycka, P. Rogala, A. Adach, and B. Barszcz, *Coord. Chem. Rev.*, **327**, 242 (2016).
16CCR(327)287	M.A. Santos, K. Chand, and S. Chaves, *Coord. Chem. Rev.*, **327**, 287 (2016).

16CCR(327)304	M. Ostrowska, I.O. Fritsky, E. Gumienna-Kontecka, and A.V. Pavlishchuk, *Coord. Chem. Rev.*, **327**, 304 (2016).
16CCR(327)349	S. Medici, M. Peana, G. Crisponi, V.M. Nurchi, J.I. Lachowicz, M. Remelli, and M.A. Zoroddu, *Coord. Chem. Rev.*, **327**, 349 (2016).
16CCR(329)16	A. Thakur and D. Mandal, *Coord. Chem. Rev.*, **329**, 16 (2016).
16CCR(329)191	W.K. Liu and R. Gust, *Coord. Chem. Rev.*, **329**, 191 (2016).
16CEJ1206	G. Rotas and N. Tagmatarchis, *Chem. Eur. J.*, **22**, 1206 (2016).
16CEJ1558	W.B. Li, L.J. Zhou, and J.L. Zhang, *Chem. Eur. J.*, **22**, 1558 (2016).
16CEJ2856	M.J. James, P. O'Brien, R.J.K. Taylor, and W.P. Unsworth, *Chem. Eur. J.*, **22**, 2856 (2016).
16CEJ3600	M. Pfaffenbach and T. Gaich, *Chem. Eur. J.*, **22**, 3600 (2016).
16CEJ3632	S. Meninno and A. Lattanzi, *Chem. Eur. J.*, **22**, 3632 (2016).
16CEJ5468	M. Ito, Y.K. Tahara, and T. Shibata, *Chem. Eur. J.*, **22**, 5468 (2016).
16CEJ6720	G. Dominguez and J. Perez-Castells, *Chem. Eur. J.*, **22**, 6720 (2016).
16CEJ7670	P. He, J.G. Zhang, X. Yin, J.T. Wu, L. Wu, Z.N. Zhou, and T.L. Zhang, *Chem. Eur. J.*, **22**, 7670 (2016).
16CEJ9456	A. Tikad, J.A. Delbrouck, and S.P. Vincent, *Chem. Eur. J.*, **22**, 9456 (2016).
16CEJ10294	P.P. de Castro, A.G. Carpanez, and G.W. Amarante, *Chem. Eur. J.*, **22**, 10294 (2016).
16CEJ10718	M.A. Shameem and A. Orthaber, *Chem. Eur. J.*, **22**, 10718 (2016).
16CEJ11918	X.W. Liang, C. Zheng, and S.L. You, *Chem. Eur. J.*, **22**, 11918 (2016).
16CEJ13430	M.D. Retamosa, E. Matador, D. Monge, J.M. Lassaletta, and R. Fernandez, *Chem. Eur. J.*, **22**, 13430 (2016).
16CEJ13764	V. Martinez and M. Henary, *Chem. Eur. J.*, **22**, 13764 (2016).
16CEJ17910	Y. Jiang, R. Sun, X.Y. Tang, and M. Shi, *Chem. Eur. J.*, **22**, 17910 (2016).
16CHIR169	F. Ghirga, D. Quaglio, P. Ghirga, S. Berardozzi, G. Zappia, B. Botta, M. Mori, and I. D'Acquarica, *Chirality*, **28**, 169 (2016).
16CHMC270	T. Weber and P.M. Selzer, *ChemMedChem*, **11**, 270 (2016).
16CHMC646	D. Song and S.T. Ma, *ChemMedChem*, **11**, 646 (2016).
16CHMC2102	J.Y. Dong, Q.J. Zhang, Q. Cui, G. Huang, X.Y. Pan, and S.S. Li, *ChemMedChem*, **11**, 2102 (2016).
16CHS266	M.E. Heckenbach, F.N. Romero, M.D. Green, and R.U. Halden, *Chemosphere*, **150**, 266 (2016).
16CJOC283	H.H. Rao, Z.J. Quan, L. Bai, and H.L. Ye, *Chin. J. Org. Chem.*, **36**, 283 (2016).
16CJOC687	X.X. Pan and L.J. Huo, *Chin. J. Org. Chem.*, **36**, 687 (2016).
16CJOC711	L.Y. Yin, L.Z. Wang, X.Q. Li, and Y.S. An, *Chin. J. Org. Chem.*, **36**, 711 (2016).
16CJOC875	A. Abdukader, Y.H. Zhang, Z.P. Zhang, and C.J. Liu, *Chin. J. Org. Chem.*, **36**, 875 (2016).
16CJOC913	Q. Miao and H.L. Sun, *Chin. J. Org. Chem.*, **36**, 913 (2016).
16CJOC939	Q.Y. Wang, H.H. Chang, W.L. Wei, Q. Liu, W.C. Gao, Y.W. Li, and X. Li, *Chin. J. Org. Chem.*, **36**, 939 (2016).
16CJOC1000	L.W. Xiao, G.X. Zhang, X.M. Jing, Q.X. Zhou, and R. Feng, *Chin. J. Org. Chem.*, **36**, 1000 (2016).
16CJOC1241	Z.R. Zhang, X.L. Zheng, and C.B. Guo, *Chin. J. Org. Chem.*, **36**, 1241 (2016).
16CJOC1484	X.W. Li, J. Zhou, and S.P. Zhuo, *Chin. J. Org. Chem.*, **36**, 1484 (2016).
16CJOC1528	Y.L. Li, M. Xue, J.L. Wang, D.L. Cao, and Z.L. Ma, *Chin. J. Org. Chem.*, **36**, 1528 (2016).

16CJOC1555	J.P. Huang, H. Zhou, and Z.Y. Chen, *Chin. J. Org. Chem.*, **36**, 1555 (2016).
16CJOC1765	X.C. Zou, K.Y. Shi, J. Li, Y. Wang, C. Wang, C.F. Deng, Y.R. Ren, J. Tan, and X.K. Fu, *Chin. J. Org. Chem.*, **36**, 1765 (2016).
16CJOC1779	Y.X. Chen, C. Zheng, X.C. Peng, Q.T. Fu, L.Y. Wu, and Q. Lin, *Chin. J. Org. Chem.*, **36**, 1779 (2016).
16CJOC1994	D.J. Li, Y.F. Wu, H.H. Chang, W.C. Gao, W.L. Wei, and X. Li, *Chin. J. Org. Chem.*, **36**, 1994 (2016).
16CJOC2024	S.S. Li, H.L. Hong, N. Zhu, L.M. Han, and J.Y. Lu, *Chin. J. Org. Chem.*, **36**, 2024 (2016).
16CJOC2284	J. Ren and M.L. Sun, *Chin. J. Org. Chem.*, **36**, 2284 (2016).
16CJOC2634	L.Q. You, J.W. Yuan, L.R. Yang, Y.M. Xiao, and P. Mao, *Chin. J. Org. Chem.*, **36**, 2634 (2016).
16CJOC2858	L.L. Lu, H. Xu, P. Zhou, and F.C. Yu, *Chin. J. Org. Chem.*, **36**, 2858 (2016).
16CL574	G. Evano, N. Blanchard, G. Compain, A. Coste, C.S. Demmer, W. Gati, C. Guissart, J. Heimburger, N. Henry, K. Jouvin, G. Karthikeyan, A. Laouiti, M. Lecomte, A. Martin-Mingot, B. Metayer, B. Michelet, A. Nitelet, C. Theunissen, S. Thibaudeau, J.J. Wang, M. Zarca, and C.Y. Zhang, *Chem. Lett.*, **45**, 574 (2016).
16CMC161	J.J. Nair, J. van Staden, and J. Bastida, *Curr. Med. Chem.*, **23**, 161 (2016).
16CMC383	M. Bu, B.B. Yang, and L.M. Hu, *Curr. Med. Chem.*, **23**, 383 (2016).
16CMC748	A. Russo, C. Aiello, P. Grieco, and D. Marasco, *Curr. Med. Chem.*, **23**, 748 (2016).
16CMC1100	R. Alvarez, L. Aramburu, P. Puebla, E. Caballero, M. Gonzalez, A. Vicente, M. Medarde, and R. Pelaez, *Curr. Med. Chem.*, **23**, 1100 (2016).
16CMC1288	M. Remesic, Y.S. Lee, and V.J. Hruby, *Curr. Med. Chem.*, **23**, 1288 (2016).
16CMC1408	S. Vincenzetti, V. Polzonetti, D. Micozzi, and S. Pucciarelli, *Curr. Med. Chem.*, **23**, 1408 (2016).
16CMC3078	B. Yang, J.L. Yang, Y.P. Zhao, H.L. Liu, and Y.M. Jiang, *Curr. Med. Chem.*, **23**, 3078 (2016).
16CMC3654	A.S. Gomes, P. Brandao, C.S.G. Fernandes, M. da Silva, M. de Sousa, and M.M.D. Pinto, *Curr. Med. Chem.*, **23**, 3654 (2016).
16CMC4151	P.B. Andrade, C. Grosso, P. Valentao, and J. Bernardo, *Curr. Med. Chem.*, **23**, 4151 (2016).
16CMC4338	D. Kumar and S.K. Jain, *Curr. Med. Chem.*, **23**, 4338 (2016).
16CMC4784	K. Muylaert, M. Jatczak, S. Mangelinckx, and C.V. Stevens, *Curr. Med. Chem.*, **23**, 4784 (2016).
16COBT155	S. Schlager and B. Drager, *Curr. Opin. Biotechnol.*, **37**, 155 (2016).
16COC4	D.M. Hodgson and S. Salik, *Curr. Org. Chem.*, **20**, 4 (2016).
16COC19	Y. Liu, *Curr. Org. Chem.*, **20**, 19 (2016).
16COC29	K.Y. Dong, L.H. Qiu, and X.F. Xu, *Curr. Org. Chem.*, **20**, 29 (2016).
16COC82	X.C. Ma, S.M. Cooper, F. Yang, W.H. Hu, and H.O. Sintim, *Curr. Org. Chem.*, **20**, 82 (2016).
16COC155	K. Arai and M. Iwaoka, *Curr. Org. Chem.*, **20**, 155 (2016).
16COC166	J. Rafique, R.F.S. Canto, S. Saba, F.A.R. Barbosa, and A.L. Braga, *Curr. Org. Chem.*, **20**, 166 (2016).
16COC580	T. Dohi and Y. Kita, *Curr. Org. Chem.*, **20**, 580 (2016).

16COC736	K.V. Wagh, K.C. Badgujar, N.M. Patil, and B.M. Bhanage, *Curr. Org. Chem.*, **20**, 736 (2016).
16COC777	Z.M. Bugarcic, M.D. Kostic, and V.M. Divac, *Curr. Org. Chem.*, **20**, 777 (2016).
16COC798	A.S. Zambare, F.A.K. Khan, S.P. Zambare, S.D. Shinde, and J.N. Sangshetti, *Curr. Org. Chem.*, **20**, 798 (2016).
16COC898	S. Tilvi and K.S. Singh, *Curr. Org. Chem.*, **20**, 898 (2016).
16COC1069	M.M. Heravi, V. Zadsirjan, and M. Dehghani, *Curr. Org. Chem.*, **20**, 1069 (2016).
16COC1359	G. Mloston, M. Jasinski, A. Wroblewska, and H. Heimgartner, *Curr. Org. Chem.*, **20**, 1359 (2016).
16COC1512	A.M. Abdelmoniem and I.A. Abdelhamid, *Curr. Org. Chem.*, **20**, 1512 (2016).
16COC1676	M.M. Heravi, V. Zadsirjan, B. Fattahi, and N. Nazari, *Curr. Org. Chem.*, **20**, 1676 (2016).
16COC1736	X.D. Jiang, S. Li, J. Guan, T. Fang, X. Liu, and L.J. Xiao, *Curr. Org. Chem.*, **20**, 1736 (2016).
16COC1851	Z.B. Wang, W.X. Hong, and J.W. Sun, *Curr. Org. Chem.*, **20**, 1851 (2016).
16COC1878	X.Y. Sun, Y.H. Sun, and Y. Rao, *Curr. Org. Chem.*, **20**, 1878 (2016).
16COC1902	F.C. Pigge, *Curr. Org. Chem.*, **20**, 1902 (2016).
16COC1955	K. Kavitha, K.S.S. Praveena, E. Ramarao, N.Y.S. Murthy, and S. Pal, *Curr. Org. Chem.*, **20**, 1955 (2016).
16COC2038	I. Szatmari, J. Sas, and F. Fulop, *Curr. Org. Chem.*, **20**, 2038 (2016).
16COC2099	R.K. Bansal, R. Gupta, P. Maheshwari, and M. Kour, *Curr. Org. Chem.*, **20**, 2099 (2016).
16COC2109	B.Y. Liu and N.X. Jin, *Curr. Org. Chem.*, **20**, 2109 (2016).
16COC2117	M.R. Iesce, F. Cermola, R. Sferruzza, and M. DellaGreca, *Curr. Org. Chem.*, **20**, 2117 (2016).
16COC2136	E. Brachet and P. Belmont, *Curr. Org. Chem.*, **20**, 2136 (2016).
16COC2161	G. Blond, M. Gulea, and V. Mamane, *Curr. Org. Chem.*, **20**, 2161 (2016).
16COC2211	D. Gaso-Sokac and M. Stivojevic, *Curr. Org. Chem.*, **20**, 2211 (2016).
16COC2222	F. Gallier, *Curr. Org. Chem.*, **20**, 2222 (2016).
16COC2284	E. Chirkin and F.H. Poree, *Curr. Org. Chem.*, **20**, 2284 (2016).
16COC2393	E. Moreno-Clavijo, A.T. Carmona, I. Robina, and A.J. Moreno-Vargas, *Curr. Org. Chem.*, **20**, 2393 (2016).
16COC2514	N. Kumbhakarna and A. Chowdhury, *Curr. Org. Chem.*, **20**, 2514 (2016).
16COC2606	M.D. Kostic, V.M. Divac, and Z.M. Bugarcic, *Curr. Org. Chem.*, **20**, 2606 (2016).
16COC2718	J.J. He, F.Y. Yan, D.P. Kong, Q.H. Ye, X.G. Zhou, and L. Chen, *Curr. Org. Chem.*, **20**, 2718 (2016).
16COC2983	B.M. Ciubuca, C.M. Saviuc, M.C. Chifiriuc, and V. Lazar, *Curr. Org. Chem.*, **20**, 2983 (2016).
16COCB22	H. Kries and S.E. O'Connor, *Curr. Opin. Chem. Biol.*, **31**, 22 (2016).
16COCB74	Y.L. Du and K.S. Ryan, *Curr. Opin. Chem. Biol.*, **31**, 74 (2016).
16COMB13	J. Janardhanan, M. Chang, and S. Mobashery, *Curr. Opin. Microbiol.*, **33**, 13 (2016).
16COS41	D. Mene and M. Kale, *Curr. Org. Synth.*, **13**, 41 (2016).
16COS58	P.T. Parvatkara and P.S. Parameswaranb, *Curr. Org. Synth.*, **13**, 58 (2016).

16COS73	W. Xu and J.X. Xu, *Curr. Org. Synth.*, **13**, 73 (2016).
16COS157	J.S.B. Forero, J. Jones, and F.M. da Silva, *Curr. Org. Synth.*, **13**, 157 (2016).
16COS220	M.M. Heravi, A. Fazeli, and Z. Faghihi, *Curr. Org. Synth.*, **13**, 220 (2016).
16COS308	M.M. Heravi, S. Asadi, N. Nazari, and B.M. Lashkariani, *Curr. Org. Synth.*, **13**, 308 (2016).
16COS372	M. Dinda, S. Maiti, S. Samanta, and P.K. Ghosh, *Curr. Org. Synth.*, **13**, 372 (2016).
16COS385	K. Singh, *Curr. Org. Synth.*, **13**, 385 (2016).
16COS408	R.M. Abdel-Rahman and H.A. Saad, *Curr. Org. Synth.*, **13**, 408 (2016).
16COS484	A. Sharma, S. Singh, and D. Utreja, *Curr. Org. Synth.*, **13**, 484 (2016).
16COS544	K. Du, W.B. Yu, C. Shen, X.Z. Chen, and P.F. Zhang, *Curr. Org. Synth.*, **13**, 544 (2016).
16COS569	M.M. Heravi, M. Ghavidel, and B. Heidari, *Curr. Org. Synth.*, **13**, 569 (2016).
16COS659	D. Pinto and A.M.S. Silva, *Curr. Org. Synth.*, **13**, 659 (2016).
16COS687	A.M.F. Phillips, *Curr. Org. Synth.*, **13**, 687 (2016).
16COS726	M.A. Chiacchio, S.V. Giofre, R. Romeo, G. Romeo, and U. Chiacchio, *Curr. Org. Synth.*, **13**, 726 (2016).
16COS780	M.M. Heravi and V. Zadsirjan, *Curr. Org. Synth.*, **13**, 780 (2016).
16COS861	J. Soloducho, D. Zajac, and J. Cabaj, *Curr. Org. Synth.*, **13**, 861 (2016).
16CPB656	K. Takasu, *Chem. Pharm. Bull.*, **64**, 656 (2016).
16CPB676	T. Ishizu, H. Tsutsumi, and T. Sato, *Chem. Pharm. Bull.*, **64**, 676 (2016).
16CPB691	N. Tanaka, T. Kusama, Y. Kashiwada, and J. Kobayashi, *Chem. Pharm. Bull.*, **64**, 691 (2016).
16CPB1079	J. Kobayashi, *Chem. Pharm. Bull.*, **64**, 1079 (2016).
16CPD3971	A. Martorana, V. Giacalone, R. Bonsignore, A. Pace, C. Gentile, I. Pibiri, S. Buscemi, A. Lauria, and A.P. Piccionello, *Curr. Pharm. Des.*, **22**, 3971 (2016).
16CPPS213	M.F. Correa and J.P.D. Fernandes, *Curr. Protein Pept. Sci.*, **17**, 213 (2016).
16CR47	E.J. Cho, *Chem. Rec.*, **16**, 47 (2016).
16CR349	J.W. Kuck, R.M. Reich, and F.E. Kuhn, *Chem. Rec.*, **16**, 349 (2016).
16CR652	H. Oguri, *Chem. Rec.*, **16**, 652 (2016).
16CR719	M. Mao and Q.H. Song, *Chem. Rec.*, **16**, 719 (2016).
16CR797	L. Li, C.M. Zhao, and H. Wang, *Chem. Rec.*, **16**, 797 (2016).
16CR1034	T. Ebata and Y. Inokuchi, *Chem. Rec.*, **16**, 1034 (2016).
16CR1106	S. Ichikawa, *Chem. Rec.*, **16**, 1106 (2016).
16CR1489	W.F. Tang and D. Du, *Chem. Rec.*, **16**, 1489 (2016).
16CR1868	A.K. Bagdi and A. Hajra, *Chem. Rec.*, **16**, 1868 (2016).
16CR1886	M. Klikar, P. Solanke, J. Tydlitat, and F. Bures, *Chem. Rec.*, **16**, 1886 (2016).
16CR2697	Y.E. Luo, Y.M. He, and Q.H. Fan, *Chem. Rec.*, **16**, 2697 (2016).
16CRF269	M. Gibis, *Compr. Rev. Food Sci. Food Saf.*, **15**, 269 (2016).
16CRSE36	P. Bhaumik and P.L. Dhepe, *Catal. Rev. Sci. Eng.*, **58**, 36 (2016).
16CRV80	L.M. De Coen, T.S.A. Heugebaert, D. Garcia, and C.V. Stevens, *Chem. Rev.*, **116**, 80 (2016).
16CRV287	L. Zhang, J.H. Dong, X.X. Xu, and Q. Liu, *Chem. Rev.*, **116**, 287 (2016).

16CRV719	S. Preshlock, M. Tredwell, and V. Gouverneur, *Chem. Rev.*, **116**, 719 (2016).
16CRV962	F. Wurthner, C.R. Saha-Moller, B. Fimmel, S. Ogi, P. Leowanawat, and D. Schmidt, *Chem. Rev.*, **116**, 962 (2016).
16CRV2170	J. Herzberger, K. Niederer, H. Pohlit, J. Seiwert, M. Worm, F.R. Wurm, and H. Frey, *Chem. Rev.*, **116**, 2170 (2016).
16CRV3328	A. Yoshimura and V.V. Zhdankin, *Chem. Rev.*, **116**, 3328 (2016).
16CRV3540	R. Improta, F. Santoro, and L. Blancafort, *Chem. Rev.*, **116**, 3540 (2016).
16CRV3919	K. De Bruycker, S. Billiet, H.A. Houck, S. Chattopadhyay, J.M. Winne, and F.E. Du Prez, *Chem. Rev.*, **116**, 3919 (2016).
16CRV4006	L. Hong, W.S. Sun, D.X. Yang, G.F. Li, and R. Wang, *Chem. Rev.*, **116**, 4006 (2016).
16CRV4441	A.K. Mailyan, J.A. Eickhoff, A.S. Minakova, Z.H. Gu, P. Lu, and A. Zakarian, *Chem. Rev.*, **116**, 4441 (2016).
16CRV4643	K. Goossens, K. Lava, C.W. Bielawski, and K. Binnemans, *Chem. Rev.*, **116**, 4643 (2016).
16CRV5689	K. Kacprzak, I. Skiera, M. Piasecka, and Z. Paryzek, *Chem. Rev.*, **116**, 5689 (2016).
16CRV5894	V.P. Boyarskiy, D.S. Ryabukhin, N.A. Bokach, and A.V. Vasilyev, *Chem. Rev.*, **116**, 5894 (2016).
16CRV6133	A.S. Amarasekara, *Chem. Rev.*, **116**, 6133 (2016).
16CRV6184	H. Lu and N. Kobayashi, *Chem. Rev.*, **116**, 6184 (2016).
16CRV6837	I. Saikia, A.J. Borah, and P. Phukan, *Chem. Rev.*, **116**, 6837 (2016).
16CRV7397	H.F. Yao, L. Ye, H. Zhang, S.S. Li, S.Q. Zhang, and J.H. Hou, *Chem. Rev.*, **116**, 7397 (2016).
16CRV7768	W. Sun, S.G. Guo, C. Hu, J.L. Fan, and X.J. Peng, *Chem. Rev.*, **116**, 7768 (2016).
16CRV7818	D.X. Hu, D.M. Withall, G.L. Challis, and R.J. Thomson, *Chem. Rev.*, **116**, 7818 (2016).
16CRV7854	B. Roy, A. Depaix, C. Perigaud, and S. Peyrottes, *Chem. Rev.*, **116**, 7854 (2016).
16CRV8256	E. Aguilar, R. Sanz, M.A. Fernandez-Rodriguez, and P. Garcia-Garcia, *Chem. Rev.*, **116**, 8256 (2016).
16CRV9162	L.N. Silva, K.R. Zimmer, A.J. Macedo, and D.S. Trentin, *Chem. Rev.*, **116**, 9162 (2016).
16CRV9375	T. Govender, P.I. Arvidsson, G.E.M. Maguire, H.G. Kruger, and T. Naicker, *Chem. Rev.*, **116**, 9375 (2016).
16CRV9485	B. Pashaei, H. Shahroosvand, M. Graetzel, and M.K. Nazeeruddin, *Chem. Rev.*, **116**, 9485 (2016).
16CRV9683	M.D. Karkas, J.A. Porco, and C.R.J. Stephenson, *Chem. Rev.*, **116**, 9683 (2016).
16CRV9816	R. Remy and C.G. Bochet, *Chem. Rev.*, **116**, 9816 (2016).
16CRV9850	D. Ravelli, S. Protti, and M. Fagnoni, *Chem. Rev.*, **116**, 9850 (2016).
16CRV9914	V. Ramamurthy and J. Sivaguru, *Chem. Rev.*, **116**, 9914 (2016).
16CRV9994	A.A. Ghogare and A. Greer, *Chem. Rev.*, **116**, 9994 (2016).
16CRV10035	K.L. Skubi, T.R. Blum, and T.P. Yoon, *Chem. Rev.*, **116**, 10035 (2016).
16CRV10075	N.A. Romero and D.A. Nicewicz, *Chem. Rev.*, **116**, 10075 (2016).
16CRV10276	D. Cambie, C. Bottecchia, N.J.W. Straathof, V. Hessel, and T. Noel, *Chem. Rev.*, **116**, 10276 (2016).
16CRV11291	J.Y. Corey, *Chem. Rev.*, **116**, 11291 (2016).

16CRV11685	M. Al Kobaisi, S.V. Bhosale, K. Latham, and A.M. Raynor, *Chem. Rev.*, **116**, 11685 (2016).
16CRV12150	J.A. Bull, R.A. Croft, O.A. Davis, R. Doran, and K.F. Morgan, *Chem. Rev.*, **116**, 12150 (2016).
16CRV12369	M. Chrzanowska, A. Grajewska, and M.D. Rozwadowska, *Chem. Rev.*, **116**, 12369 (2016).
16CRV12564	P. Ruiz-Castillo and S.L. Buchwald, *Chem. Rev.*, **116**, 12564 (2016).
16CRV13752	E. Yashima, N. Ousaka, D. Taura, K. Shimomura, T. Ikai, and K. Maeda, *Chem. Rev.*, **116**, 13752 (2016).
16CRV14225	J.R. Pouliot, F. Grenier, J.T. Blaskovits, S. Beaupre, and M. Leclerc, *Chem. Rev.*, **116**, 14225 (2016).
16CRV14307	R. Nunez, M. Tarres, A. Ferrer-Ugalde, F.F. de Biani, and F. Teixidor, *Chem. Rev.*, **116**, 14307 (2016).
16CRV14379	J. Shelton, X. Lu, J.A. Hollenbaugh, J.H. Cho, F. Amblard, and R.F. Schinazi, *Chem. Rev.*, **116**, 14379 (2016).
16CRV14620	A. Gorczynski, J.M. Harrowfield, V. Patroniak, and A.R. Stefankiewicz, *Chem. Rev.*, **116**, 14620 (2016).
16CRV14726	J.R. Johansson, T. Beke-Somfai, A.S. Stalsmeden, and N. Kann, *Chem. Rev.*, **116**, 14726 (2016).
16CRV14769	Z.F. Zhang, N.A. Butt, and W.B. Zhang, *Chem. Rev.*, **116**, 14769 (2016).
16CRV15035	B.M. Trost, J.D. Knopf, and C.S. Brindle, *Chem. Rev.*, **116**, 15035 (2016).
16CRV15167	J.M. Longo, M.J. Sanford, and G.W. Coates, *Chem. Rev.*, **116**, 15167 (2016).
16CRV15235	M. Berthet, T. Cheviett, G. Dujardin, I. Parrot, and J. Martinez, *Chem. Rev.*, **116**, 15235 (2016).
16CS3482	R.A. Manderville and S.D. Wetmore, *Chem. Sci.*, **7**, 3482 (2016).
16CS6298	A. Marrocchi, A. Facchetti, D. Lanari, S. Santoro, and L. Vaccaro, *Chem. Sci.*, **7**, 6298 (2016).
16CS6591	L.P. Tong and R.P. Thummel, *Chem. Sci.*, **7**, 6591 (2016).
16CSR494	T. Castanheiro, J. Suffert, M. Donnard, and M. Gulea, *Chem. Soc. Rev.*, **45**, 494 (2016).
16CSR546	T. Cernak, K.D. Dykstra, S. Tyagarajan, P. Vachal, and S.W. Krska, *Chem. Soc. Rev.*, **45**, 546 (2016).
16CSR1257	Q.J. Wang, Y.J. Su, L.X. Li, and H.M. Huang, *Chem. Soc. Rev.*, **45**, 1257 (2016).
16CSR1331	D. Pflasterer and A.S.K. Hashmi, *Chem. Soc. Rev.*, **45**, 1331 (2016).
16CSR1513	M.M. Sun, K. Mullen, and M.Z. Yin, *Chem. Soc. Rev.*, **45**, 1513 (2016).
16CSR1557	R. Ardkhean, D.F.J. Caputo, S.M. Morrow, H. Shi, Y. Xiong, and E.A. Anderson, *Chem. Soc. Rev.*, **45**, 1557 (2016).
16CSR2044	J.R. Chen, X.Q. Hu, L.Q. Lu, and W.J. Xiao, *Chem. Soc. Rev.*, **45**, 2044 (2016).
16CSR2212	B.M. Trost and J.T. Masters, *Chem. Soc. Rev.*, **45**, 2212 (2016).
16CSR2637	D. Gillingham, S. Geigle, and O.A. von Lilienfeld, *Chem. Soc. Rev.*, **45**, 2637 (2016).
16CSR2825	L. Levi and T.J.J. Muller, *Chem. Soc. Rev.*, **45**, 2825 (2016).
16CSR2900	T. Gensch, M.N. Hopkinson, F. Glorius, and J. Wencel-Delord, *Chem. Soc. Rev.*, **45**, 2900 (2016).
16CSR3069	T. Xiong and Q. Zhang, *Chem. Soc. Rev.*, **45**, 3069 (2016).
16CSR3188	G.M. Peters and J.T. Davis, *Chem. Soc. Rev.*, **45**, 3188 (2016).

16CSR3766	X.S. Hou, C.F. Ke, and J.F. Stoddart, *Chem. Soc. Rev.*, **45**, 3766 (2016).
16CSR3846	Y. Ge and D.F. O'Shea, *Chem. Soc. Rev.*, **45**, 3846 (2016).
16CSR4307	G.K. Cui, J.J. Wang, and S.J. Zhang, *Chem. Soc. Rev.*, **45**, 4307 (2016).
16CSR4364	C. Raviola, S. Protti, D. Ravelli, and M. Fagnoni, *Chem. Soc. Rev.*, **45**, 4364 (2016).
16CSR4448	Z.T. Zheng, Z.X. Wang, Y.L. Wang, and L.M. Zhang, *Chem. Soc. Rev.*, **45**, 4448 (2016).
16CSR4471	A.M. Asiri and A.S.K. Hashmi, *Chem. Soc. Rev.*, **45**, 4471 (2016).
16CSR4524	K. Sekine and T. Yamada, *Chem. Soc. Rev.*, **45**, 4524 (2016).
16CSR4825	C. Liu, K. Wang, X. Gong, and A.J. Heeger, *Chem. Soc. Rev.*, **45**, 4825 (2016).
16CSR5147	R. Nunez, I. Romero, F. Teixidor, and C. Vinas, *Chem. Soc. Rev.*, **45**, 5147 (2016).
16CSR5474	A. Borissov, T.Q. Davies, S.R. Ellis, T.A. Fleming, M.S.W. Richardson, and D.J. Dixon, *Chem. Soc. Rev.*, **45**, 5474 (2016).
16CSR6048	Z.M. Kaskova, A.S. Tsarkova, and I.V. Yampolsky, *Chem. Soc. Rev.*, **45**, 6048 (2016).
16CSR6327	T. Ochiai, D. Franz, and S. Inoue, *Chem. Soc. Rev.*, **45**, 6327 (2016).
16CSR6270	S.V. Kohlhepp and T. Gulder, *Chem. Soc. Rev.*, **45**, 6270 (2016).
16CSS156	Z.H. Zhang, *ChemSusChem*, **9**, 156 (2016).
16CSS2015	B. Liu and Z.H. Zhang, *ChemSusChem*, **9**, 2015 (2016).
16CSS2037	U. Omoruyi, S. Page, J. Hallett, and P.W. Miller, *ChemSusChem*, **9**, 2037 (2016).
16CST923	S. Chassaing, V. Beneteau, and P. Pale, *Catal. Sci. Technol.*, **6**, 923 (2016).
16CST2005	C.B. Bheeter, L. Chen, J.F. Soule, and H. Doucet, *Catal. Sci. Technol.*, **6**, 2005 (2016).
16CST3694	P. Zhou and Z.H. Zhang, *Catal. Sci. Technol.*, **6**, 3694 (2016).
16CSTR8	R.A. Shastri, *Chem. Sci. Trans.*, **5**, 8 (2016).
16CTMC343	R. De Marco, G. Mazzotti, A. Greco, and L. Gentilucci, *Curr. Top. Med. Chem.*, **16**, 343 (2016).
16CTMC1200	M.H. Feng, B.Q. Tang, S.H. Liang, and X.F. Jiang, *Curr. Top. Med. Chem.*, **16**, 1200 (2016).
16CTMC1262	R.A. Rane, S. Karunanidhi, K. Jain, M. Shaikh, G. Hampannavar, and R. Karpoormath, *Curr. Top. Med. Chem.*, **16**, 1262 (2016).
16CTMC1290	L.S. Tsutsumi, D. Gundisch, and D.Q. Sun, *Curr. Top. Med. Chem.*, **16**, 1290 (2016).
16CTMC1362	Z.Y. Wang, X.L. Chen, C.L. Wu, H.W. Xu, and H.M. Liu, *Curr. Top. Med. Chem.*, **16**, 1362 (2016).
16CTMC1392	L.H. Shan, Y. Liu, Y.H. Li, H.M. Liu, and Y. Ke, *Curr. Top. Med. Chem.*, **16**, 1392 (2016).
16CTMC1717	G.L. Tang, M.C. Tang, L.Q. Song, and Y. Zhang, *Curr. Top. Med. Chem.*, **16**, 1717 (2016).
16CTMC1727	G.T. Zhang, W.J. Zhang, S. Saha, and C.S. Zhang, *Curr. Top. Med. Chem.*, **16**, 1727 (2016).
16CTMC2125	A.C. Lele, D.A. Mishra, T.K. Kamil, S. Bhakta, and M.S. Degani, *Curr. Top. Med. Chem.*, **16**, 2125 (2016).
16CTMC2245	A.S. Grewal, D. Pandita, S. Bhardwaj, and V. Lather, *Curr. Top. Med. Chem.*, **16**, 2245 (2016).
16CTMC2648	M. Kiyama, R. Saito, S. Iwano, R. Obata, H. Niwa, and S.A. Maki, *Curr. Top. Med. Chem.*, **16**, 2648 (2016).

16CTMC2841	M.T. Chhabria, S. Patel, P. Modi, and P.S. Brahmkshatriya, *Curr. Top. Med. Chem.*, **16**, 2841 (2016).
16CTMC2863	M.A. Barmade, P.R. Murumkar, M.K. Sharma, and M.R. Yadav, *Curr. Top. Med. Chem.*, **16**, 2863 (2016).
16CTMC2884	J. Dwivedi, N. Kaur, D. Kishore, S. Kumari, and S. Sharma, *Curr. Top. Med. Chem.*, **16**, 2884 (2016).
16CTMC3133	K.S. Jain, N. Arya, N.N. Inamdar, P.B. Auti, S.A. Unawane, H.H. Puranik, M.S. Sanap, A.D. Inamke, V.J. Mahale, C.S. Prajapati, and C.J. Shishoo, *Curr. Top. Med. Chem.*, **16**, 3133 (2016).
16CTMC3175	G. Joshi, H. Nayyar, J.M. Alex, G.S. Vishwakarma, S. Mittal, and R. Kumar, *Curr. Top. Med. Chem.*, **16**, 3175 (2016).
16CTMC3258	R.K. Rawal, J. Bariwal, and V. Singh, *Curr. Top. Med. Chem.*, **16**, 3258 (2016).
16CTMC3274	T.L.S. Kishbaugh, *Curr. Top. Med. Chem.*, **16**, 3274 (2016).
16CTMC3365	J.L. Huang and H. Xu, *Curr. Top. Med. Chem.*, **16**, 3365 (2016).
16CTMC3590	R. Goel, V. Luxami, and K. Paul, *Curr. Top. Med. Chem.*, **16**, 3590 (2016).
16CTMC3617	N.K. Downer-Riley and Y.A. Jackson, *Curr. Top. Med. Chem.*, **16**, 3617 (2016).
16DDT1814	S. Genovese, S. Fiorito, V.A. Taddeo, and F. Epifano, *Drug Discov. Today*, **21**, 1814 (2016).
16DP100	J.F. Li, C.X. Yin, and F.J. Huo, *Dyes Pigm.*, **131**, 100 (2016).
16DP311	A. Gök, E.B. Orman, Ü. Salan, A.R. Özkaya, and M. Bulut, *Dyes Pigm.*, **133**, 311 (2016).
16DT15	C. Hille and F.E. Kuhn, *Dalton Trans.*, **45**, 15 (2016).
16DT32	V.T. Annibale and D.T. Song, *Dalton Trans.*, **45**, 32 (2016).
16DT440	C. Hemmert and H. Gornitzka, *Dalton Trans.*, **45**, 440 (2016).
16DT1810	A.M. Caminade, A. Hameau, and J.P. Majoral, *Dalton Trans.*, **45**, 1810 (2016).
16DT1823	P. Bagi, V. Ujj, M. Czugler, E. Fogassy, and G. Keglevich, *Dalton Trans.*, **45**, 1823 (2016).
16DT2338	N. Zabarska, A. Stumper, and S. Rau, *Dalton Trans.*, **45**, 2338 (2016).
16DT3210	M. Ishikawa, A. Naka, and K. Yoshizawa, *Dalton Trans.*, **45**, 3210 (2016).
16DT4457	P.M. Kozlowski, B.D. Garabato, P. Lodowski, and M. Jaworska, *Dalton Trans.*, **45**, 4457 (2016).
16DT5880	S. Wurtemberger-Pietsch, U. Radius, and T.B. Marder, *Dalton Trans.*, **45**, 5880 (2016).
16DT5896	D. Gudat, *Dalton Trans.*, **45**, 5896 (2016).
16DT11233	N.N. Adarsh, F. Novio, and D. Ruiz-Molina, *Dalton Trans.*, **45**, 11233 (2016).
16DT12252	M.S. Balakrishna, *Dalton Trans.*, **45**, 12252 (2016).
16DT13261	G.Y. Li, L.L. Sun, L.N. Ji, and H. Chao, *Dalton Trans.*, **45**, 13261 (2016).
16DT13631	C. Kreitner and K. Heinze, *Dalton Trans.*, **45**, 13631 (2016).
16DT13996	Y. Ren and F. Jakle, *Dalton Trans.*, **45**, 13996 (2016).
16DT16081	R.S. Ghadwal, *Dalton Trans.*, **45**, 16081 (2016).
16DT17557	B.H. Huang, C.Y. Tsai, C.T. Chen, and B.T. Ko, *Dalton Trans.*, **45**, 17557 (2016).
16EA2672	S. Majeed, W.Y. Gao, Y. Zholudov, K. Muzyka, and G.B. Xu, *Electroanalysis*, **28**, 2672 (2016).

16EJIC300	B. Wrackmeyer and E. Khan, *Eur. J. Inorg. Chem.*, 300 (2016).
16EJIC575	J.W. Heinicke, *Eur. J. Inorg. Chem.*, 575 (2016).
16EJIC595	J.A.W. Sklorz and C. Muller, *Eur. J. Inorg. Chem.*, 595 (2016).
16EJIC1306	A. Galvan, F.J. Fananas, and F. Rodriguez, *Eur. J. Inorg. Chem.*, 1306 (2016).
16EJIC1448	B. Ramasamy and P. Ghosh, *Eur. J. Inorg. Chem.*, 1448 (2016).
16EJIC2214	C. Kammerer and G. Rapenne, *Eur. J. Inorg. Chem.*, 2214 (2016).
16EJIC2227	G.M. Pawar, J.B. Sheridan, and F. Jakle, *Eur. J. Inorg. Chem.*, 2227 (2016).
16EJIC2253	R.F. Semeniuc and D.L. Reger, *Eur. J. Inorg. Chem.*, 2253 (2016).
16EJIC2296	B.A. McKeown, J.P. Lee, J. Mei, T.R. Cundari, and T.B. Gunnoe, *Eur. J. Inorg. Chem.*, 2296 (2016).
16EJMC(107)63	M. Maqbool, M. Mobashir, and N. Hoda, *Eur. J. Med. Chem.*, **107**, 63 (2016).
16EJMC(108)455	M.M. Bastos, C.C.P. Costa, T.C. Bezerra, F.D. da Silva, and N. Boechat, *Eur. J. Med. Chem.*, **108**, 455 (2016).
16EJMC(109)89	D. Das, P. Sikdar, and M. Bairagi, *Eur. J. Med. Chem.*, **109**, 89 (2016).
16EJMC(110)13	S.S. Gholap, *Eur. J. Med. Chem.*, **110**, 13 (2016).
16EJMC(112)347	A. Unzue, K. Lafleur, H.T. Zhao, T. Zhou, J. Dong, P. Kolb, J. Liebl, S. Zahler, A. Caflisch, and C. Nevado, *Eur. J. Med. Chem.*, **112**, 347 (2016).
16EJMC(113)145	D. Havrylyuk, O. Roman, and R. Lesyk, *Eur. J. Med. Chem.*, **113**, 145 (2016).
16EJMC(113)198	B. Kuppast and H. Fahmy, *Eur. J. Med. Chem.*, **113**, 198 (2016).
16EJMC(114)65	H. Zhang, D.W. Kang, B.S. Huang, N. Liu, F.B. Zhao, P. Zhan, and X.Y. Liu, *Eur. J. Med. Chem.*, **114**, 65 (2016).
16EJMC(114)257	A.M. Gouda and A.H. Abdelazeem, *Eur. J. Med. Chem.*, **114**, 257 (2016).
16EJMC(116)267	Shagufta and I. Ahmad, *Eur. J. Med. Chem.*, **116**, 267 (2016).
16EJMC(116)290	A. Saeed, *Eur. J. Med. Chem.*, **116**, 290 (2016).
16EJMC(119)141	J. Dandriyal, R. Singla, M. Kumar, and V. Jaitak, *Eur. J. Med. Chem.*, **119**, 141 (2016).
16EJMC(121)47	L.P. Guan and B.Y. Liu, *Eur. J. Med. Chem.*, **121**, 47 (2016).
16EJMC(121)671	N.R. Patel, D.V. Patel, P.R. Murumkar, and M.R. Yadav, *Eur. J. Med. Chem.*, **121**, 671 (2016).
16EJMC(122)366	H.S. Ibrahim, S.M. Abou-Seri, and H.A. Abdel-Aziz, *Eur. J. Med. Chem.*, **122**, 366 (2016).
16EJMC(123)236	M.Z. Hassan, H. Osman, M.A. Ali, and M.J. Ahsan, *Eur. J. Med. Chem.*, **123**, 236 (2016).
16EJMC(123)858	M. Kaur, M. Singh, N. Chadha, and O. Silakari, *Eur. J. Med. Chem.*, **123**, 858 (2016).
16EJMC(124)500	Shaveta, S. Mishra, and P. Singh, *Eur. J. Med. Chem.*, **124**, 500 (2016).
16EJOC233	G. Reginato, A. Mordini, L. Zani, M. Calamante, and A. Dessi, *Eur. J. Org. Chem.*, 233 (2016).
16EJOC612	D.A. Alonso, A. Baeza, R. Chinchilla, G. Guillena, I.M. Pastor, and D.J. Ramon, *Eur. J. Org. Chem.*, 612 (2016).
16EJOC860	S. Patil and A. Bugarin, *Eur. J. Org. Chem.*, 860 (2016).
16EJOC1459	S. Motevalli, Y. Sokeirik, and A. Ghanem, *Eur. J. Org. Chem.*, 1459 (2016).
16EJOC1630	P.G. McCaw, N.M. Buckley, S.G. Collins, and A.R. Maguire, *Eur. J. Org. Chem.*, 1630 (2016).
16EJOC1937	R. Zhou and Z.J. He, *Eur. J. Org. Chem.*, 1937 (2016).

16EJOC2231	A.J. Clark, *Eur. J. Org. Chem.*, 2231 (2016).
16EJOC2244	D. Saccone, S. Galliano, N. Barbero, P. Quagliotto, G. Viscardi, and C. Barolo, *Eur. J. Org. Chem.*, 2244 (2016).
16EJOC2907	L. Levi and T.J.J. Muller, *Eur. J. Org. Chem.*, 2907 (2016).
16EJOC3143	A. Gualandi, S. Grilli, and D. Savoia, *Eur. J. Org. Chem.*, 3143 (2016).
16EJOC3264	L.A. Jouanno, K. Renault, C. Sabot, and P.Y. Renard, *Eur. J. Org. Chem.*, 3264 (2016).
16EJOC3282	S. Moghimi, M. Mahdavi, A. Shafiee, and A. Foroumadi, *Eur. J. Org. Chem.*, 3282 (2016).
16EJOC3671	J.B. Jiang, M. Artola, T.J.M. Beenakker, S.P. Schroder, R. Petracca, C. de Boer, J. Aerts, G.A. van der Marel, J.D.C. Codee, and H.S. Overkleeft, *Eur. J. Org. Chem.*, 3671 (2016).
16EJOC4124	U.V.S. Reddy, M. Chennapuram, C. Seki, E. Kwon, Y. Okuyama, and H. Nakano, *Eur. J. Org. Chem.*, 4124 (2016).
16EJOC4249	P. Das and J.T. Njardarson, *Eur. J. Org. Chem.*, 4249 (2016).
16EJOC5060	M. Neumeyer and R. Bruckner, *Eur. J. Org. Chem.*, 5060 (2016).
16EJOC5194	B. Cecconi, N. Manfredi, T. Montini, P. Fornasiero, and A. Abbotto, *Eur. J. Org. Chem.*, 5194 (2016).
16EJOC5446	E. Bunuel and D.J. Cardenas, *Eur. J. Org. Chem.*, 5446 (2016).
16EJOC5589	J. Merad, J.M. Pons, O. Chuzel, and C. Bressy, *Eur. J. Org. Chem.*, 5589 (2016).
16EJOC5778	R.A. Fernandes, P.H. Patil, and D.A. Chaudhari, *Eur. J. Org. Chem.*, 5778 (2016).
16EJOC5908	Q. Zhao, H. Abou-Hamdan, and L. Desaubry, *Eur. J. Org. Chem.*, 5908 (2016).
16EJPS(88)101	B. Testa, G. Vistoli, and A. Pedretti, *Eur. J. Pharm. Sci.*, **88**, 101 (2016).
16EJPS(91)1	T.V. Sravanthi and S.L. Manju, *Eur. J. Pharm. Sci.*, **91**, 1 (2016).
16EPJ202	G.Z. Papageorgiou, D.G. Papageorgiou, Z. Terzopoulou, and D.N. Bikiaris, *Eur. Polym. J.*, **83**, 202 (2016).
16EPJ478	H. Al-Kutubi, H.R. Zafarani, L. Rassaei, and K. Mathwig, *Eur. Polym. J.*, **83**, 478 (2016).
16FC19	Z. Jahnz-Wechmann, G.R. Framski, P.A. Januszczyk, and J. Boryski, *Front. Chem.*, **4**, 19 (2016).
16FCT23	T. Herraiz, *Food Chem. Toxicol.*, **97**, 23 (2016).
16FMC227	H. Turner, *Future Med. Chem.*, **8**, 227 (2016).
16FMC1291	R. Patil, S.A. Patil, and K.D. Beaman, *Future Med. Chem.*, **8**, 1291 (2016).
16FMC2113	M.A. Santos, K. Chand, and S. Chaves, *Future Med. Chem.*, **8**, 2113 (2016).
16FRR(S)102	C.S. Gloora, F. Denes, and P. Renaud, *Free Radic. Res.*, **50**, S102 (2016).
16GC5701	H. Li, S. Yang, A. Riisager, A. Pandey, R.S. Sangwan, S. Saravanamurugan, and R. Luque, *Green Chem.*, **18**, 5701 (2016).
16HI461	M. Krstic, S. Sovilj, S. Borozan, M. Rancic, J. Poljarevic, and S.R. Grguric-Sipka, *Hem. Ind.*, **70**, 461 (2016).
16H(92)185	D. Moderhack, *Heterocycles*, **92**, 185 (2016).
16H(92)1013	N. Kubodera, *Heterocycles*, **92**, 1013 (2016).
16H(92)1171	X.X. Xie, L.H. Li, X. Wu, C. Ma, and J.S. Zhang, *Heterocycles*, **92**, 1171 (2016).
16H(92)1761	D.I.A. Othman and M. Kitamura, *Heterocycles*, **92**, 1761 (2016).
16H(92)1931	M.A. Baashen, B.F. Abdel-Wahab, and G.A. El-Hiti, *Heterocycles*, **92**, 1931 (2016).

16HC175	M.L. Zhang, D. Liu, G.Q. Fan, R.X. Wang, X.H. Lu, Y.C. Gu, and Q.W. Shi, *Heterocycl. Commun.*, **22**, 175 (2016).
16HL817	M. Asif, *Heterocycl. Lett.*, **6**, 817 (2016).
16IAN321	N.A. Bumagin and V.I. Potkin, *Izv. Akad. Nauk Ser. Khim.*, **321**, (2016) [*Russ. Chem. Bull. Int. Ed.* (Engl. Version), **66**, 321 (2016)].
16IAN333	M.A. Grin and A.F. Mironov, *Izv. Akad. Nauk Ser. Khim.*, **333**, (2016) [*Russ. Chem. Bull. Int. Ed.* (Engl. Version), **66**, 333 (2016)].
16IAN384	D.S. Druzhilovskii, A.V. Rudik, D.A. Filimonov, A.A. Lagunin, T.A. Gloriozova, and V.V. Poroikov, *Izv. Akad. Nauk Ser. Khim.*, **384**, (2016) [*Russ. Chem. Bull. Int. Ed.* (Engl. Version), **66**, 384 (2016)].
16IAN923	L.V. Myznikov, S.V. Vorona, T.V. Artamonova, and Y.E. Zevatskii, *Izv. Akad. Nauk Ser. Khim.*, **923**, (2016) [*Russ. Chem. Bull. Int. Ed.* (Engl. Version), **66**, 923 (2016)].
16IAN1418	M.V. Orlov, *Izv. Akad. Nauk Ser. Khim.*, **1418**, (2016) [*Russ. Chem. Bull. Int. Ed.* (Engl. Version), **66**, 1418 (2016)].
16IAN1441	L.I. Belen'kii and N.D. Chuvylkin, *Izv. Akad. Nauk Ser. Khim.*, **1441**, (2016) [*Russ. Chem. Bull. Int. Ed.* (Engl. Version), **66**, 1441 (2016)].
16IAN1687	A.Yu. Rulev, *Izv. Akad. Nauk Ser. Khim.*, **1687**, (2016) [*Russ. Chem. Bull. Int. Ed.* (Engl. Version), **66**, 1687 (2016)].
16IAN1700	V.I. Saloutin, M.V. Goryaeva, O.L. Khudina, A.E. Ivanov, and Ya.V. Burgart, *Izv. Akad. Nauk Ser. Khim.*, **1700**, (2016) [*Russ. Chem. Bull. Int. Ed.* (Engl. Version), **66**, 1700 (2016)].
16IAN1709	N.M. Przheval'skii, R.A. Laipanov, G.P. Tokmakov, and N.L. Nam, *Izv. Akad. Nauk Ser. Khim.*, **1709**, (2016) [*Russ. Chem. Bull. Int. Ed.* (Engl. Version), **66**, 1709 (2016)].
16IAN2163	K.V. Shcherbakov, Ya.V. Burgart, V.I. Saloutin, and O.N. Chupakhin, *Izv. Akad. Nauk Ser. Khim.*, **2163**, (2016) [*Russ. Chem. Bull. Int. Ed.* (Engl. Version), **66**, 2163 (2016)].
16IAN2172	G.A. Gazieva, T.B. Karpova, T.V. Nechaeva, and A.N. Kravchenko, *Izv. Akad. Nauk Ser. Khim.*, **2172**, (2016) [*Russ. Chem. Bull. Int. Ed.* (Engl. Version), **66**, 2172 (2016)].
16IAN2183	A. Padva, *Izv. Akad. Nauk Ser. Khim.*, **2183**, (2016) [*Russ. Chem. Bull. Int. Ed.* (Engl. Version), **66**, 2183 (2016)].
16IAN2315	A.S. Smirnov, S.P. Smirnov, T.S. Pivina, D.B. Lempert, and L.K. Maslova, *Izv. Akad. Nauk Ser. Khim.*, **2315**, (2016) [*Russ. Chem. Bull. Int. Ed.* (Engl. Version), **66**, 2315 (2016)].
16IAN2559	A.Zh. Kasanova, E.A. Krasnokutskaya, and V.D. Filimonov, *Izv. Akad. Nauk Ser. Khim.*, **2559**, (2016) [*Russ. Chem. Bull. Int. Ed.* (Engl. Version), **66**, 2559 (2016)].
16IJC119	Z.K. Zhou and Z. Shen, *Isr. J. Chem.*, **56**, 119 (2016).
16IJC431	N.R. O'Connor, J.L. Wood, and B.M. Stoltz, *Isr. J. Chem.*, **56**, 431 (2016).
16IJC445	R. Talukdar, A. Saha, and M.K. Ghorai, *Isr. J. Chem.*, **56**, 445 (2016).
16IJC463	L.J. Wang and Y. Tang, *Isr. J. Chem.*, **56**, 463 (2016).
16IJC499	M.C. Martin, R. Shenje, and S. France, *Isr. J. Chem.*, **56**, 499 (2016).
16IJC512	A.K. Pandey, A. Ghosh, and P. Banerjee, *Isr. J. Chem.*, **56**, 512 (2016).
16IJC531	O. Reiser, *Isr. J. Chem.*, **56**, 531 (2016).
16IJCT1	R.K. Singh, S. Sharma, A. Kaur, M. Saini, and S. Kumar, *Iran. J. Catal.*, **6**, 1 (2016).
16IJPS482	D. Singh, D.P. Pathak, and Anjali, *Int. J. Pharm. Sci. Res.*, **7**, 482 (2016).

16IJPS1375	G. Achaiah, N.S. Goud, K.P. Kumar, and P. Mayuri, *Int. J. Pharm. Sci. Res.*, **7**, 1375 (2016).
16JA10056	T. Bura, J.T. Blaskovits, and M. Leclerc, *J. Am. Chem. Soc.*, **138**, 10056 (2016).
16JA12692	M. Yan, J.C. Lo, J.T. Edwards, and P.S. Baran, *J. Am. Chem. Soc.*, **138**, 12692 (2016).
16JAB500	A. Minami and H. Oikawa, *J. Antibiot.*, **69**, 500 (2016).
16JAFC6659	Y.M. Ma, X.A. Liang, Y. Kong, and B. Jia, *J. Agric. Food Chem.*, **64**, 6659 (2016).
16JCC3131	D.X. Bao, S. Xiang, J. Wang, Y.C. Li, and X.Q. Zhao, *J. Coord. Chem.*, **69**, 3131 (2016).
16JCR1	M.T.W. Lee, A. Maruani, and V. Chudasama, *J. Chem. Res.*, 1 (2016).
16JEIMC173	S. Kumar, R.K. Singh, B. Patial, S. Goyal, and T.R. Bhardwaj, *J. Enzyme Inhib. Med. Chem.*, **31**, 173 (2016).
16JFC(185)118	B.I. Usachev, *J. Fluor. Chem.*, **185**, 118 (2016).
16JFC(192)1	G.N. Lipunova, E.V. Nosova, V.N. Charushin, and O.N. Chupakhin, *J. Fluor. Chem.*, **192**, 1 (2016).
16JHC13	R.E. Khidre, B.F. Abdel-Wahab, A.A. Farahat, and H.A. Mohamed, *J. Heterocycl. Chem.*, **53**, 13 (2016).
16JHC345	A. Zula, D. Kikelj, and J. Ilas, *J. Heterocycl. Chem.*, **53**, 345 (2016).
16JHC356	M.A. Gouda, *J. Heterocycl. Chem.*, **53**, 356 (2016).
16JHC685	N. Monakhova, S. Ryabova, and V. Makarov, *J. Heterocycl. Chem.*, **53**, 685 (2016).
16JHC1687	R. Gupta, *J. Heterocycl. Chem.*, **53**, 1687 (2016).
16JHC1697	A.K. Pathak, C. Ameta, R. Ameta, and P.B. Punjabi, *J. Heterocycl. Chem.*, **53**, 1697 (2016).
16JIPMC(85)1	M. Yamada, M.R. Gandhi, U.M.R. Kunda, and F. Hamada, *J. Incl. Phenom. Macrocycl. Chem.*, **85**, 1 (2016).
16JMC16	R. Cannalire, M.L. Barreca, G. Manfroni, and V. Cecchetti, *J. Med. Chem.*, **59**, 16 (2016).
16JMC504	A.J. Roecker, C.D. Cox, and P.J. Colemant, *J. Med. Chem.*, **59**, 504 (2016).
16JMC1271	F.A. Romero, A.M. Taylor, T.D. Crawford, V. Tsui, A. Cote, and S. Magnuson, *J. Med. Chem.*, **59**, 1271 (2016).
16JMC1308	T.E. McAllister, K.S. England, R.J. Hopkinson, P.E. Brennan, A. Kawamura, and C.J. Schofield, *J. Med. Chem.*, **59**, 1308 (2016).
16JMC2269	Y.B. Feng, P.V. LoGrasso, O. Defert, and R.S. Li, *J. Med. Chem.*, **59**, 2269 (2016).
16JMC2849	P. Zhan, C. Pannecouque, E. De Clercq, and X.Y. Liu, *J. Med. Chem.*, **59**, 2849 (2016).
16JMC2894	S.P. Andrews and R.J. Cox, *J. Med. Chem.*, **59**, 2894 (2016).
16JMC3579	K. Ritter, C. Buning, N. Halland, C. Poverlein, and L. Schwink, *J. Med. Chem.*, **59**, 3579 (2016).
16JMC3593	S.H. Myers, V.G. Brunton, and A. Unciti-Broceta, *J. Med. Chem.*, **59**, 3593 (2016).
16JMC3609	M. Fiore, S. Forli, and F. Manetti, *J. Med. Chem.*, **59**, 3609 (2016).
16JMC4062	Y. Hu, D. Stumpfe, and J. Bajorath, *J. Med. Chem.*, **59**, 4062 (2016).
16JMC4087	B. Kuhn, W. Guba, J. Hert, D. Banner, C. Bissantz, S. Ceccarelli, W. Haap, M. Korner, A. Kuglstatter, C. Lerner, P. Mattei, W. Neidhart, E. Pinard, M.G. Rudolph, T. Schulz-Gasch, T. Wokering, and M. Stahl, *J. Med. Chem.*, **59**, 4087 (2016).
16JMC4428	J.A. Gonzalez-Delgado, P.J. Kennedy, M. Ferreira, J.P.C. Tome, and B. Sarmento, *J. Med. Chem.*, **59**, 4428 (2016).

16JMC5587	E. Fernandez-Alvaro, W.D. Hong, G.L. Nixon, P.M. O'Neil, and F. Calderon, *J. Med. Chem.*, **59**, 5587 (2016).
16JMC5622	M.A.M. Behnam, C. Nitsche, V. Boldescu, and C.D. Klein, *J. Med. Chem.*, **59**, 5622 (2016).
16JMC5987	S. Neidle, *J. Med. Chem.*, **59**, 5987 (2016).
16JMC6553	Y.P. Xu, *J. Med. Chem.*, **59**, 6553 (2016).
16JMC6580	Z.D. Song, Y. Ge, C.Y. Wang, S.S. Huang, X.H. Shu, K.X. Liu, Y.W. Zhou, and X.D. Ma, *J. Med. Chem.*, **59**, 6580 (2016).
16JMC6629	R.M. Franzini and C. Randolph, *J. Med. Chem.*, **59**, 6629 (2016).
16JMC7011	N.T. Zaveri, *J. Med. Chem.*, **59**, 7011 (2016).
16JMC7029	C. Jansen, A.J. Kooistra, G.K. Kanev, R. Leurs, I.J.P. de Esch, and C. de Graaf, *J. Med. Chem.*, **59**, 7029 (2016).
16JMC7311	N.A. Meanwell, *J. Med. Chem.*, **59**, 7311 (2016).
16JMC7360	T. Frohlich, A.C. Karagoz, C. Reiter, and S.B. Tsogoeva, *J. Med. Chem.*, **59**, 7360 (2016).
16JMC7738	Z.J. Lesnikowski, *J. Med. Chem.*, **59**, 7738 (2016).
16JMC8168	J.D. Aubert and L. Juillerat-Jeanneret, *J. Med. Chem.*, **59**, 8168 (2016).
16JMC8667	Y.A. Sonawane, M.A. Taylor, J.V. Napoleon, S. Rana, J.I. Contreras, and A. Natarajan, *J. Med. Chem.*, **59**, 8667 (2016).
16JMC9277	M. Guccione, R. Ettari, S. Taliani, F. Da Settimo, M. Zappala, and S. Grasso, *J. Med. Chem.*, **59**, 9277 (2016).
16JMC9295	B.L. DeCorte, *J. Med. Chem.*, **59**, 9295 (2016).
16JMC9305	L. Carlino and G. Rastelli, *J. Med. Chem.*, **59**, 9305 (2016).
16JMC9575	Y.Q. Wang, P.Y. Wang, Y.T. Wang, G.F. Yang, A. Zhang, and Z.H. Miao, *J. Med. Chem.*, **59**, 9575 (2016).
16JMC9645	W.B. Wan and P.P. Seth, *J. Med. Chem.*, **59**, 9645 (2016).
16JMC9981	S. Conroy, N. Kindon, B. Kellam, and M.J. Stocks, *J. Med. Chem.*, **59**, 9981 (2016).
16JMC10006	M.J.P. de Vega, I. Gomez-Monterrey, A. Ferrer-Montiel, and R. Gonzalez-Muniz, *J. Med. Chem.*, **59**, 10006 (2016).
16JMC10030	T.P. Heffron, *J. Med. Chem.*, **59**, 10030 (2016).
16JMC10343	M. Serpi, V. Ferrari, and F. Pertusati, *J. Med. Chem.*, **59**, 10343 (2016).
16JMC10807	M.A.T. Blaskovich, *J. Med. Chem.*, **59**, 10807 (2016).
16JMC10865	O. Van der Poorten, A. Knuhtsen, D.S. Pedersen, S. Ballet, and D. Tourwe, *J. Med. Chem.*, **59**, 10865 (2016).
16JMCA15771	J. Du, M.C. Biewer, and M.C. Stefan, *J. Mater. Chem. A*, **4**, 15771 (2016).
16JMCA16288	P. Puthiaraj, Y.R. Lee, S.Q. Zhang, and W.S. Ahn, *J. Mater. Chem. A*, **4**, 16288 (2016).
16JMCB4952	X. Strakosas, B. Wei, D.C. Martin, and R.M. Owens, *J. Mater. Chem. B*, **4**, 4952 (2016).
16JMCB7060	P.Y. Gu, Z.L. Wang, and Q.C. Zhang, *J. Mater. Chem. B*, **4**, 7060 (2016).
16JMCC2404	Z.S. Huang, H. Meier, and D.R. Cao, *J. Mater. Chem. C*, **4**, 2404 (2016).
16JMCC6200	Y. Wang and T. Michinobu, *J. Mater. Chem. C*, **4**, 6200 (2016).
16JMCC8596	A. Paun, N.D. Hadade, C.C. Paraschivescu, and M. Matache, *J. Mater. Chem. C*, **4**, 8596 (2016).
16JMCC10574	X.C. Li, C.Y. Wang, W.Y. Lai, and W. Huang, *J. Mater. Chem. C*, **4**, 10574 (2016).
16JNP662	S.A. Rasmussen, A.J.C. Andersen, N.G. Andersen, K.F. Nielsen, P.J. Hansen, and T.O. Larsen, *J. Nat. Prod.*, **79**, 662 (2016).
16JOC4421	D. Kaiser and N. Maulide, *J. Org. Chem.*, **81**, 4421 (2016).

16JOC9521	L.L. Anderson, M.A. Kroc, T.W. Reidl, and J. Son, *J. Org. Chem.*, **81**, 9521 (2016).
16JOC10109	C. Cabrele and O. Reiser, *J. Org. Chem.*, **81**, 10109 (2016).
16JOC10126	D. Sucunza, A.M. Cuadro, J. Alvarez-Builla, and J.J. Vaquero, *J. Org. Chem.*, **81**, 10126 (2016).
16JOC10136	C.M. Gober and M.M. Joullie, *J. Org. Chem.*, **81**, 10136 (2016).
16JOC10145	A.A. Jaworski and K.A. Scheidt, *J. Org. Chem.*, **81**, 10145 (2016).
16JOMC(801)139	T.M. Shaikh and F.E. Hong, *J. Organomet. Chem.*, **801**, 139 (2016).
16JOMC(811)48	B. Minaev, A. Baryshnikova, and W.H. Sun, *J. Organomet. Chem.*, **811**, 48 (2016).
16JOMC(823)50	I. Omae, *J. Organomet. Chem.*, **823**, 50 (2016).
16JPB203	Y.X. Huang, H.X. Tan, Z.Y. Guo, X.X. Wu, Q.L. Zhang, L. Zhang, and Y. Diao, *J. Plant Biol.*, **59**, 203 (2016).
16JPC(28)116	A.J. Stasyuk, P.J. Cywinski, and D.T. Gryko, *J. Photochem. Photobiol. C Photochem. Rev.*, **28**, 116 (2016).
16JPC(29)29	M.L.P. Reddy and K.S. Bejoymohandas, *J. Photochem. Photobiol. C Photochem. Rev.*, **29**, 29 (2016).
16JPC(29)73	A. Mukhopadhyay and J.N. Moorthy, *J. Photochem. Photobiol. C Photochem. Rev.*, **29**, 73 (2016).
16JPP1	G.W. Craig, *J. Porphyr. Phthalocyanines*, **20**, 1 (2016).
16JPP96	M. Konig, F. Faschinger, L.M. Reith, and W. Schofberger, *J. Porphyr. Phthalocyanines*, **20**, 96 (2016).
16JPP117	S. Le Gac and B. Boitrel, *J. Porphyr. Phthalocyanines*, **20**, 117 (2016).
16JPP134	P. Ascenzi and M. Brunori, *J. Porphyr. Phthalocyanines*, **20**, 134 (2016).
16JPP571	F. Dumoulin, D. Topkaya, S. Yasar, V. Ahsen, and U. Isci, *J. Porphyr. Phthalocyanines*, **20**, 571 (2016).
16JPP843	C.A. Henriques, S.M.A. Pinto, J. Canotilho, M.E.S. Eusebio, and M.J.F. Calvete, *J. Porphyr. Phthalocyanines*, **20**, 843 (2016).
16JPP855	T.D. Lash, *J. Porphyr. Phthalocyanines*, **20**, 855 (2016).
16JPP889	M.L. Rodriguez-Mendez, C. Garcia-Hernandez, C. Medina-Plaza, C. Garcia-Cabezon, and J.A. de Saja, *J. Porphyr. Phthalocyanines*, **20**, 889 (2016).
16JPP895	R. Sharma and M. Ravikanth, *J. Porphyr. Phthalocyanines*, **20**, 895 (2016).
16JPP935	K. Mase, S. Aoi, K. Ohkubo, and S. Fukuzumi, *J. Porphyr. Phthalocyanines*, **20**, 935 (2016).
16JPP950	F. Figueira, J.M.M. Rodrigues, A.A.S. Farinha, J.A.S. Cavaleiro, and J.P.C. Tome, *J. Porphyr. Phthalocyanines*, **20**, 950 (2016).
16JPP966	G. Bottari, A. de la Escosura, D. Gonzalez-Rodriguez, and G. de la Torre, *J. Porphyr. Phthalocyanines*, **20**, 966 (2016).
16JSSE2599	M. Ates and N. Uludag, *J. Solid State Electrochem.*, **20**, 2599 (2016).
16KGS71	M.Y. Kornev and V.Y. Sosnovskikh, *Khim. Geterotsikl. Soedin.*, **52**, 71 (2016) [*Chem. Heterocycl. Comp.* (Engl. Version), **52**, 71 (2016)].
16KGS427	V.Y. Sosnovskikh, *Khim. Geterotsikl. Soedin.*, **52**, 427 (2016) [*Chem. Heterocycl. Comp.* (Engl. Version), **52**, 427 (2016)].
16KGS509	I. Mierina, M. Jure, and A. Stikute, *Khim. Geterotsikl. Soedin.*, **52**, 509 (2016) [*Chem. Heterocycl. Comp.* (Engl. Version), **52**, 509 (2016)].
16KGS616	A. Padwa and S. Bur, *Khim. Geterotsikl. Soedin.*, **52**, 616 (2016) [*Chem. Heterocycl. Comp.* (Engl. Version), **52**, 616 (2016)].
16KGS627	N.P. Belskaya, V.A. Bakulev, and Z.J. Fan, *Khim. Geterotsikl. Soedin.*, **52**, 627 (2016) [*Chem. Heterocycl. Comp.* (Engl. Version), **52**, 627 (2016)].

16KGS651	T.V. Beryozkina, N.N. Kolos, and V.A. Bakulev, *Khim. Geterotsikl. Soedin.*, **52**, 651 (2016) [*Chem. Heterocycl. Comp.* (Engl. Version), **52**, 651 (2016)].
16KGS658	A.G. Lvov and V.Z. Shirinyan, *Khim. Geterotsikl. Soedin.*, **52**, 658 (2016) [*Chem. Heterocycl. Comp.* (Engl. Version), **52**, 658 (2016)].
16KGS666	E.V. Babaev, *Khim. Geterotsikl. Soedin.*, **52**, 666 (2016) [*Chem. Heterocycl. Comp.* (Engl. Version), **52**, 666 (2016)].
16KGS675	N.V. Vasil'ev, T.S. Kostryukova, G.V. Zatonsky, and S.Z. Vatsadze, *Khim. Geterotsikl. Soedin.*, **52**, 675 (2016) [*Chem. Heterocycl. Comp.* (Engl. Version), **52**, 675 (2016)].
16KGS753	A.V. Smolobochkin, A.S. Gazizov, A.R. Burilov, and M.A. Pudovik, *Khim. Geterotsikl. Soedin.*, **52**, 753 (2016) [*Chem. Heterocycl. Comp.* (Engl. Version), **52**, 753 (2016)].
16KGS766	M.S. Bekheit, A.A. Farahat, and B.F. Abdel-Wahab, *Khim. Geterotsikl. Soedin.*, **52**, 766 (2016) [*Chem. Heterocycl. Comp.* (Engl. Version), **52**, 766 (2016)].
16KGS887	L.V. Myznikov, S.V. Vorona, T.V. Artamonova, and Y.E. Zevatskii, *Khim. Geterotsikl. Soedin.*, **52**, 887 (2016) [*Chem. Heterocycl. Comp.* (Engl. Version), **52**, 887 (2016)].
16KGS973	V.T. Abaev, I.V. Trushkov, and M.G. Uchuskin, *Khim. Geterotsikl. Soedin.*, **52**, 973 (2016) [*Chem. Heterocycl. Comp.* (Engl. Version), **52**, 973 (2016)].
16MI1	J. Mortier, editor: *Arene Chemistry: Reaction Mechanisms and Methods for Aromatic Compounds*, Wiley: (2016).
16MI2	S. Sadjadi, editor: *Organic Nanoreactors. From Molecular to Supramolecular Organic Compounds*, Elsevier: (2016).
16MI3	N.E. Jacobsen, *NMR Data Interpretation Explained: Understanding 1D and 2D NMR Spectra of Organic Compounds and Natural Products*, Wiley: (2016).
16MI4	T. Nishinaga, editor: *Organic REDOX Systems: Synthesis, Properties, and Applications*, Wiley: (2016).
16MI5	H. Hopf and M.S. Sherburn, editors: *Cross Conjugation: Modern Dendralene, Radialene and Fulvene Chemistry*, Wiley: (2016).
16MI6	H.-J. Schneider, editor: *Applications of Supramolecular Chemistry*, CRC Press: (2016).
16MI7	Y. Voloshin, I. Belaya, and R. Krämer, *The Encapsulation Phenomenon*, Springer: (2016).
16MI8	G. Keglevich, editor: *Milestones in Microwave Chemistry*, Springer: (2016).
16MI9	M. Vrabel and T. Carell, *Cycloadditions in Bioorthogonal Chemistry*, Springer: (2016).
16MI10	J.H. Zagal and F. Bedioui, editors: *Electrochemistry of N4 Macrocyclic Metal Complexes. Vol. 1: Energy; Vol. 2: Biomimesis, Electroanalysis and Electrosynthesis of MN4 Metal Complexes*, Springer: (2016).
16MI11	S.-L. You, editor: *Asymmetric Dearomatization Reactions*, Wiley: (2016).
16MI12	R.J. Sharpe, *Stereoselective Desymmetrization Methods in the Assembly of Complex Natural Molecules*, Springer: (2016).
16MI13	M. North, editor: *Sustainable Catalysis: Without Metals or Other Endangered Elements*, Royal Society of Chemistry: (2016), Pt. 2.
16MI14	S.L. McDonald, *Copper-Catalyzed Electrophilic Amination of sp^2 and sp^3 C–H Bonds*, Springer: (2016).

16MI15	M. Harmata, editor: *Strategies and Tactics in Organic Synthesis*, Elsevier: (2016), **Vol. 12**.
16MI16	S. Chandrasekaran, editor: *Click Reactions in Organic Synthesis*, Wiley: (2016).
16MI17	J.J. Li, *Carbocation Chemistry: Applications in Organic Synthesis*, CRC Press: (2016).
16MI18	V. Šunjić and V. Petrović Peroković, *Organic Chemistry From Retrosynthesis to Asymmetric Synthesis*, Springer: (2016).
16MI19	M. Rueping, D. Parmar, and E. Sugiono, *Asymmetric Brønsted Acid Catalysis*, Wiley: (2016).
16MI20	J. Whittall, P.W. Sutton, and W. Kroutil, editors: *Practical Methods for Biocatalysis and Biotransformations 3*, Wiley: (2016).
16MI21	X.-F. Wu, editor: *Transition Metal-Catalyzed Heterocycle Synthesis via C–H Activation*, Wiley: (2016).
16MI22	M. Stradiotto and R.J. Lundgren, editors: *Ligand Design in Metal Chemistry: Reactivity and Catalysis*, S. L. Buchwald (foreword by), D. Milstein (foreword by), Wiley: 2016.
16MI23	N.V. Plechkova and K.R. Seddon, editors: *Ionic Liquids Completely UnCOILed: Critical Expert Overviews*, Wiley: (2016).
16MI24	Y. Marcus, *Ionic Liquid Properties. From Molten Salts to Room Temperature Ionic Liquids (RTILs)*, Springer: (2016).
16MI25	Atta-ur-Rahman, editor: *Studies in Natural Products Chemistry*, (2016).
16MI26	Atta-ur-Rahman, editor: *Studies in Natural Products Chemistry*, (2016).
16MI27	Atta-ur-Rahman, editor: *Studies in Natural Products Chemistry*, (2016).
16MI28	Atta-ur-Rahman, editor: *Studies in Natural Products Chemistry*, (2016).
16MI29	K.J. Powell, *Synthetic and Biophysical Studies on the Tridachiahydropyrone Family of Natural Products*, Springer: (2016).
16MI30	A.L. Zografos, editor: *From Biosynthesis to Total Synthesis: Strategies and Tactics for Natural Products*, Wiley: (2016).
16MI31	M.K. Bhattacharjee, *Chemistry of Antibiotics and Related Drugs*, Springer: (2016).
16MI32	J.J. Li and D.S. Johnson, *Innovative Drug Synthesis*, Wiley: (2016).
16MI33	D.C. Blakemore, P.M. Doyle, and Y.M. Fobian, editors: *Synthetic Methods in Drug Discovery*, Royal Society of Chemistry: (2016), **Vol. 2**.
16MI34	R. Vardanyan and V. Hruby, *Synthesis of Best-Seller Drugs*, Elsevier: (2016).
16MI35	H. Brittain, editor: *Profiles of Drug Substances, Excipients and Related Methodology*, Elsevier: (2016), **Vol. 41**.
16MI36	T. Zou, *Anti-cancer N-Heterocyclic Carbene Complexes of Gold(III), Gold(I) and Platinum(II). Thiol "Switch-on" Fluorescent Probes, Thioredoxin Reductase Inhibitors and Endoplasmic Reticulum Targeting Agents*, Springer: (2016).
16MI37	R.S. Thomas and M.D. Waters, editors: *Toxicogenomics in Predictive Carcinogenicity*, Royal Society of Chemistry: (2016).
16MI38	B.A. Trofimov, A.I. Mikhaleva, E.Y. Schmidt, and L.N. Sobenina, *Chemistry of Pyrroles*, CRC Press: (2016).
16MI39	G.W. Gribble, *Indole Ring Synthesis: From Natural Products to Drug Discovery*, Wiley: (2016).
16MI40	X.-F. Wu, *Transition Metal-Catalyzed Pyridine Synthesis*, Elsevier: (2016).
16MI41	J.P.E. Spencer and A. Crozier, editors: *Flavonoids and Related Compounds: Bioavailability and Function*, CRC Press: (2016).

16MI42	S.H. Lone and M.A. Khuroo, *Biflavanoids. Chemical and Pharmacological Aspects*, 1st ed. Elsevier: (2016).
16MI43	X.F. Wu and Z. Wang, *Transition Metal Catalyzed Pyrimidine, Pyrazine, Pyridazine and Triazine Synthesis*, Elsevier: (2016).
16MI44	V.A. Mamedov, *Quinoxalines. Synthesis, Reactions, Mechanisms and Structure*, Springer: (2016).
16MI45	R.M. Izatt, editor: *Macrocyclic and Supramolecular Chemistry: How Izatt-Christensen Award Winners Shaped the Field*, Wiley: (2016).
16MI46	P. Neri, J.L. Sessler, and M.-X. Wang, editors: *Calixarenes and Beyond*, Springer: (2016).
16MI47	In D.W. Allen, D. Loakes, and J.C. Tebby, editors: *Organophosphorus Chemistry*, Royal Society of Chemistry: (2016), **Vol. 45**.
16MI48	D. Zhao, *Functionalization of Carborane via Carboryne Intermediates*, Springer: (2016).
16MI49	H.W. Roesky, *Efficient Methods for Preparing Silicon Compounds*, Elsevier: (2016).
16NPR382	J.W. Blunt, B.R. Copp, R.A. Keyzers, M.H.G. Munro, and M.R. Prinsep, *Nat. Prod. Rep.*, **33**, 382 (2016).
16NPR562	Y.Y. Qiu and S.H. Gao, *Nat. Prod. Rep.*, **33**, 562 (2016).
16NPR602	C. Li, A.X. Jones, and X.G. Lei, *Nat. Prod. Rep.*, **33**, 602 (2016).
16NPR801	M. Millot, A. Dieu, and S. Tomasi, *Nat. Prod. Rep.*, **33**, 801 (2016).
16NPR820	M.A. Beniddir, L. Evanno, D. Joseph, A. Skiredj, and E. Poupon, *Nat. Prod. Rep.*, **33**, 820 (2016).
16NPR861	M.D. Norris and M.V. Perkins, *Nat. Prod. Rep.*, **33**, 861 (2016).
16NPR1146	J. Liu, X.J. Zhu, S.J. Kim, and W.J. Zhang, *Nat. Prod. Rep.*, **33**, 1146 (2016).
16NPR1268	Z. Jin, *Nat. Prod. Rep.*, **33**, 1268 (2016).
16NPR1318	Z. Jin, *Nat. Prod. Rep.*, **33**, 1318 (2016).
16NPR1425	S.O. Simonetti, E.L. Larghi, and T.S. Kaufman, *Nat. Prod. Rep.*, **33**, 1425 (2016).
16OBC409	H. Ishikawa and S. Shiomi, *Org. Biomol. Chem.*, **14**, 409 (2016).
16OBC2593	J. Lei, J.B. Huang, and Q. Zhu, *Org. Biomol. Chem.*, **14**, 2593 (2016).
16OBC4008	A.S. Henderson, J.F. Bower, and M.C. Galan, *Org. Biomol. Chem.*, **14**, 4008 (2016).
16OBC4170	G. Clixby and L. Twyman, *Org. Biomol. Chem.*, **14**, 4170 (2016).
16OBC4335	H. Cong, X.L. Ni, X. Xiao, Y. Huang, Q.J. Zhu, S.F. Xue, Z. Tao, L.F. Lindoy, and G. Wei, *Org. Biomol. Chem.*, **14**, 4335 (2016).
16OBC4365	C.C. Li and S.D. Yang, *Org. Biomol. Chem.*, **14**, 4365 (2016).
16OBC4554	S.S. Li, L. Qin, and L. Dong, *Org. Biomol. Chem.*, **14**, 4554 (2016).
16OBC4970	R.K. Chambers, T.A. Khan, D.B. Olsen, and B.E. Sleebs, *Org. Biomol. Chem.*, **14**, 4970 (2016).
16OBC4986	J.D. Neuhaus and M.C. Willis, *Org. Biomol. Chem.*, **14**, 4986 (2016).
16OBC5367	S. Kaneko, Y. Kumatabara, and S. Shirakawa, *Org. Biomol. Chem.*, **14**, 5367 (2016).
16OBC5390	X.B. Ding, M.A. Brimble, and D.P. Furkert, *Org. Biomol. Chem.*, **14**, 5390 (2016).
16OBC5402	B. Wu and N. Yoshikai, *Org. Biomol. Chem.*, **14**, 5402 (2016).
16OBC5875	R. Jacques, R. Pal, N.A. Parker, C.E. Sear, P.W. Smith, A. Ribaucourt, and D.M. Hodgson, *Org. Biomol. Chem.*, **14**, 5875 (2016).
16OBC5894	S.R. McCabe and P. Wipf, *Org. Biomol. Chem.*, **14**, 5894 (2016).

16OBC6611	A.P. Taylor, R.P. Robinson, Y.M. Fobian, D.C. Blakemore, L.H. Jones, and O. Fadeyi, *Org. Biomol. Chem.*, **14**, 6611 (2016).
16OBC7136	N.N. Zhou, H.T. Zhu, D.S. Yang, and Z.H. Guan, *Org. Biomol. Chem.*, **14**, 7136 (2016).
16OBC7639	T. Aggarwal, S. Kumar, and A.K. Verma, *Org. Biomol. Chem.*, **14**, 7639 (2016).
16OBC7804	B. Sadowski, J. Klajn, and D.T. Gryko, *Org. Biomol. Chem.*, **14**, 7804 (2016).
16OBC8014	T.M.M. Maiden and J.P.A. Harrity, *Org. Biomol. Chem.*, **14**, 8014 (2016).
16OBC8911	A.H. Harkiss and A. Sutherland, *Org. Biomol. Chem.*, **14**, 8911 (2016).
16OBC9184	A. Blanc, V. Beneteau, J.M. Weibel, and P. Pale, *Org. Biomol. Chem.*, **14**, 9184 (2016).
16OBC10134	J. Caruano, G.G. Muccioli, and R. Robiette, *Org. Biomol. Chem.*, **14**, 10134 (2016).
16OBC10547	L. Mertens and R.M. Koenigs, *Org. Biomol. Chem.*, **14**, 10547 (2016).
16OBC11301	H.M. Hugel, A.T. Smith, and M.A. Rizzacasa, *Org. Biomol. Chem.*, **14**, 11301 (2016).
16OBC11317	X.R. Song, Y.F. Qiu, X.Y. Liu, and Y.M. Liang, *Org. Biomol. Chem.*, **14**, 11317 (2016).
16OPPI223	S.T. Martinez, C. Belouezzane, A.C. Pinto, and T. Glasnov, *Org. Prep. Proced. Int.*, **48**, 223 (2016).
16OPPI254	F. Rivas and T.T. Ling, *Org. Prep. Proced. Int.*, **48**, 254 (2016).
16OPPI425	A.S. Pashev, N.T. Burdzhiev, and E.R. Stanoeva, *Org. Prep. Proced. Int.*, **48**, 425 (2016).
16PCR121	I. Mawlong, M.S.S. Kumar, and D. Singh, *Phytochem. Rev.*, **15**, 121 (2016).
16PCR147	J.P. de Andrade, R.B. Giordani, L. Torras-Claveria, N.B. Pigni, S. Berkov, M. Font-Bardia, T. Calvet, E. Konrath, K. Bueno, L.G. Sachett, J.H. Dutilh, W.D. Borges, F. Viladomat, A.T. Henriques, J.J. Nair, J.A.S. Zuanazzi, and J. Bastida, *Phytochem. Rev.*, **15**, 147 (2016).
16PCR197	S.R.M. Ibrahim, H.M. Abdallah, A.M. El-Halawany, and G.A. Mohamed, *Phytochem. Rev.*, **15**, 197 (2016).
16PCR221	Q.F. Pan, N.R. Mustafa, K.X. Tang, Y.H. Choi, and R. Verpoorte, *Phytochem. Rev.*, **15**, 221 (2016).
16PCR317	M.B. Kilgore and T.M. Kutchan, *Phytochem. Rev.*, **15**, 317 (2016).
16PCR339	A.M.K. Thamm, Y. Qu, and V. De Luca, *Phytochem. Rev.*, **15**, 339 (2016).
16PCR363	A. Berim and D.R. Gang, *Phytochem. Rev.*, **15**, 363 (2016).
16PHC1	C.A. Ramsden, *Prog. Heterocycl. Chem.*, **28**, 1 (2016).
16PHC27	N. Piens, N. De Kimpe, and M. D'hooghe, *Prog. Heterocycl. Chem.*, **28**, 27 (2016).
16PHC57	J.C. Badenock, *Prog. Heterocycl. Chem.*, **28**, 57 (2016).
16PHC95	J. Anaya and R.M. Sánchez, *Prog. Heterocycl. Chem.*, **28**, 95 (2016).
16PHC121	E.R. Biehl, *Prog. Heterocycl. Chem.*, **28**, 121 (2016).
16PHC165	M. Lopchuk, *Prog. Heterocycl. Chem.*, **28**, 165 (2016).
16PHC219	Z. Ke, G.C. Tsui, X.S. Peng, and Y.Y. Yeung, *Prog. Heterocycl. Chem.*, **28**, 219 (2016).
16PHC275	L. Yet, *Prog. Heterocycl. Chem.*, **28**, 275 (2016).
16PHC317	Y.J. Wu, *Prog. Heterocycl. Chem.*, **28**, 317 (2016).
16PHC341	R.A. Aitken and G.-I. Dragomir, *Prog. Heterocycl. Chem.*, **28**, 341 (2016).

16PHC361	F.M. Cordero, D. Giomi, and L. Lascialfari, *Prog. Heterocycl. Chem.*, **28**, 361 (2016).
16PHC391	G.W. Gribble and T.L.S. Kishbaugh, *Prog. Heterocycl. Chem.*, **28**, 391 (2016).
16PHC439	K.A. Rinderspacher, *Prog. Heterocycl. Chem.*, **28**, 439 (2016).
16PHC493	P. Audebert, G. Clavier, and C. Allain, *Prog. Heterocycl. Chem.*, **28**, 493 (2016).
16PHC523	C.M.M. Santos and A.M.S. Silva, *Prog. Heterocycl. Chem.*, **28**, 523 (2016).
16PHC579	A.G. Meyer, J.A. Smith, C. Hyland, C.C. Williams, A.C. Bissember, and T.P. Nicholls, *Prog. Heterocycl. Chem.*, **28**, 579 (2016).
16PHC623	G.R. Newkome, *Prog. Heterocycl. Chem.*, **28**, 623 (2016).
16RCI2119	M.A. Gouda, A.A. Abu-Hashem, H.H. Saad, and K.M. Elattar, *Res. Chem. Intermed.*, **42**, 2119 (2016).
16RCI2731	S. Nayak, S. Chakroborty, S. Bhakta, P. Panda, and S. Mohapatra, *Res. Chem. Intermed.*, **42**, 2731 (2016).
16RCI2749	S.M. Roopan, S.M. Patil, and J. Palaniraja, *Res. Chem. Intermed.*, **42**, 2749 (2016).
16RCI5147	I. Khan, A. Ibrar, N. Abbas, and A. Saeed, *Res. Chem. Intermed.*, **42**, 5147 (2016).
16RCI5617	P.H. Yang, *Res. Chem. Intermed.*, **42**, 5617 (2016).
16RCI6143	M.A. Gouda, W.S. Hamama, H.A.K. El-din, and H.H. Zoorob, *Res. Chem. Intermed.*, **42**, 6143 (2016).
16RCI6777	J. Rani, S. Kumar, M. Saini, J. Mundlia, and P.K. Verma, *Res. Chem. Intermed.*, **42**, 6777 (2016).
16RCI6805	A. Saeed, F.A. Larik, and P.A. Channar, *Res. Chem. Intermed.*, **42**, 6805 (2016).
16RSCA1984	A.Y. Rulev and A.R. Romanov, *RSC Adv.*, **6**, 1984 (2016).
16RSCA2119	H. Vite-Caritino, O. Mendez-Lucio, H. Reyes, A. Cabrera, D. Chavez, and J.L. Medina-Franco, *RSC Adv.*, **6**, 2119 (2016).
16RSCA7575	S. Qian, M. Zhang, Q.L. Chen, Y.Y. He, W. Wang, and Z.Y. Wang, *RSC Adv.*, **6**, 7575 (2016).
16RSCA9194	J. Nowicki, M. Muszynski, and J.P. Mikkola, *RSC Adv.*, **6**, 9194 (2016).
16RSCA14778	H. Mahmud, A.K.O. Huq, and R.B. Yahya, *RSC Adv.*, **6**, 14778 (2016).
16RSCA17955	T.A. Farghaly, S.M. Gomha, K.M. Dawood, and M.R. Shaaban, *RSC Adv.*, **6**, 17955 (2016).
16RSCA18002	M.S. Alehashem, C.G. Lim, and N.F. Thomas, *RSC Adv.*, **6**, 18002 (2016).
16RSCA18419	A. Gualandi and D. Savoia, *RSC Adv.*, **6**, 18419 (2016).
16RSCA18619	E. Vessally, *RSC Adv.*, **6**, 18619 (2016).
16RSCA21979	G.M. Ziarani, Z. Hassanzadeh, P. Gholamzadeh, S. Asadi, and A. Badiei, *RSC Adv.*, **6**, 21979 (2016).
16RSCA23169	P. Alreja and N. Kaur, *RSC Adv.*, **6**, 23169 (2016).
16RSCA23760	R.K. Pai, T.N. Ahipa, and B. Hemavathi, *RSC Adv.*, **6**, 23760 (2016).
16RSCA24010	W.S. Hamama, G.G. El-Bana, S. Shaaban, O.M.O. Habib, and H.H. Zoorob, *RSC Adv.*, **6**, 24010 (2016).
16RSCA30498	M.M. Heravi, V. Zadsirjan, and B. Farajpour, *RSC Adv.*, **6**, 30498 (2016).
16RSCA37039	R. Khajuria, S. Dham, and K.K. Kapoor, *RSC Adv.*, **6**, 37039 (2016).
16RSCA37286	K.M. Elattar, B.D. Mert, M.A. Abozeid, and A. El-Mekabaty, *RSC Adv.*, **6**, 37286 (2016).

16RSCA37784	S. Ponra and K.C. Majumdar, *RSC Adv.*, **6**, 37784 (2016).
16RSCA38827	G.M. Ziarani, N.H. Nasab, and N. Lashgari, *RSC Adv.*, **6**, 38827 (2016).
16RSCA42045	M.M. Khan, S. Khan, Saigal, and S. Iqbal, *RSC Adv.*, **6**, 42045 (2016).
16RSCA42132	V.A. Mamedov, *RSC Adv.*, **6**, 42132 (2016).
16RSCA42660	B. Kumar, Sheetal, A.K. Mantha, and V. Kumar, *RSC Adv.*, **6**, 42660 (2016).
16RSCA46394	E.C. Gaudino, S. Tagliapietra, K. Martina, G. Palmisano, and G. Cravotto, *RSC Adv.*, **6**, 46394 (2016).
16RSCA49730	E. Vessally, L. Edjlali, A. Hosseinian, A. Bekhradnia, and M.D. Esrafili, *RSC Adv.*, **6**, 49730 (2016).
16RSCA50384	B. Maitiand and K. Chanda, *RSC Adv.*, **6**, 50384 (2016).
16RSCA63991	A. Chinnappan, C. Baskar, and H. Kim, *RSC Adv.*, **6**, 63991 (2016).
16RSCA71662	E. Vessally, A. Hosseinian, L. Edjlali, A. Bekhradnia, and M.D. Esrafili, *RSC Adv.*, **6**, 71662 (2016).
16RSCA71827	K.M. Elattar and B.D. Mert, *RSC Adv.*, **6**, 71827 (2016).
16RSCA77603	P. Mahbub and P.N. Nesterenko, *RSC Adv.*, **6**, 77603 (2016).
16RSCA93016	A. Ibrar, I. Khan, N. Abbas, U. Farooq, and A. Khan, *RSC Adv.*, **6**, 93016 (2016).
16RSCA93932	R. Abro, M. Abro, S.R. Gao, A.W. Bhutto, Z.M. Ali, A. Shah, X.C. Chen, and G.R. Yu, *RSC Adv.*, **6**, 93932 (2016).
16RSCA96154	S. Nayak, P. Panda, S. Bhakta, S.K. Mishra, and S. Mohapatra, *RSC Adv.*, **6**, 96154 (2016).
16RSCA99220	Y. Kumar, P. Singh, and G. Bhargava, *RSC Adv.*, **6**, 99220 (2016).
16RSCA99781	E. Vessally, A. Hosseinian, L. Edjlali, A. Bekhradnia, and M.D. Esrafili, *RSC Adv.*, **6**, 99781 (2016).
16RSCA111564	A.K. Ghosh and M. Brindisi, *RSC Adv.*, **6**, 111564 (2016).
16RSCA114758	K. Bujnowski, L. Synoradzki, R.C. Darlak, T.A. Zevaco, and E. Dinjus, *RSC Adv.*, **6**, 114758 (2016).
16S1	M.G. Brant and J.E. Wulff, *Synthesis*, **48**, 1 (2016).
16S18	X.D. Jia, *Synthesis*, **48**, 18 (2016).
16S184	D. Tilly, F. Chevallier, and F. Mongin, *Synthesis*, **48**, 184 (2016).
16S615	L.G. Voskressensky, N.E. Golantsov, and A.M. Maharramov, *Synthesis*, **48**, 615 (2016).
16S1253	R.S. Rohokale and U.A. Kshirsagar, *Synthesis*, **48**, 1253 (2016).
16S1269	S. Das, S. Santra, P. Mondal, A. Majee, and A. Hajra, *Synthesis*, **48**, 1269 (2016).
16S1437	H. Erhardt and S.F. Kirsch, *Synthesis*, **48**, 1437 (2016).
16S1457	S. Collins, S. Bartlett, F.L. Nie, H.F. Sore, and D.R. Spring, *Synthesis*, **48**, 1457 (2016).
16S1769	Z.P. Chen and Y.G. Zhou, *Synthesis*, **48**, 1769 (2016).
16S1782	N. Hoffmann, *Synthesis*, **48**, 1782 (2016).
16S1974	I. Sasaki, *Synthesis*, **48**, 1974 (2016).
16S1993	S. Florio and A. Salomone, *Synthesis*, **48**, 1993 (2016).
16S2009	G.J. Tanoury, *Synthesis*, **48**, 2009 (2016).
16S2151	M. Montesinos-Magraner, C. Vila, G. Blay, and J.R. Pedro, *Synthesis*, **48**, 2151 (2016).
16S2303	Priyanka, R.K. Sharma, and D. Katiyar, *Synthesis*, **48**, 2303 (2016).
16S2323	A.L. Garcia-Cabeza, F.J. Moreno-Dorado, M.J. Ortega, and F.M. Guerra, *Synthesis*, **48**, 2323 (2016).
16S2681	J. Gmach, L. Joachimiak, and K.M. Blazewska, *Synthesis*, **48**, 2681 (2016).

16S2705	M. Amat, N. Llor, G. Guignard, and J. Bosch, *Synthesis*, **48**, 2705 (2016).
16S2721	M. Giustiniano, E. Novellino, and G.C. Tron, *Synthesis*, **48**, 2721 (2016).
16S2969	N.L. Sloan and A. Sutherland, *Synthesis*, **48**, 2969 (2016).
16S3470	J.A. Varela and C. Saa, *Synthesis*, **48**, 3470 (2016).
16S3625	J.F. Vincent-Rocan and A.M. Beauchemin, *Synthesis*, **48**, 3625 (2016).
16S3646	L. Zani, A. Mordini, M. Calamante, and G. Reginato, *Synthesis*, **48**, 3646 (2016).
16S3684	V. Fulopova and M. Soural, *Synthesis*, **48**, 3684 (2016).
16S3821	R. Rossi, M. Lessi, C. Manzini, G. Marianetti, and F. Bellina, *Synthesis*, **48**, 3821 (2016).
16S3863	J. Rintjema and A.W. Kleij, *Synthesis*, **48**, 3863 (2016).
16S3879	T. Besson and C. Fruit, *Synthesis*, **48**, 3879 (2016).
16S4050	L. Moni, L. Banfi, R. Riva, and A. Basso, *Synthesis*, **48**, 4050 (2016).
16S4305	R. Sakhuja, K. Pericherla, K. Bajaj, B. Khungar, and A. Kumar, *Synthesis*, **48**, 4305 (2016).
16SC93	K.M. Elattar, M.A. Abozeid, and H.A. Etman, *Synth. Commun.*, **46**, 93 (2016).
16SC569	S. Govori, *Synth. Commun.*, **46**, 569 (2016).
16SC645	S.M. Roopan and R. Sompalle, *Synth. Commun.*, **46**, 645 (2016).
16SC719	K.M. Elattar, I. Youssef, and A.A. Fadda, *Synth. Commun.*, **46**, 719 (2016).
16SC831	F.A. Saddique, A.F. Zahoor, S. Faiz, S.A.R. Naqvi, M. Usman, and M. Ahmad, *Synth. Commun.*, **46**, 831 (2016).
16SC993	S.Y. Abbas, K.A.M. El-Bayouki, and W.M. Basyouni, *Synth. Commun.*, **46**, 993 (2016).
16SC1081	S. Bondock, O. Albormani, A.M. Fouda, and K.A. Abu Safieh, *Synth. Commun.*, **46**, 1081 (2016).
16SC1155	S. Jain, J. Dwivedi, P. Jain, and D. Kishore, *Synth. Commun.*, **46**, 1155 (2016).
16SC1397	S. Ahmad, M. Yousaf, A. Mansha, N. Rasool, A.F. Zahoor, F. Hafeez, and S.M.A. Rizvi, *Synth. Commun.*, **46**, 1397 (2016).
16SC1477	K.M. Elattar, R. Rabie, and M.M. Hammouda, *Synth. Commun.*, **46**, 1477 (2016).
16SC1567	K.M. Elattar and A.A. Fadda, *Synth. Commun.*, **46**, 1567 (2016).
16SC1643	P. Saraswat, G. Jeyabalan, M.Z. Hassan, M.U. Rahman, and N.K. Nyola, *Synth. Commun.*, **46**, 1643 (2016).
16SC1781	K. Tanemura and T. Suzuki, *Synth. Commun.*, **46**, 1781 (2016).
16SC1849	R. Akhtar, M. Yousaf, S.A.R. Naqvi, M. Irfan, A.F. Zahoor, A.I. Hussain, and S.A.S. Chatha, *Synth. Commun.*, **46**, 1849 (2016).
16SL477	H. Song, Y. Kim, J. Park, K. Kim, and E. Lee, *Synlett*, **27**, 477 (2016).
16SL486	Y.C. Mu, C.D. Zhu, and Z.Z. Shi, *Synlett*, **27**, 486 (2016).
16SL493	B. Zhou, L. Li, and L.W. Ye, *Synlett*, **27**, 493 (2016).
16SL498	T. Hensel, N.N. Andersen, M. Plesner, and M. Pittelkow, *Synlett*, **27**, 498 (2016).
16SL526	A. Gini and O.G. Mancheno, *Synlett*, **27**, 526 (2016).
16SL656	L. Meng and J. Wang, *Synlett*, **27**, 656 (2016).
16SL1145	K.K. Wan and R.A. Shenvi, *Synlett*, **27**, 1145 (2016).
16SL1318	W.P. Dong, Z.K. Chen, J.F. Xu, M.Z. Miao, and H.J. Ren, *Synlett*, **27**, 1318 (2016).

16SL1486	J. Dolfen, N. De Kimpe, and M. D'Hooghe, *Synlett*, **27**, 1486 (2016).
16SL1765	H. Yorimitsu, D. Vasu, M. Bhanuchandra, K. Murakami, and A. Osuka, *Synlett*, **27**, 1765 (2016).
16SL1775	A. Granzhan and H. Ihmels, *Synlett*, **27**, 1775 (2016).
16SL2051	W.P. Unsworth and R.J.K. Taylor, *Synlett*, **27**, 2051 (2016).
16SL2171	H. Zhou, M.H. Shen, and H.D. Xu, *Synlett*, **27**, 2171 (2016).
16SL2293	P. Hindenberg and C. Romero-Nieto, *Synlett*, **27**, 2293 (2016).
16SL2659	X.Y. Sun and S.Y. Yu, *Synlett*, **27**, 2659 (2016).
16T1	K.K. Krishnan, A.M. Thomas, K.S. Sindhu, and G. Anilkumar, *Tetrahedron*, **72**, 1 (2016).
16T355	M.R. Fructos and A. Prieto, *Tetrahedron*, **72**, 355 (2016).
16T1603	M.S. Singh, S. Chowdhury, and S. Koley, *Tetrahedron*, **72**, 1603 (2016).
16T3133	H. Ellissier, *Tetrahedron*, **72**, 3133 (2016).
16T3345	N. Hill, K. Paruch, and J. Svenda, *Tetrahedron*, **72**, 3345 (2016).
16T3379	A. Ghorai, B. Achari, and P. Chattopadhyay, *Tetrahedron*, **72**, 3379 (2016).
16T3389	D. Mandala, W.A. Thompson, and P. Watts, *Tetrahedron*, **72**, 3389 (2016).
16T4989	N. Veerasamy and R.G. Carter, *Tetrahedron*, **72**, 4989 (2016).
16T5003	A. de la Torre, C. Cuyamendous, V. Bultel-Ponce, T. Durand, J.M. Galano, and C. Oger, *Tetrahedron*, **72**, 5003 (2016).
16T5027	B.A. Shainyan, *Tetrahedron*, **72**, 5027 (2016).
16T5257	M.S. Singh, S. Chowdhury, and S. Koley, *Tetrahedron*, **72**, 5257 (2016).
16T6175	S.M. Ujwaldev, K.S. Sindhu, A.P. Thankachan, and G. Anilkumar, *Tetrahedron*, **72**, 6175 (2016).
16T6191	A.Y. Sukhorukov, A.A. Sukhanova, and S.G. Zlotin, *Tetrahedron*, **72**, 6191 (2016).
16T6711	S. El Kazzouli and G. Guillaumet, *Tetrahedron*, **72**, 6711 (2016).
16T7093	N.F. O'Rourke, M.J. Kier, and G.C. Micalizio, *Tetrahedron*, **72**, 7093 (2016).
16TCC(372)1	T. Kawabata, editor: *Site-Selective Catalysis. Topics in Current Chemistry*, **Vol. 372**, (2016), p. 236.
16THC(41)1	M. D'hooghe and H.-J. Ha, editors: *Topics in Heterocyclic Chemistry*, **Vol. 41**, (2016), p. 367.
16THC(42)1	X.-F. Wu and M. Beller, editors: *Topics in Heterocyclic Chemistry*, **Vol. 42**, (2016), p. 171.
16THC(43)1	J.-C.M. Monbaliu, editor: *Topics in Heterocyclic Chemistry*, **Vol. 43**, (2016), p. 289.
16THC(44)1	Z. Casar, editor: *Topics in Heterocyclic Chemistry*, **Vol. 44**, (2016), p. 273.
16THC(45)1	T. Patonay and K. Kónya, editors: *Topics in Heterocyclic Chemistry*, **Vol. 45**, (2016), p. 362.
16THC(46)1	M. Bandini, editor: *Topics in Heterocyclic Chemistry*, **Vol. 46**, (2016), p. 289.
16TJC1	O. Mazimba, *Turk. J. Chem.*, **40**, 1 (2016).
16TJC225	R.E. Khidre and W.M. Abdou, *Turk. J. Chem.*, **40**, 225 (2016).
16TL243	G. Sathiyan, E.K.T. Sivakumar, R. Ganesamoorthy, R. Thangamuthu, and P. Sakthivel, *Tetrahedron Lett.*, **57**, 243 (2016).
16TL2665	O.N. Chupakhin and V.N. Charushin, *Tetrahedron Lett.*, **57**, 2665 (2016).

16TL2683	E. Bogdan, N.D. Hadade, A. Terec, and I. Grosu, *Tetrahedron Lett.*, **57**, 2683 (2016).
16TL3575	M.A. Kuznetsov, L.M. Kuznetsova, and A.S. Pankova, *Tetrahedron Lett.*, **57**, 3575 (2016).
16TL3586	V. Franckevicius, *Tetrahedron Lett.*, **57**, 3586 (2016).
16TL5135	H. Nagarajaiah, A. Mukhopadhyay, and J.N. Moorthy, *Tetrahedron Lett.*, **57**, 5135 (2016).
16TL5416	V.A. Azov, *Tetrahedron Lett.*, **57**, 5416 (2016).
16TL5519	J.L. Hu, M. Bian, and H.F. Ding, *Tetrahedron Lett.*, **57**, 5519 (2016).
16TOC(54)1	G. van Koten and R.A. Gossage, editors: *The Privileged Pincer-Metal Platform: Coordination Chemistry and Applications, Topics in Organometallic Chemistry*, (2016).
16TOC(55)1	P.H. Dixneuf, H. Doucet, editors: *C–H Bond Activation and Catalytic Functionalization I. Topics in Organometallic Chemistry*, **Vol. 55**, (2016), p. 264.
16TOC(56)1	P.H. Dixneuf, H. Doucet, editors: *C–H Bond Activation and Catalytic Functionalization II Topics in Organometallic Chemistry*, **Vol. 56**, (2016), p. 212.
16TOC(59)1	In P. Kalck, editor: *Homo- and Heterobimetallic Complexes in Catalysis. Cooperative Catalysis, Topics in Organometallic Chemistry*, (2016).
16UK81	A.Y. Pulyalina, G.A. Polotskaya, and A.M. Toikka, *Usp. Khim.*, **85**, 81 (2016) [*Russ. Chem. Rev.* (Engl. Transl.), **85**, 81 (2016)].
16UK172	S.M. Pluzhnik-Gladyr, *Usp. Khim.*, **85**, 172 (2016) [*Russ. Chem. Rev.* (Engl. Transl.), **85**, 172 (2016)].
16UK226	K.I. Galkin and V.P. Ananikov, *Usp. Khim.*, **85**, 226 (2016) [*Russ. Chem. Rev.* (Engl. Transl.), **85**, 226 (2016)].
16UK308	D.A. Babkov, A.N. Geisman, A.L. Khandazhinskaya, and M.S. Novikov, *Usp. Khim.*, **85**, 308 (2016) [*Russ. Chem. Rev.* (Engl. Transl.), **85**, 308 (2016)].
16UK477	N.A. Bragina, K.A. Zhdanova, and A.F. Mironov, *Usp. Khim.*, **85**, 477 (2016) [*Russ. Chem. Rev.* (Engl. Transl.), **85**, 477 (2016)].
16UK637	D.S. Ryabukhin and A.V. Vasilyev, *Usp. Khim.*, **85**, 637 (2016) [*Russ. Chem. Rev.* (Engl. Transl.), **85**, 637 (2016)].
16UK667	N.S. Goulioukina, N.N. Makukhin, and I.P. Beletskaya, *Usp. Khim.*, **85**, 667 (2016) [*Russ. Chem. Rev.* (Engl. Transl.), **85**, 667 (2016)].
16UK700	S.V. Dudkin, E.A. Makarova, and E.A. Lukyanets, *Usp. Khim.*, **85**, 700 (2016) [*Russ. Chem. Rev.* (Engl. Transl.), **85**, 700 (2016)].
16UK759	G.N. Lipunova, E.V. Nosova, V.N. Charushin, and O.N. Chupakhin, *Usp. Khim.*, **85**, 759 (2016) [*Russ. Chem. Rev.* (Engl. Transl.), **85**, 759 (2016)].
16UK795	O.A. Gerasko, E.A. Kovalenko, and V.P. Fedin, *Usp. Khim.*, **85**, 795 (2016) [*Russ. Chem. Rev.* (Engl. Transl.), **85**, 795 (2016)].
16UK817	O.A. Omelchuk, A.S. Tikhomirov, and A.E. Shchekotikhin, *Usp. Khim.*, **85**, 817 (2016) [*Russ. Chem. Rev.* (Engl. Transl.), **85**, 817 (2016)].
16UK917	Y.A. Ustynyuk, M.Y. Alyapyshev, V.A. Babain, and N.A. Ustynyuk, *Usp. Khim.*, **85**, 917 (2016) [*Russ. Chem. Rev.* (Engl. Transl.), **85**, 917 (2016)].
16UK1056	O.P. Shkurko, T.G. Tolstikova, and V.F. Sedova, *Usp. Khim.*, **85**, 1056 (2016) [*Russ. Chem. Rev.* (Engl. Transl.), **85**, 1056 (2016)].
16UK1097	L.L. Fershtat and N.N. Makhova, *Usp. Khim.*, **85**, 1097 (2016) [*Russ. Chem. Rev.* (Engl. Transl.), **85**, 1097 (2016)].

16YGK56	H. Tabata, T. Yoneda, H. Takahashi, and H. Natsugari, *J. Synth. Org. Chem. Jpn.*, **74**, 56 (2016).
16YGK104	H. Ishikawa, *J. Synth. Org. Chem. Jpn.*, **74**, 104 (2016).
16YGK117	T. Abe and M. Ishikura, *J. Synth. Org. Chem. Jpn.*, **74**, 117 (2016).
16YGK130	H. Asahara, S.T. Le, and N. Nishiwaki, *J. Synth. Org. Chem. Jpn.*, **74**, 130 (2016).
16YGK154	S. Mori and N. Shibata, *J. Synth. Org. Chem. Jpn.*, **74**, 154 (2016).
16YGK243	Y. Haketa, R. Yamakado, and H. Maeda, *J. Synth. Org. Chem. Jpn.*, **74**, 243 (2016).
16YGK316	T. Hamura, *J. Synth. Org. Chem. Jpn.*, **74**, 316 (2016).
16YGK326	K. Okuma, *J. Synth. Org. Chem. Jpn.*, **74**, 326 (2016).
16YGK426	T. Chiba, T. Nakaya, K. Katayama, A. Matsuda, and S. Ichikawa, *J. Synth. Org. Chem. Jpn.*, **74**, 426 (2016).
16YGK599	T. Sato and N. Chida, *J. Synth. Org. Chem. Jpn.*, **74**, 599 (2016).
16YGK676	Y. Le and Y. Aso, *J. Synth. Org. Chem. Jpn.*, **74**, 676 (2016).
16YGK700	A.R. Pradipta, A. Tsutsui, and K. Tanaka, *J. Synth. Org. Chem. Jpn.*, **74**, 700 (2016).
16YGK760	Y. Ooyama and J. Ohshita, *J. Synth. Org. Chem. Jpn.*, **74**, 760 (2016).
16YGK781	T. Ishi-i, *J. Synth. Org. Chem. Jpn.*, **74**, 781 (2016).
16YGK854	H. Mizoguchi and H. Oguri, *J. Synth. Org. Chem. Jpn.*, **74**, 854 (2016).
16YZ607	S. Hibino, *J. Pharm. Soc. Jpn.*, **136**, 607 (2016).
16YZ841	H. Sashida, *J. Pharm. Soc. Jpn.*, **136**, 841 (2016).
16YZ1517	Y. Yamaoka, *J. Pharm. Soc. Jpn.*, **136**, 1517 (2016).
16ZAK3	A.Z. Temerdashev, A.M. Grigoriev, and I.V. Rybalchenko, *Zh. Anal. Khim.*, **71**, 3 (2016) [*Russ. J. Anal. Chem.* (Engl. Version), **71**, 1 (2016)].
16ZOK159	E.V. Koroleva, Z.I. Ignatovich, Y.V. Sinyutich, and K.N. Gusak, *Zh. Org. Khim.*, **52**, 159 (2016) [Engl. Transl.: *Russ. J. Org. Chem.*, **52**, 139 (2016)].
16ZOK1087	E.S. Balenkova, A.V. Shastin, V.M. Muzalevskii, and V.G. Nenaidenko, *Zh. Org. Khim.*, **52**, 1087 (2016) [Engl. Transl.: *Russ. J. Org. Chem.*, **52**, No. 8 (2016)].
16ZOK1239	R.R. Gataullin, *Zh. Org. Khim.*, **52**, 1239–1275 (2016) [Engl. Transl.: *Russ. J. Org. Chem.*, **52**, 1227 (2016)].
16ZOK1551	I.D. Grishin and D.F. Grishin, *Zh. Org. Khim.*, **52**, 1551 (2016) [Engl. Transl.: *Russ. J. Org. Chem.*, **52**, 1551 (2016)].
17AHC(122)245	L.I. Belen'kii and Y.B. Evdokimenkova, *Adv. Heterocycl. Chem.*, **122**, 245 (2017).
18AHC(124)121	L.I. Belen'kii and Y.B. Evdokimenkova, *Adv. Heterocycl. Chem.*, **124**, 121 (2018).

INDEX

Note: Page numbers followed by "*f*" indicate figures, and "*s*" indicate schemes.

A

1-Acetylazulenes, Friedlander reaction, 5, 5*s*
Acetyl-coenzyme A carboxylase (ACCase) inhibitor, 95–96
Acylhydrazines, rhodium-catalyzed pyrazole synthesis using, 91*s*
Agriculture, 1,2,3-thiadiazole in, 162–163
Aldehydes, 3-component reaction of, 70*s*
Aliphatic hydrazines, 65
Aliphatic hydrazones, 57
Alkaloids
   individual groups of, 199–200
   synthesis, 199
Alkenes, diazo compounds reaction with, 73*s*
Alkyl groups, 65
N-Alkyl hydrazines, 63–64
Alkynes
   cycloaddition of sydnones with, 80*s*
   intramolecular cyclization of, 22
   N-substituted pyrazoles synthesis from, 79*s*
   Pd-catalyzed synthesis of pyrazoles from, 66*s*
   pyrazole synthesis from, 65*s*
   tosylhydrazones reaction with, 79*s*
Alkynethiolates, 156–158
Alkynylboranes, cycloaddition of sydnones with, 81, 82*s*
Alkynyltrifluoroborates, Lewis acid/base-directed cycloaddition of sydnones with, 83, 83*s*
Allenic esters, pyrazoles and pyrazolines formation from, 63*s*
Allylic carbonates, 77
   mechanism of reaction, 78*s*
   phosphine-catalyzed [3+2] cycloaddition of, 77*s*
Alzheimer's disease, 94–95
Amino acids, 206
2-Aminoazulene, Sandmeyer reaction, 13
2-Amino-6-bromoazulene derivatives, cross-coupling reaction, 35
2-Aminofurans synthesis, 41*s*
Ammonium acetate, cyclization reaction with, 36*s*
Anionic *N*-heterocyclic carbenes, 178
Antibiotics, 200–201
Apixaban, 93–94
Aromatic *N*-sulfonylhydrazones, 58–59
Arylation, palladium-catalyzed direct, 35
Arylazasulfones
   mechanism of reaction of allylic carbonates with, 78*s*
   phosphine-catalyzed [3+2] cycloaddition of allylic carbonates with, 77*s*
Aryldiazonium salts, 90*s*
Arylhydrazines, 65
   Pd-catalyzed synthesis of pyrazoles from, 66*s*
   Sonogashira-carbonylation of, 65
N-Arylhydrazones, Ru-catalyzed oxidative cyclization of, 89*s*
2-Arylhydrazonothiocarbonyl compound, oxidative cyclization of, 121–125
N-Arylpyrazoles, 60
   synthesis, 61*s*
   transition metal-catalyzed functionalization of, 97*s*
2-Aryl-1,2,3-thiadiazoline-5-imines synthesis, 121–122
Asymmetric copper-catalyzed azide-alkyne cycloadditions, 177
Asymmetric dearomatization reactions, 177
Azodicarboxylates, pyrazoles and pyrazolines formation from, 63*s*
Azoles, azulene derivatives with, 46–50
Azolyl- and benzoazolylazulenes synthesis and reactions, 48–50
1-(Azolyl)- and 1-(Benzoazolyl) azulenes, synthesis and reactions, 46–48

Azulene. See also *specific types of azulene*
  derivatives with
    azoles and benzoazoles, 46–50
    6-membered ring heterocycle, 4–22
    oxygen-containing 6-membered
      heterocycles, 20–22
    numbering and resonance structure,
      2, 2s
    1-(4-pyridyl)azulenes, pyridine
      moiety, 8
    reaction with N-fluoro-3,5-
      dichloropyridinium triflate, 5s
    substituted isocoumarin derivatives, 22
Azulene synthesis, 2–4, 3s
  azulene-fused porphyrin synthesis, 44, 45s
  azulene-substituted
    isocoumarins synthesis, 22s
    quinoxaline synthesis, 15s
  1-(2,6-diphenyl- and 2,6-
    dimethyl-4-pyridyl)azulene, 6–7, 6s
  1-(4-pyridyl- and 3-pyridyl)azulenes,
    5–6, 6s
  1-(4-pyridyl)azulenes, 7–8, 7s
  1-(2-quinolyl)azulenes, 5, 5s
Azulenopyranones synthesis, 21s
1- and 2-Azulenyl porphyrins
  synthesis, 44s
Azulenylporphyrins, 43–44

B

B-cell antigen receptor (BCR), 93–94
Benzoazoles, azulene derivatives with,
  46–50
6-(Benzoazol-2-yl)- and
  6-(1H-perimidin-2-yl)azulenes
  synthesis, 50s
1-(2-Benzoazolyl)azulenes synthesis, 47, 47s
2-(2-Benzo[b]furyl)azulenes synthesis, 41s
1-(2-Benzofuryl)azulene synthesis, 40s
Benzoin, cyclization reaction with, 36s
5-(2-Benzolyl) and 5,7-di
  (2-benzothiazolyl)azulenes
  synthesis, 49s
Bestmann–Ohira reagent
  for construction of
    phosphoryl-pyrazoles, 66–67, 67s, 70s
    vinyl-substituted phosphoryl-
      pyrazoles, 68, 68s

Bestmann–Ohira reagent
  ynones reaction with, 69s
Betaines, 178
Bifunctional sydnone, 98–99
1,3-Bis(2-benzo[b]thienyl)azulene
  synthesis, 26–27
N-Bochydrazones, 58–59
Boron heterocycles, 222–223
Bowl-shaped conjugated polycycles, 176
Bronsted acid-mediated intramolecular
  nucleophilic reaction, 22s
Bruton's tyrosine kinase, 93–94

C

Carbazoles, 210–211
Carbene reactivity, 181
Carbohydrates, 177
Catalysis, pyrazole-based ligands for, 97s
Catalytic enantioselective Friedel–Crafts
  reactions, 180
Celecoxib, 94
Chalcone analogues, 5–6
C–H bond
  activation reaction, palladium-catalyzed,
    29–30s, 35s
  functionalization, 96–97
Chiral phase-transfer catalysts, 178
2-Chloroazulene derivative,
  29–30
Claisen–Schmidt reaction, 12–13
Coenzymes, 205–206
Condensation reaction, pyrazole synthesis,
  57–65, 60–61s
Cooperative capture synthesis, 179
Coordination chemistry, pyrazole in, 96–99,
  97–99s
Copper acetate, 85
Copper-catalyzed azide-alkyne
  cycloadditions, 177
Copper salt, on cycloaddition
  regioselectivity, 85, 85s
Copper triflate, 85
Corey–Fuchs reaction, 49–50
Cornforth rearrangement,
  1,2,3-thiadiazole, 149
Coumarin derivatives, 20–21
Crop protection, pyrazoles in,
  95–96, 96s

Cross-coupling reaction, 181–182
  palladium-catalyzed, 31s
  Suzuki–Miyaura, 8, 14, 14s, 19, 20s, 27, 27s, 29s, 32, 33s, 42, 42–43s, 45s
Cu-catalyzed sydnone alkyne cycloaddition (CuSAC), 84–85, 84s
  mechanism of reaction, 84s
Cu-catalyzed synthesis, of pyrazoles from iodoenynes, 87s
CuSAC. *See* Cu-catalyzed sydnone alkyne cycloaddition (CuSAC)
Cushing's syndrome, 95s
Cyclization reaction, 21
  AlCl$_3$-mediated, 21s
  of 1-azulenyl ketones, 36–37
  with benzoin and ammonium acetate, 36s
  of α-diazo thiocarbonyl compounds, 116–121
Cycloaddition, 186–187
  alkynylboronates
    with C4-substituted sydnones, 81, 82s
    with C4-unsubstituted sydnones, 81, 81s
  of diazoalkanes, 116
  regioselectivity, effect of copper salt, 85, 85s
  of sydnone
    with alkynes, 80–81, 80s
    with alkynylboranes, 81, 82s
    with alkynyltrifluoroborates, 83, 83s
    and sydnone imines with cyclooctanes, 85, 86s
Cyclooctanes, cycloaddition of sydnones and sydnone imines with, 85, 86s
Cyclopentadiene (CPD), 8
Cyclopenta-fused heterocycles, 176

D

Dearomatization reactions, 177
Dehydrofluorination reaction, 4
Dehydrogenation, troublesome, 2
Density functional theory (DFT), 118
Diarylhydrazones
  Fe-catalyzed synthesis of pyrazoles from, 89s
  pyrazole synthesis from, 89s
Diazoalkanes, cycloaddition of, 116
Diazo compounds
  indium-catalyzed cycloaddition of alkynes with, 75s
  in pyrazole synthesis, 66–77
  reaction with alkenes, 73s
α-Diazo thiocarbonyl compounds, cyclization of, 116–121
4,7-Di(2-thienyl)azulene derivative synthesis, 32s
5,6-Di(3-thienyl)azulene derivative synthesis, 33s
4,7-Di(2-furyl)azulene preparation, 43s
1,3-Di(4-pyridyl)azulenes synthesis, 10–11, 11s
1,3-Di(azolyl)azulenes synthesis, 47–48, 48s
1,3-Di(2-pyrrolyl)azulene synthesis, 36s
2,3-Di(1-azulenyl)benzofurans synthesis, 40s
1,3-Dicarbonyl compounds, 56
2,3-Dichloro-5,6-dicyanobenzoquinone (DDQ), 24–25
Diels–Alder reaction, 176
1,3-Diethyl carboxylate pyrazolide anions, 96
Di(1-azulenyl)ketone reaction with Lawesson's reagent, 24–25, 25s
N,N-Dimethylformamide dimethylacetal (DMFDMA), 44
Dimroth rearrangements, 1,2,3-thiadiazole, 145–148
1,3-Dioxane derivatives, 220
1-(2,6-Diphenyl- and 2,6-dimethyl-4-pyridyl)azulene synthesis, 6–7, 6s
1,3-Dipolar cycloaddition, 68–69
Dipyrrolyl-substituted pyrazoles, anion binding of, 99s
1,6-Di-*tert*-butylazulene reaction with pyridinium salt, 8–9, 9s
Dithienylethenes synthesis, 27s
Di, tri, and tetra(4-pyridyl)biazulenes structure, 11f
Drugs, heterocyclic chemistry, 201–205
  definite types of activity
    antibacterial activity, 202
    anti-HIV activity, 203–204
    antimalarial activity, 204
    antitumor activity, 203

Drugs, heterocyclic chemistry (Continued)
  CNS-targeted drugs, 203
  enzyme inhibitors and activators, 205
  receptor antagonisting activities, 204–205
Dye, 188–189
  pH-dependent near-infrared, 100s

E

Electron-deficient alkynes, rhodium-catalyzed pyrazole synthesis using, 91s
Electron–neutral aromatics, 65
Electron-withdrawing group (EWG), 74–75
Electrophile reactivity, 180
Electrophilic substitution, 2, 7–9, 13, 15–17, 23–24
Enamine reaction of diazoacetates, 76–77, 76s
Enaminones, copper-mediated oxidative coupling of, 92s
Enantioconvergent catalysis, 179
Enzymes, 205–206
ESIPT. See Excited-state intramolecular proton transfer (ESIPT)
3,4-Ethylenedioxythiophene (EDOT), 29–30
Excited-state intramolecular proton transfer (ESIPT), 100, 101s

F

Fe-catalyzed synthesis of pyrazole, 89s
Fipronil®, 58
Five-membered lactones, 212
Five-membered rings, heterocyclic chemistry
  four heteroatom, 215
  one heteroatom, 208–212
  three heteroatom, 214
  two heteroatom, 212–214
Fluorination reagent, 4
N-Fluoro-3,5-dichloropyridinium triflate, azulene reaction with, 5s
Free radical reaction, 181
Friedlander reaction, 1-acetylazulenes, 5, 5s
Furans, 211
1-(2-Furyl)azulenes, 39, 39f
Furylazulene, synthesis and reactions, 39–42

3-(2-Furyl) guaiazulene, phosphine-mediated synthesis, 39s
Fused 1,2,3-thiadiazoles, 126–132

H

α-Halohydrazones, 58, 58s
Hantzsch synthesis, 140
Hartwig–Buchwald reaction, 23–24, 24s
Heteroarylazulenes, 16
  1-heteroarylazulenes, 50
  2-heteroarylazulenes, 50
    synthesis, 15–16
  6-heteroarylazulenes, 13, 50
    synthesis, 4
Heterocycle
  5-membered, 83
  1,2-heterocycles, 213, 217
  1,3-heterocycles, 213, 217–218
  as intermediates, 182
  nitrogen containing, 86
  into position 4 of 1,2,3-thiadiazole, 135–136
  into position 5 of 1,2,3-thiadiazole, 138–143
  synthesis, 56
  to 1,2,3-thiadiazole, annelation, 131–132
  1,2,3-thiadiazoles linearly connected to another, 133–143
N-Heterocyclic carbenes, 178
Heterocyclic chemistry
  alkaloids, 199–200
  antibiotics, 200–201
  applications, 188–194
  boron heterocycles, 222–223
  coordination compounds, 191
  drugs, 201–205
  dyes, 188–189
  five-membered rings
    four heteroatom, 215
    one heteroatom, 208–212
    three heteroatom, 214
    two heteroatom, 212–214
  four-membered rings, 208
  history and biographies, 176
  intermediates, 188–189
  ionic liquids, 193
  large rings, 221–222

luminescent substances, 189–190
metallacycles, 223
miscellaneous, 193–194, 205–207
monographs, 176
natural bioactive heterocycles, 195–207
nitrogen heterocycles, 194
nonconventional synthetic
 methodologies, 183–185
organic conductors, 190
organic synthesis, 182
organocatalysts, 182–183
oxygen heterocycles, 195
phosphorus heterocycles, 222
photovoltaics, 190
polymers, 191–193
properties, 188–194
reaction types, 176–194
reactivity, 178–180
  carbenes, 181
  cross-coupling reactions, 181–182
  electrophiles, 180
  free radicals, 181
  nucleophiles, 180
  oxidants, 180
  reducing agents, 180
ring synthesis from nonheterocyclic
 compounds, 188
seven-membered rings, 220–221
silicon heterocycles, 223
six-membered rings
  one heteroatom, 215–217
  three heteroatom, 220
  two heteroatom, 217–220
structure and stereochemistry
  betaines and unusual structures, 178
  stereochemical aspects, 177–178
sulfur heterocycles, 195
synthetic bioactive heterocycles, 195–207
synthetic strategies, 185
  cycloadditions, 186–187
  metal-catalyzed reactions, 186
  miscellaneous methods, 187
  multicomponent reactions, 186–187
  photoreactions, 185–186
three-membered rings, 207–208
versatile synthons, 188
vitamins, 201
Heterocyclic tautomerism, 177

Hetero-Diels–Alder reaction, 178
Highest occupied molecular orbital
 (HOMO) energy level, 24
HMR, 111–114, 129–130
1-(1H-perimidin-2-yl)azulene, 47
Huisgen zwitterion, 62–63
Hurd–Mori synthesis, 111–116
Hydrazines, pyrazole synthesis from, 65s
Hydrazones
  aliphatic, 57
  cyclization with thionyl chloride,
   111–116
  as pyrazole precursors, 58
  reaction with isocyanides, 59s
Hydropyrans, 216
Hydropyrazines, 219
Hydropyridines, 215
Hydropyrroles, 209
Hypervalent heterocycles, 178

I

Ibrutinib, 93–94
ICT. *See* Intramolecular charge transfer
 (ICT)
Imidazoles, 213
Indoles, 210–211
Inhibitors
  acetyl-coenzyme A carboxylase, 95–96
  enzyme, 205
  polo-like kinase-2, 136
Intramolecular charge transfer (ICT), 2
Intramolecular cyclization of alkynes, 22
Iodine/*tert*-butyl peroxybenzoate (TBPB)
 oxidant system, 58–59
 reaction of hdrazones with
  isocyanides, 59s
2-Iodoazulene synthesis, 13
Iodocyclization reaction, acid-mediated
 and, 22, 22s
4-Iodoisocoumarin, 22
Ionic cyclopentadienide, 2
Ionic liquids, 176, 193
Isocoumarin derivatives, 20–21
Isocyanides
 hydrazones reaction with, 59s
 reaction, 58s
Isoindoles, 211

K

β-Ketoesters, 57
  reaction of trichloroacetylhydrazones with, 57, 57s
Krohnke type reaction, 12–13, 12s
Kumada–Tamao–Corriu coupling reaction, 25–26

L

L'abbe-type rearrangement, 1,2,3-thiadiazole, 150
Lawesson's reagent, di(1-azulenyl)ketone reaction with, 24–25, 25s
Lewis acid
  base-directed cycloaddition of sydnones with alkynyltrifluoroborates, 83, 83s
  trisalkynylboranes, 82
Ligand, pyrazole, 97–98s
Luminescent substance, 189–190

M

Malononitriles, 3-component reaction of, 70s
Marine organisms, heterocycles, 206
Medicine, 1,2,3-thiadiazole in, 162–163
5-Mercapto-1,3,4-triazoles synthesis, 142
Metal catalysis and metal-promoted synthesis, 86–93, 87–92s
Metal-catalyzed reactions, 186
Metallacycles, 223
Metal–organic frameworks (MOFs), 98, 98s
Mitsunobu reaction, 62–63
Monocyclic pyrroles, 208–209
Monocyclic 1,2,3-thiadiazoles synthesis, 110–125
Monosubstituted pyrazoles, 78
Monothio-1,3-diketones, pyrazole synthesis from, 61s
Multicomponent reaction, 186–187

N

Natural bioactive heterocycles
  biological functions, 195–196
  syntheses of, 196–198
N-bromosuccinimide (NBS), 21
Negishi coupling reaction, 9–10, 16, 16–17s
Neutral tris(azolyl)phosphanes, 177

N-iodosuccinimide (NIS), 22
Nitriles, copper-mediated oxidative coupling of, 92s
Nitrogen heterocycles, 194
Nitro-Michael reactions, 176
Nonconventional synthetic methodologies, 183–185
Nonsteroidal antiinflammatory drug (NSAID), 94
N-substituted pyrazoles, 99–100
  regiochemistry of, 56
  synthesis from tosylhydrazones and alkynes, 79s
Nucleic acids, 219
Nucleophile reactivity, 180
Nucleosides, 218–219
Nucleotides, 218–219
N-unsubstituted pyrazoles, 99–100

O

OHR, 122–123
Organic conductors, 190
Organic field-effect transistors (OFETs), 4
Organic light-emitting diodes (OLEDs), 4
Organic light-emitting (OLE) properties, 24
Organic synthesis, 182
Organocatalysts, 182–183
Organocatalytic enantioselective desymmetrization, 178
Oxidant reactivity, 180
Oxidation of thiadiazoles, 132
Oxidative cyclization, of 2-arylhydrazonothiocarbonyl compounds, 121–125
Oxygen heterocycles, 195

P

Palladium-catalyzed C–H bond activation reaction, 29–30s, 35s
Palladium-catalyzed cross-coupling reaction, 21, 31s
Palladium-catalyzed direct arylation, 35
Pechmann synthesis, 116
Peptides, 206
Pericyclic pyrazole formation, 73–74
Phenanthroline, 83–84
Phenazone, 56s
Phenazone, 56

Phosphine-mediated synthesis, 3-(2-furyl) guaiazulene, 39s
Phosphorus heterocycles, 222
Phosphoryl-pyrazoles, Bestmann-Ohira reagent for construction of, 66–67, 67s, 70s
Photocyclization reaction, 27
Photoreaction, 185–186
Photovoltaics, 190
Plant metabolites, 206
Polo-like kinase-2 (PLK2) inhibitors, 136
Poly[1,3-bis(3-alkyl-2-thienyl) azulene]s preparation, 25s
Polymers, 191–193
Porphyrins, 209–210
Pt-catalyzed [3,3] sigmatropic rearrangement/cyclization of propargylhydrazones, 88s
Pyrans, 216
Pyrazines, 219
Pyrazole, 56, 212
   applications, 99–100
   cocatalyzed formation of, 90s
   in coordination chemistry, 96–99, 97–99s
   in crop protection, 95–96, 96s
   excited-state intramolecular proton transfer, 101s
   high performance thermosets preparation, 100s
   ligand, 97–98s
     for catalysis, 97s
   medicinal chemistry applications of, 93–95, 94–95s
   metal–organic framework, 98, 98s
   monosubstituted, 78
   3H-pyrazole, [1,3]-Sigmatropic benzyl shift of, 75s
   sp^2 hybridization of, 96
   3,4,5-trisubstituted, 71s
Pyrazole precursors
   hydrazones as, 58
   tosylhydrazones as, 77–80, 77s
Pyrazole synthesis
   from alkynes
     and arylhydrazines, 66s
     and hydrazines, 65s
   from allenic esters and azodicarboxylates, 63s

4-component coupling, 66s
   condensation reactions, 57–65, 60–61s
   cycloaddition reactions, 66–85
   from diarylhydrazones and vicinal diols, 89s
   from α-diazo-β-ketoesters, 71s
   diazo compounds in, 66–77
   from 3-methylthio-2-propenones, 62s
   from monothio-1,3-diketones, 61s
   Ru-catalyzed hydrogen transfer for, 90s
   sydnones in, 80–85, 80–86s
   by tandem crosscoupling/ electrocyclization of diazoacetates and enol triflates, 74s
   Ynones in, 63–65, 64s
Pyrazole trifluoroborates synthesis, 64s
Pyrazolide anions, 96
Pyrazoline synthesis, from allenic esters and azodicarboxylates, 63s
1-(3-Pyrazolyl)azulenes synthesis, 47s
Pyridazines, 217
Pyridine, 215–216
   N-oxides, 215
Pyridinium
   compounds, 215
   salt, 7–8
     1,6-di-tert-butylazulene reaction with, 8–9, 9s
     reaction with 2-(4-pyridyl)azulene, 16s
1-(4-Pyridyl- and 3-pyridyl)azulenes synthesis, 5–6, 6s
2-(3-Pyridyl and pyrimidinyl)azulenes synthesis, 17s
2-(Pyridyl)- and 2-(quinolyl)azulene derivatives synthesis, 13
1-(2-pyridyl)- and 1-(3-quinolyl) azulenes synthesis, 12s
1-(Pyridyl)azulene, 12
1-(4-Pyridyl)azulene
   pyridine moiety of, 8
   synthesis and reaction, 7–8, 7s
Pyridylazulene derivatives, synthesis and reactions, 4–20
1-(2-Pyridyl)azulene synthesis, 9–10, 10s
2-(Pyridyl)azulene synthesis, 14, 14s
2-(3-Pyridyl)azulene synthesis, 20s
2-(4-Pyridyl)azulene synthesis, 16s

6-(4-Pyridyl)azulene synthesis, 19s
1-(Pyridyl)-, 1-(quinolyl)-, and
    1-(isoquinolyl)azulene derivatives,
    4–13, 5s
2-(Pyridyl)-, 2-(quinolyl)-, and
    2-(isoquinolyl) azulene derivatives,
    13–17
5-(Pyridyl)-, 5-(quinolyl)-, and
    5-(isoquinolyl) azulene derivatives,
    17–18
6-(Pyridyl)-, 6-(quinolyl)-, and
    6-(isoquinolyl)azulene derivatives,
    synthesis and reactivity, 18–20
Pyrimidine nucleoside, 218
Pyrone-fused azulene derivative formation,
    21–22, 22s
Pyrrolylazulene, synthesis and reactions,
    35–38
Pyrylium salt, 6–7

Q

5-(2-Quinolyl) and 5,7-bis(2-quinolyl)
    azulenes synthesis, 18s
1-(2-Quinolyl)azulenes
    synthesis, 5, 5s
5-(2-Quinolyl)azulene synthesis, 17, 18s

R

Reducing agents, 180
Regioselectivity, copper salt on
    cycloaddition, 85, 85s
Rhodium-catalyzed pyrazole synthesis,
    using acylhydrazines and
    electron-deficient alkynes, 91s
Rhodium-catalyzed transannulation
    of 1,2,3-thiadiazole, 159–161
Ru-catalyzed hydrogen transfer, for
    synthesis of pyrazoles, 90s
Ru-catalyzed oxidative cyclization, of
    N-arylhydrazones, 89s
Ruthenium-mediated hydrogen transfer,
    89–90

S

Sandmeyer reaction, 2-aminoazulene, 13
Scholl-type oxidative dehydrogenation
    reaction, 15
Sildenafil, 93–94

Silicon heterocycles, 223
Silyl-substituted ynones, 68
Six-membered rings, heterocyclic chemistry
    one heteroatom, 215–217
    three heteroatom, 220
    two heteroatom, 217–220
Sodium ascorbate, 83–84
Solar cells, 4
Sonogashira–Hagihara reaction, 21–22,
    39–40
Stille coupling reaction, 26–27, 26s,
    31, 32s, 36s
Sulfonylpyrazoles, diazosulfones reaction for
    construction of, 72s
Sulfur heterocycles, 195
Suzuki–Miyaura cross-coupling reactions, 8,
    12, 14, 14s, 19, 20s, 27, 27s, 29s, 32,
    33s, 42, 42–43s, 45s
Sydnones, in pyrazole synthesis, 80–85,
    80–86s
Synthetic bioactive heterocycles
    biological functions, 195–196
    syntheses of, 196–198

T

Tert-butyl hydroperoxide (TBHP), 72–73
Tetraarylpyrroles synthesis, 36s
Tetrabutyl ammonium iodide (TBAI),
    72–73
Tetrahydroquinoline, 56
Tetrathiafulvalene (TTF) derivatives, 30
Thermal decomposition reaction, of
    1,2,3-thiadiazole, 154–155
Thiadiazole-4-carboxylic acids, 136
Thiadiazole oxidation, 132
1,2,3-Thiadiazoles
    cleavage of ring, 151
        photochemical, 151–154
    Cornforth rearrangement, 149
    Dimroth rearrangements, 145–148
    5H-1,2,3-thiadiazoles transformations,
        155–159
    heterocycle into position 4 of, 135–136
    heterocycle into position 5 of, 138–143
    L'abbe-type rearrangements, 150
    linearly connected to another heterocycle,
        133–143
    in medicine and agriculture, 162–163

rearrangement of 5-vinyl-1,2,3-
thiadiazoles to 4,5-
dihydropyridazine-4-thioles,
150–151
rhodium-catalyzed transannulations of,
159–161
ring transformations of, 144–161
thermal decomposition reactions of,
154–155
1,2,3-Thiadiazoles synthesis
fused 1,2,3-thiadiazoles, 126–132
monocyclic 1,2,3-thiadiazoles, 110–125
1,2,3-Thiadiazole to heterocycle, annelation
of, 126–131
1,2,3-Thiadiazole-5-yl to heterocycle,
136–137
(1,2,3-Thiadiazol-5-yl)-benzoxazinone,
142
1,2,3-Thiadiazo-4-yl to heterocycle,
133–135
1,3-Thiazine derivatives, 219
1,4-Thiazine derivatives, 219
1-(4-Thiazolyl)azulenes synthesis, 46s
1-(2-Thienyl) and 1,3-di(2-thienyl)
azulenes synthesis, 23s
4-Thienyl- and 5-thienylazulene derivatives
and compounds, 31–33
2-Thienylazulene
derivatives and related compounds, 27–31
synthesis and sulfanylation, 28s, 34s
6-Thienylazulene, 35
1-Thienylazulene derivatives and
compounds, 23–27
6-Thienylazulene derivatives and
compounds, 33–35
2-(2-Thienyl)azulene derivatives
synthesis, 29s
6-(2-Thienyl)azulene derivatives
synthesis, 35s
4-(2-Thienyl)azulenes synthesis, 31–32, 31s
6-(2-Thienyl)azulene synthesis, 33–34s
Thionyl chloride, cyclization of hydrazones
with, 111–116

Thiopyran compounds, 217
Tosylhydrazone
alternative cycloaddition partners for, 80s
N-substituted pyrazoles synthesis
from, 79s
as pyrazole precursors, 77–80, 77s
reaction with alkynes, 79s
Transition metal catalysis, 56–57
functionalization of N-arylpyrazoles, 97s
Trichloroacetylhydrazones reaction with
β-Ketoesters, 57, 57s
3,4,5-Trisubstituted pyrazoles, 71s
Tropolones, azulene derivatives from, 3s
Tropylium, 2
Troublesome dehydrogenation, 2

V

Vicinal diols
Fe-catalyzed synthesis of pyrazoles
from, 89s
pyrazole synthesis from, 89s
Vilsmeier–Haack-type reaction, 37, 37s
Vinyl diazoacetates, cocatalyzed formation
of pyrazoles from, 90s
Vitamins, 201

W

Withasomnine synthesis, 73s
Wolff synthesis, 116–121

Y

Ylides, 215
Ynones
in pyrazole synthesis, 63–65, 64s
reaction with Bestmann–Ohira
reagent, 69s
silyl-substituted, 68

Z

Ziegler–Hafner's azulene synthesis, 2–3, 8,
19, 33–34, 33–34s, 37–38, 38s, 44s

Printed in the United States
By Bookmasters